AF683436

Student Solutions Manual

for
Kaseberg's

Intermediate Algebra: A Just-in-Time Approach

Third Edition

Cindy Rubash
Southern Oregon State College

THOMSON
BROOKS/COLE
Australia • Canada • Mexico • Singapore • Spain • United Kingdom • United States

COPYRIGHT © 2004 Brooks/Cole, a division of Thomson Learning, Inc. Thomson Learning™ is a trademark used herein under license.

ALL RIGHTS RESERVED. No part of this work covered by the copyright hereon may be reproduced or used in any form or by any means—graphic, electronic, or mechanical, including but not limited to photocopying, recording, taping, Web distribution, information networks, or information storage and retrieval systems—without the written permission of the publisher.

Printed in the United States of America
1 2 3 4 5 6 7 07 06 05 04 03

Printer: West Group

0-534-38725-X

For more information about our products, contact us at:
Thomson Learning Academic Resource Center
1-800-423-0563

For permission to use material from this text, contact us by:
Phone: 1-800-730-2214
Fax: 1-800-730-2215
Web: http://www.thomsonrights.com

Brooks/Cole—Thomson Learning, Inc.
10 Davis Drive
Belmont, CA 94002-3098
USA

Asia
Thomson Learning
5 Shenton Way #01-01
UIC Building
Singapore 068808

Australia/New Zealand
Thomson Learning
102 Dodds Street
Southbank, Victoria 3006
Australia

Canada
Nelson
1120 Birchmount Road
Toronto, Ontario M1K 5G4
Canada

Europe/Middle East/South Africa
Thomson Learning
High Holborn House
50/51 Bedford Row
London WC1R 4LR
United Kingdom

Latin America
Thomson Learning
Seneca, 53
Colonia Polanco
11560 Mexico D.F.
Mexico

Spain/Portugal
Paraninfo
Calle/Magallanes, 25
28015 Madrid, Spain

CONTENTS

Exercises 1.1

3. **a.** real numbers **b.** integers **c.** irrational numbers
d. multiplicative inverses **e.** additive inverses

5. **a.** distributive property of multiplication over addition
b. factors **c.** commutative property for multiplication
d. factoring **e.** associative property for addition

7. **a.** negative **b.** one **c.** positive
d. one **e.** zero

9. $\frac{1}{2}$ is not an integer

11. $\sqrt{9}$ is not an irrational number

13. 0 is not a natural number

15. natural numbers

17. a.
$4 + (-12) = -(12 - 4) = -8$
$4 - (-12) = 4 + (+12) = 16$
$(4)(-12) = -48$
$4 \div (-12) = \frac{4}{-12} = \frac{4}{-3 \cdot 4} = -\frac{1}{3}$

b.
$\frac{1}{2} + \frac{1}{3} = \frac{3 \cdot 1}{3 \cdot 2} + \frac{2 \cdot 1}{2 \cdot 3} = \frac{3}{6} + \frac{2}{6} = \frac{5}{6}$
$\frac{1}{2} - \frac{1}{3} = \frac{3}{6} - \frac{2}{6} = \frac{1}{6}$
$\frac{1}{2} \cdot \frac{1}{3} = \frac{1 \cdot 1}{2 \cdot 3} = \frac{1}{6}$
$\frac{1}{2} \div \frac{1}{3} = \frac{1}{2} \cdot \frac{3}{1} = \frac{3}{2}$

c.
$-\frac{2}{3} + \frac{1}{6} = -(\frac{4}{6} - \frac{1}{6}) = -\frac{3}{6} = -\frac{1}{2}$
$-\frac{2}{3} - \frac{1}{6} = -\frac{4}{6} + (-\frac{1}{6}) = -\frac{5}{6}$
$(-\frac{2}{3})(\frac{1}{6}) = -\frac{2 \cdot 1}{3 \cdot 6} = -\frac{2}{18} = -\frac{1}{9}$
$(-\frac{2}{3}) \div (\frac{1}{6}) = (-\frac{2}{3}) \cdot \frac{6}{1} = -\frac{12}{3} = 4$

d.
$0.75 + (-0.2) = +(0.75 - 0.2) = 0.55$
$0.75 - (-0.2) = 0.75 + (+0.2) = 0.95$
$(0.75)(-0.2) = (\frac{75}{100})(-\frac{2}{10}) = -\frac{150}{1000} = -0.15$
$(0.75) \div (-0.2) = (\frac{75}{100})(-\frac{10}{2}) = -\frac{375}{100} = -3.75$

e.
$(-0.45) + (-1.5) = -(0.45 + 1.5) = -1.95$
$-0.45 - (-1.5) = -0.45 + 1.5 = (1.5 - 0.45) = 1.05$
$(-0.45)(-1.5) = (\frac{45}{100})(\frac{15}{10}) = \frac{675}{1000} = 0.675$
$(-0.45) \div (-1.5) = (\frac{45}{100})(\frac{10}{15}) = \frac{3}{10} = 0.3$

f.
$(-\frac{5}{8}) + (-\frac{10}{9}) = -(\frac{45}{72} + \frac{80}{72}) = -\frac{125}{72} = -1\frac{53}{72}$
$(-\frac{5}{8}) - (-\frac{10}{9}) = -\frac{45}{72} + \frac{80}{72} = \frac{35}{72}$
$(-\frac{5}{8})(-\frac{10}{9}) = \frac{50}{72} = \frac{25}{36}$
$(-\frac{5}{8}) \div (-\frac{10}{9}) = (\frac{5}{8})(\frac{9}{10}) = \frac{45}{80} = \frac{9}{16}$

19. a. $8 - (-6) + (-2) = (8 + 6) + (-2) = 14 + (-2) = 14 - 2 = 12$

b. $15 + (-9) - (-8) = 15 + (-9) + 8 = 15 + [8 + (-9)] = 15 + (-1) = 15 - 1 = 14$

c. $-3.6 - (-4.8) + (4.3) = -3.6 + 4.8 + 4.3 = (4.8 - 3.6) + 4.3 = 1.2 + 4.3 = 5.5$

d. $-8.3 + (-3.8) - (-2.1) = -8.3 + [(-3.8) + 2.1] = -8.3 + (-1.7) = -(8.3 + 1.7) = -10$

e. $4\frac{1}{4} - 5\frac{1}{2} + 2\frac{1}{4} = 4\frac{1}{4} + 2\frac{1}{4} - 5\frac{1}{2} = 6\frac{1}{2} - 5\frac{1}{2} = 1$

f. $6\frac{1}{4} + 4\frac{1}{2} - 3\frac{1}{2} = 6\frac{1}{4} + 1 = 7\frac{1}{4}$

21. a. $\sqrt{-9}$ is undefined in the set of real numbers.

b. $\sqrt{9} = 3$ **c.** $-\sqrt{9} = -3$

23. a. $-(-3)$; the opposite of negative three **b.** -2; negative two

c. $5 - (-3)$; five subtract negative three

25. a. $4 + (-2) + (-8) + 6 = 4 + 6 + (-2) + (-8)$ *commutative property for addition*

$= (4 + 6) + [(-2) + (-8)]$ *associative property for addition*

$= 10 + (-10) = 0$

b. $-5(7)(-2)(3) = (-5)(-2)(7)(3)$ *commutative property for multiplication*

$= [(-5)(-2)][(7)(3)]$ *associative property for multiplication*

$= (10)(21) = 210$

c. $5 + (-2) + 15 + 12 = 5 + 15 + 12 + (-2)$ *commutative property for addition*

$= (5 + 15) + (12 - 2)$ *associative property for addition*

$= 20 + 10 = 30$

d. $\frac{1}{2}(19)(20) = \frac{1}{2}(20)(19)$ *commutative property for multiplication*

$= [\frac{1}{2}(20)](19)$ *associative property for multiplication*

$= (10)(19) = 190$

27. a. $19 \cdot 92 + 19 \cdot 8 = 19(92 + 8) = 19(100) = 1900$ **b.** $20(\frac{1}{5} + 0.5) = 4 + 10 = 14$

c. $299 \cdot 5 = (300 - 1)(5) = 1500 - 5 = 1495$

d. $107 \cdot 29 - 7 \cdot 29 = 29(107 - 7) = 29(100) = 2900$ **e.** $8(4\frac{1}{4}) = 8(4 + \frac{1}{4}) = 32 + 2 = 34$

29. a.
$A = \pi(122 \div 2)^2$
$A = \pi(61)^2$
$A = \pi(3721)$
$A \approx 11{,}689.9 \text{ cm}^2$

b.
$S = 2\pi \cdot 2^2 + 2\pi(2)(4.5)$
$S = 2\pi(4) + 2\pi(9)$
$S = \pi(8 + 18)$
$S = \pi(26)$
$S \approx 81.7 \text{ in}^2$

29. c. $d = (-5)^2 - 4(6)(-7)$

$d = 25 - 4(-42)$

$d = 25 + 168$

$d = 193$

d. $x = 9 - 4(9 - 2) + 8 \div 4 - 2 \cdot 5$

$x = 9 - 4(7) + 2 - 10$

$x = 9 - 28 + 2 - 10$

$x = -27$

31. a. $m = \frac{4-8}{6-(-2)}$

$m = \frac{-4}{8}$

$m = -\frac{1}{2}$

b. $c = \sqrt{(3^2 + 4^2)}$

$c = \sqrt{(9 + 16)}$

$c = \sqrt{25} = 5$

c. $F = \frac{9}{5}(100) + 32$

$F = 180 + 32$

$F = 212$

d. $C = \frac{5}{9}(212 - 32)$

$C = \frac{5}{9}(180)$

$C = 100$

e. $h = \frac{4}{125}(-20)^2$

$h = \frac{4}{125}(400)$

$h = \frac{1600}{125} = 12.8$

33. a. $x = \frac{3-12}{3^2-6}$

$x = \frac{-9}{9-6}$

$x = \frac{-9}{3} = -3$

b. $d = \sqrt{(6-(-2))^2 + (3-(-3))^2}$

$d = \sqrt{(8^2 + 6^2)}$

$d = \sqrt{64 + 36}$

$d = \sqrt{100} = 10$

c. $m = \frac{3-(-3)}{6-(-2)}$

$m = \frac{6}{8}$

$m = \frac{3}{4}$

d. $x = \frac{-(-2)+\sqrt{(-2)^2-4(3)(-5)}}{2(3)}$

$x = \frac{2+\sqrt{4+60}}{6}$

$x = \frac{2+8}{6},\ x = \frac{10}{6} = 1\frac{2}{3}$

e. $d = \{\,|4.05 - 7.12| + |7.51 - 7.12| + |9.80 - 7.12|\,\} \div 3$

$d = (3.07 + 0.39 + 2.68) \div 3$

$d = 6.14 \div 3$

$d = 2.04\overline{6}$

35. Possible answer: Subtraction can be restated as addition of the opposite.

37. 31a - slope; 31b - Pythagorean theorem; 31c - Celsius to Fahrenheit temperatures; 31d - Fahrenheit to Celsius temperatures; 33b - distance formula between two coordinates; 33c - slope; 33d - quadratic formula; 29c - discriminate for quadratic formula

Exercises 1.2

1. **a.** In $2x + 3$, the variable is x, the numerical coefficient is 2, and the constant term is 3

b. The variable is r and the numerical coefficient is π. There is no constant term

c. The variable is x, the numerical coefficient is -1 (implied by the – in front of x), the constant term is 4.

d. The constant is x, there are two numerical coefficients both are 1, the constant term is -1.

3. Visualize x columns of 2 dots with one dot missing from the last column.

5. **a.**

Input, x	1	2	3	4
Output, y	3	6	9	12

b. $6 - 3 = 3$; $9 - 6 = 3$; $12 - 9 = 3$;

c. The rule is $y = 3x$

7. **a.**

Input, x	1	2	3	4	5
Output, y	2	5	8	11	14

b. $5 - 2 = 3$; $8 - 5 = 3$; $11 - 8 = 3$; $14 - 11 = 3$

c. We can estimate the rule as $y = 3x + ?$. When $x = 1$, $y = 3(1) - 1$ or $y = 2$.
The rule is $y = 3x - 1$

9. **a.** Four subtract the opposite of c.

b. The opposite of the cube of z (or z-cubed).

c. Negative z-squared or the opposite of z-squared.

d. The output maybe positive or negative

Input, z	-2	-1	0	1	2
Output, $-z^3$	$-(-2)^3 = 8$	$-(-1)^3 = 1$	$-(0)^3 = 0$	$-(1)^3 = -1$	$-(2)^3 = -8$

e. $-z^2$ will always be negative because z^2 is always positive for any real number z.

11. **a.** $(-3)^2 = (-3)(-3) = 9$

b. $(-2)^3 = (-2)(-2)(-2) = -8$

c. $-3^2 = -(3)(3) = -9$

d. $(3x)^2 = (3x)(3x) = (3 \cdot 3)(x \cdot x) = 9x^2$

e. $3x^2 = (3)(x)(x) = 3x^2$

f. $0 - 3x^2 = -3x^2$

13. 11a. $(-3)^2$, base is -3 11b. $(-2)^3$, base is -2 11c. -3^2, base is 3
11d. $(3x)^2$, base is 3x 11e. $3x^2$, base is x 11f. $0 - 3x^2$, base is x

15. a. $(-2x)^3 = (-2x)(-2x)(-2x) = -8x^3$ **b.** $-(2x)^3 = -(2x)(2x)(2x) = -8x^3$

c. $(-3y)^4 = (-3y)(-3y)(-3y)(-3y) = 81y^4$ **d.** $-(xy)^4 = -(xy)(xy)(xy)(xy) = -x^4y^4$

e. $(-xy)^4 = (-xy)(-xy)(-xy)(-xy) = x^4y^4$

f. $0 -(-2y)^3 = -(-2y)(-2y)(-2y) = -(-8)y^3 = 8y^3$

17. a.

Multiply	x	–2
3x	$3x \cdot x = 3x^2$	$3x \cdot (-2) = -6x$

b.

Factor	a	–b
2a	$\mathbf{2a^2 = 2a(a)}$	$\mathbf{-2ab = 2a(-b)}$

19. a. $3(2x + 4) = 3(2x) + 3(4) = 6x + 12$ **b.** $15y + 10 = 5(3y) + 5(2) = 5(3y + 2)$

21. a. $x(x + 5) = x(x) + x(5) = x^2 + 5x$ **b.** $ab + bc = b(a) + b(c) = b(a + c)$

23. a. $-2(3 - 4x) = (-2)(3) + (-2)(-4x) = -6 + 8x$

b. $-a(b - c) = (-a)(b) + (-a)(-c) = -ab + ac$

c. $x(x^2 - 2x - 3) = (x)(x^2) + (x)(-2x) + (x)(-3) = x^3 - 2x^2 - 3x$

25. a. $6x - 54 = (6)(x) - (6)(9)$; greatest common factor (gcf) is 6, $6(x - 9)$

b. $15x - 225 = (15)(x) - (15)(15)$; gcf is 15, $15(x - 15)$

c. $6a^2 - 8a = (2a)(3a) - (2a)(4)$; gcf is 2a, $2a(3a - 4)$

27. a. $ab - b^2 = (a)(b) - (b)(b)$; gcf is b, $b(a - b)$

b. $xy - x^2y = (xy)(1) - (xy)(x)$; gcf is xy, $xy(1 - x)$

c. $(xy)^2 - xy^2 = (xy^2)(x) - (xy^2)(1)$; gcf is xy^2, $xy^2(x - 1)$

29. a. $6ab - 12a + 3 = (3)(2ab) - (3)(4a) + (3)(1)$; gcf is 3, $3(2ab - 4a + 1)$

b. $4ac + 12a - 16 = (4)(ac) + (4)(3a) - (4)(4)$; greatest common factor is 4, $4(ac + 3a - 4)$

31. There are 3 (x^2) tiles, 5 (x) tiles and 6 (1) tiles, total is $3x^2 + 5x + 6$

33. There are 4 (a^2) tiles, 3 (ab) tiles and 2 (b^2) tiles, total is $4a^2 + 3ab + 2b^2$

35. a. $3a - 2b + 4b - 2a = 3a - 2a - 2b + 4b = a(3 - 2) + b(-2 + 4) = a + 2b$

b. $6x^2 + 2x - 4x^2 - 3 = 6x^2 - 4x^2 + 2x - 3 = x^2(6 - 4) + 2x - 3 = 2x^2 + 2x - 3$

37. a. $6ac - 3ab + 2ca + 7bc = 6ac + 2ac - 3ab + 7bc = ac(6 + 2) - 3ab + 7bc = 8ac - 3ab + 7bc$

b. $x(4 + 2x) - x(x - 4) = 4x + 2x^2 - x^2 + 4x = 2x^2 - x^2 + 4x + 4x = x^2 + 8x$

39. a. gcf is 15, $\frac{15}{45} = \frac{15\cdot 1}{15\cdot 3} = \frac{1}{3}$

b. gcf is 16, $\frac{48}{160} = \frac{16\cdot 3}{16\cdot 10} = \frac{3}{10}$

c. gcf is 240, $\frac{5280}{3600} = \frac{240\cdot 22}{240\cdot 15} = \frac{22}{15}$

41. a. gcf is be, $\frac{abe}{ben} = \frac{a\cdot be}{be\cdot n} = \frac{a}{n}$

b. gcf is b, $\frac{ab+bc}{b} = \frac{b(a+c)}{b} = a+c$

c. $\frac{x-2}{2}$ cannot be simplified

43. a. gcf is x, $\frac{xy+xz}{xy} = \frac{x(y+z)}{x\cdot y} = \frac{y+z}{y}$

b. gcf is x, $\frac{x^2+2x}{2x} = \frac{x(x+2)}{x\cdot 2} = \frac{x+2}{x}$

c. gcf is xy, $\frac{xy^2-2x^2y}{2xy} = \frac{xy(y-2x)}{xy\cdot 2} = \frac{y-2x}{2}$

45. a. $(7)^2 + 2(7) + 1 = 49 + 14 + 1 = 64$

b. $(7+1)^2 = 8^2 = 64$

47. a. $(9)^2 - (7)^2 = 81 - 49 = 32$

b. $(9-7)^2 = 2^2 = 4$

49. a. $9 - 3[(8) - 2] = 9 - 3(6) = 9 - 18 = -9$

b. $6[(8) - 2] = 6(6) = 36$

51. a. $C = \frac{5}{9}[(32) - 32]$

$C = \frac{5}{9}(0)$

$C = 0$

b. $C = \frac{5}{9}[(-40) - 32]$

$C = \frac{5}{9}(-72)$

$C = -40$

c. $C = \frac{5}{9}[(98.6) - 32]$

$C = \frac{5}{9}(66.6)$

$C = 37$

53. a. $V = \frac{4}{3}\pi(1.5\text{ cm})^3$

$V = \frac{4}{3}\pi(3.375)\text{ cm}^3$

$V = 4.5\pi\text{ cm}^3$

$V = 14.1\text{ cm}^3$

b. $V = \frac{4}{3}\pi(3\text{ cm})^3$

$V = \frac{4}{3}\pi(27)\text{ cm}^3$

$V = 36\pi\text{ cm}^3$

$V = 113.1\text{ cm}^3$

c. $V = \frac{4}{3}\pi(6\text{ cm})^3$

$V = \frac{4}{3}\pi(216)\text{ cm}^3$

$V = 288\pi\text{ cm}^3$

$V = 904.8\text{ cm}^3$

55. a. $S = \frac{100}{2}\{2(1) + [(100) - 1](2)\}$

$S = 50[2 + (99)2]$

$S = 50(2 + 198)$

$S = 50(200)$

$S = 10,000$

b. $S = \frac{100}{2}\{2(2) + [(100) - 1](2)\}$

$S = 50[4 + (99)2]$

$S = 50(4 + 198)$

$S = 50(202)$

$S = 10,100$

c. $S = \frac{100}{2}\{2(5) + [(100) - 1](5)\}$

$S = 50[10 + (99)5]$

$S = 50(10 + 495)$

$S = 50(505)$

$S = 25,250$

57. a. $D = \sqrt{[6-(-2)]^2 + [3-(-3)]^2}$

$D = \sqrt{(8)^2 + (6)^2}$

$D = \sqrt{64 + 36}$

$D = \sqrt{100}$

$D = 10$

b. $D = \sqrt{(14-2)^2 + [2-(-3)]^2}$

$D = \sqrt{(12)^2 + (5)^2}$

$D = \sqrt{144 + 25}$

$D = \sqrt{169}$

$D = 13$

c. $D = \sqrt{[(-4) - 11)]^2 + [3-(-5)]^2}$

$D = \sqrt{(-15)^2 + (8)^2}$

$D = \sqrt{225 + 64}$

$D = \sqrt{289}$

$D = 17$

59. When a is > 0, $a > -a$; when $a < 0$, $a < -a$; when $a = 0$, $a = -a$

Exercises 1.3

1.

Change in x		1	1	1
x	1	2	3	4
y	2	5	8	11
Change in y		3	3	3

The changes in x and y are constant. Working backwards in the table, when $x = 0$, $y = 2 - 3 = -1$. Using the change in y as the coefficient on x: the equation is $y = 3x - 1$

3.

Change in x		1	1	1
x	1	2	3	4
y	3	5	7	9
Change in y		2	2	2

The changes in x and y are constant. Working backwards in the table, when x = 0, y = 3 – 2 = 1. Using the change in y as the coefficient on x: the equation is y = 2x + 1

5.

Change in x		1	1	1
x	1	2	3	4
y	2	4	6	8
Change in y		2	2	2

The changes is x and y are constant. Working backwards in the table, when x = 0, y = 2 – 2 = 0. Using the change in y as the coefficient on x: the equation is y = 2x.

7.

Change in x		1	1	1
x	1	2	3	4
y	2	5	10	17
Change in y		3	5	7

The change in y is not constant. We can observe that y is close to the value of x^2 but something more is needed. When x = 1, y = 2; 1 more than 1^2. Checking a different combination; when x = 3, y = 10; again 1 more than 3^2. The equation is $y = x^2 + 1$.

9.

Input	0	2	4	5	n
Output	0	1	**2**	10	$\frac{1}{2}$**n**

11.

Input	0	1	2	3	n
Output	0	1	**8**	27	$\mathbf{n^3}$

13. y – 2 = x + 5, y – 2 + 2 = x + 5 + 2, y = x + 7
x + y = 7, x + y – x = 7 – x, y = 7 – x
x – y = –7, x – y + y = –7 + y, x + 7 = –7 + y + 7, x + 7 = y, y = x + 7
x + y = 7 is not equivalent to the other equations.

15. $x + y = 8$, $x + y - x = 8 - x$, $y = 8 - x$, $y = -x + 8$
$y - x = -8$, $y - x + x = -8 + x$, $y = x - 8$
$y = -x + 8$
$y - x = -8$ is not equivalent to the other equations.

17. The difference between x and seven. **19.** The quotient of x and seven.

21. The difference between seven and x. **23.** The quotient of x and five.

25. The quotient of seven and the sum of x and two

27 The quotient of the sum of x and two and the difference between x and two.

29. $10 - 2x = y$ **31.** $y = \frac{1}{2}x + 10$

33. $\frac{x}{15} + 2 = y$ **35.** $y = 7x - 14$

37. The output is five less than the product of 2 and the input.

39. The sum of the input and the output is eleven.

41. The difference between two times the input and the output is 5.

43. The output is the quotient of the input and 8.

45. In $A = s^2$, s is the independent variable, A is the dependent variable.

47. In $A = \frac{s^2\sqrt{3}}{4}$, s is the independent variable, A is the dependent variable.

49. A is the independent variable, E is the dependent variable.

51. Independent variable, weight of package; dependent variable, cost of shipping.

53. Independent variable, temperature (a cooler day might bring more kittens); dependent variable, number of kittens.

55. $y = 1.29x - 0.45$ **57.** $y = 1,25 - 0.05x$

59. $y = 125x + 85$

61. a. $\frac{1}{6} = 0.167, 0.167 \cdot 100 = 16.7\%$ **b.** $\frac{3}{5}$ $0.60, 0.60 \cdot 100 = 60\%$

c. $20 \cdot 100 = 2000\%$ **d.** $0.05 \cdot 100 = 5\%$

63. a. $20\% = \frac{20}{100} = 0.20$ **b.** $12.5\% = \frac{12.5}{100} = 0.125$

c. $0.005\% = \frac{0.005}{100} = 0.00005$ **d.** $500\% = \frac{500}{100} = 5$

65. x = season ticket cost, y = daily cost, 5% = 0.05; y = 0.05x

67. x = wages, y = amount of tax, 6.2% = 0.062; y = 0.062x

69. x = amount of loan, y = one cost, 1% = 0.01; y = 0.01x

71. x = amount of wages, y = net pay, 7.65% = 0.0765, 20% = 0.20, $n\% = \frac{n}{100}$;
$y = x - (0.0765x + 0.20x + \frac{n}{100}x)$

73. The group is a woman with her mother and her daughter.

Mid-Chapter 1 Test

1. a. $-3 + (-5) - (-6) = -3 + (-5) + 6 = -8 + 6 = -2$

b. $-\frac{1}{2} + \frac{1}{4} - \frac{3}{4} - (-\frac{1}{4}) = -\frac{1}{2} + \frac{1}{4} + \frac{1}{4} - \frac{3}{4} = -\frac{1}{2} + \frac{1}{2} - \frac{3}{4} = -\frac{3}{4}$

c. $2.25 + 8.50 - 3.75 = 10.75 - 3.75 = 7.0$

d. $6.2 - 8.6 + 1.8 = 6.2 + 1.8 - 8.6 = 8.0 - 8.6 = -0.6$

2. a. $2 \cdot 17 \cdot 50 = 2 \cdot 50 \cdot 17 = 100 \cdot 17 = 1700$ **b.** $\frac{1}{2}(-7)(-6) = \frac{1}{2}(42) = 21$

c. $-4(-8)(-5) = -4(40) = -160$ **d.** $\frac{3}{5}(\frac{11}{6})(\frac{15}{121}) = \frac{3(11)(3)(5)}{5(3)(2)(11)(11)} = \frac{3}{2(11)} = \frac{3}{22}$

3. a. $-4^2 = -(4)(4) = -16$ **b.** $(-4)^2 = (-4)(-4) = 16$

c. $0 - 4^2 = -(4)(4) = -16$

d. $8 - 6 \div 3 + 3 \cdot 4 + \frac{1}{2}(8 - 3) = 8 - 2 + 12 + \frac{1}{2}(5) = 6 + 12 + 2\frac{1}{2} = 20\frac{1}{2}$

e. $\frac{3-6}{2-(-7)} = \frac{-3}{2+7} = -\frac{3}{9} = -\frac{1}{3}$ **f.** $\frac{-(-2)-\sqrt{(-2)^2-4(5)(-3)}}{2(5)} = \frac{2-\sqrt{64}}{10} = \frac{2-8}{10} = \frac{-6}{10} = -\frac{3}{5}$

g. $\sqrt{(2-(-7))^2 + (3-6)^2} = \sqrt{9^2 + (-3)^2} = \sqrt{81+9} = \sqrt{90} = 3\sqrt{10} \approx 9.5$

h. $\frac{|1.25-2.83|+|2.56-2.83|+|4.68-2.83|}{3} = \frac{1.58+0.27+1.85}{3} = \frac{3.7}{3} = \frac{37}{30} \approx 1.2$

4. a. $\frac{1}{2} + 96.43 + 3 \div 6 = \frac{1}{2} + 96.43 + \frac{1}{2} = 96.43 + 1 = 97.43$

b. $9 \cdot 59 + 59 = 59(9 + 1) = 59(10) = 590$ **c.** $8(8\frac{3}{8}) = 8(8) + 8(\frac{3}{8}) = 64 + 3 = 67$

d. $3.57(39) + 6.43(39) = 39(3.57 + 6.43) = 39(10) = 390$

e. $0.0333333\overline{3} + \frac{3}{10} = 0.033333\overline{3} + 0.300 = 0.33333\overline{3} = \frac{1}{3}$

f. $(225 + ?) + (576 - 127) = 225 + 576 - 127 + ? = 225 + 576 + 0$; ? must be 127

5. a. $-4x^2$ will be negative because $4x^2$ is always positive when x is a real number.

b. 4x could be positive or negative **c.** x could be positive or negative

d. $-|x|$ will be negative because $|x|$ is always positive.

6. a. $3(x + 2) = 3x + 3(2) = 3x + 6$ **b.** $-3(x - 2) = -3x + (-3)(-2) = -3x + 6$

c. $7 - 4(x + 5) = 7 - 4x + (-4)(5) = 7 - 4x - 20 = -4x - 20 + 7 = -4x - 13$

d. $8 - 2(4 - x) = 8 + (-2)(4) + (-2)(-x) = 8 - 8 + 2x = 2x$

7. a. gcf is 2; $4x - 18 = 2(2x) - (2)(9) = 2(2x - 9)$ **b.** gcf is x; $x^2 + 5x = x(x + 5)$

c. gcf is n; $mn^2 - np^2 = n(mn - p^2)$ **b.** gcf is 7xy; $63x^2y - 49xy^2 = 7xy(9x - 7y)$

8. a. The variable is x, the constant term is –4, the numerical coefficient is –1.

b. The variable is x, there is no constant term, the numerical coefficients are 2 and –1.

9. a. $3(x^2 + 3x - 4) - x(x^2 + 3x - 4) = 3x^2 + 9x - 12 - x^3 - 3x^2 + 4x = -x^3 + 13x - 12$

b. $a(a^2 - 2ab + b^2) - b(a^2 + 2ab - b^2) = a^3 - 2a^2b + ab^2 - a^2b - 2ab^2 + b^3$

$= a^3 - 3a^2b - ab^2 + b^3$

10. a. $\frac{x^2yz}{xy^2z} = \frac{xxyz}{xyyz} = \frac{x}{y}$ **b.** $\frac{x^2yz}{(xy)^2z} = \frac{xxyz}{xyxyz} = \frac{1}{y}$ **c.** $\frac{xy-yz}{xy} = \frac{y(x-z)}{xy} = \frac{x-z}{x}$

11. a. s is the independent variable, h is the dependent variable; $h = \frac{4\sqrt{3}}{2} = 2\sqrt{3}$ yd

b. d is the independent variable, A is the dependent variable; $A = \pi(\frac{5}{2})^2$, $A = \frac{25}{4}\pi$ cm^2 or $A = 6.25\pi$ cm^2

12. a. Independent variable is the number of quarters, dependent variable is parking time.

b. Independent variable is speed, dependent variable is distance traveled.

c. Independent variable is size of luggage, dependent variable is cost.

13. a. $C = x + 0.0875x$ or $C = 1.0875x$ **b.** $A = x + 0.03x$ or $A = 1.03x$

c. $C = 1.49x - 0.30$

14. a.

# of pairs of pens	1	2	3	4	10	50	100
Total # of panels	7	**12**	**17**	**22**	52	**252**	**502**

b. The change in y is constant at 5. Working backwards in the table, when x = 0, y = 2. The equation is y = 5x + 2

15. j, 4 + 2x

16. a, 5x + 3

17. g, 3x – 5

18. h, 0.5x + 3

19. i, 2x – 4

20. b, 4x – 2

21. e, 5 – 3x

22. $\frac{1}{2} - 4x$

Exercises 1.4

1.

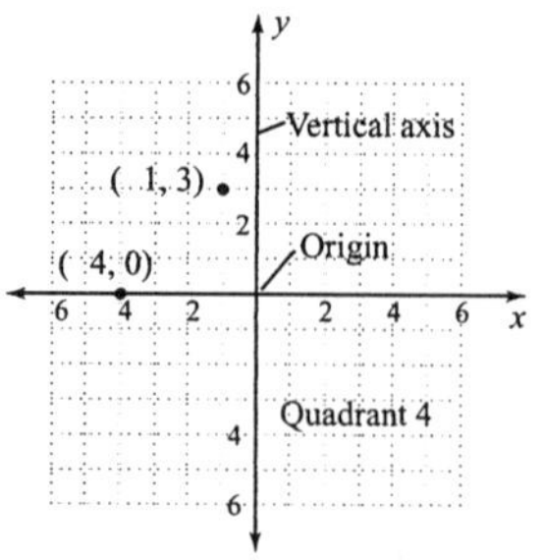

3. a. Solution set **b.** evaluate **c.** scale
d. dependent variable **e.** independent variable

5. a. (75, 167), J **b.** (10, 50), E
c. (37, 98.6), G **d.** (100, 212), K

7. a. B = (–18, 0) **b.** A, (–40, –40)

9.

x	y = 3x–2
–3	3(–3) – 2 = –11
–2	3(–2) – 2 = –8
–1	3(–1) – 2 = –5
0	3(0) – 2 = –2
1	3(1) – 2 = 1
2	3(2) – 2 = 4
3	3(3) – 2 = 7

11.

x	y = 4 – 2x
–3	4 – 2(–3) = 10
–2	4 – 2(–2) = 8
–1	4 – 2(–1) = 6
0	4 – 2(0) = 4
1	4 – 2(1) = 2
2	4 – 2(2) = 0
3	4 – 2(3) = –2

13.

x	y = –2 – x
–3	–2 – (–3) = 1
–2	–2 – (–2) = 0
–1	–2 – (–1) = –1
0	–2 – (0) = –2
1	–2 – (1) = –3
2	–2 – (2) = –4
3	–2 – (3) = –5

15.

x	$y = 2x - x^2$
–3	$2(-3) - (-3)^2 = -15$
–2	$2(-2) - (-2)^2 = -8$
–1	$2(-1) - (-1)^2 = -3$
0	$2(0) - (0)^2 = 0$
1	$2(1) - (1)^2 = 1$
2	$2(2) - (2)^2 = 0$
3	$2(3) - (3)^2 = -3$

17.

x	$y = x^2 + 2$
–3	$(-3)^2 + 2 = 11$
–2	$(-2)^2 + 2 = 6$
–1	$(1)^2 + 2 = 3$
0	$(0)^2 + 2 = 2$
1	$(1)^2 + 2 = 3$
2	$(2)^2 + 2 = 6$
3	$(3)^2 + 2 = 11$

19.

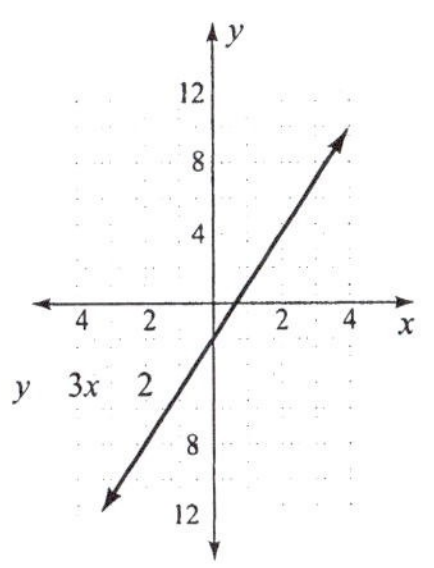

21.

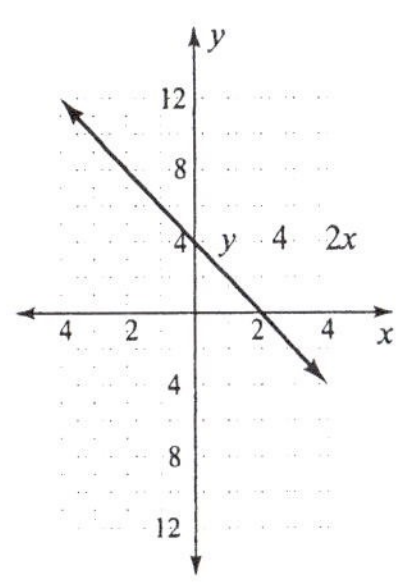

23.

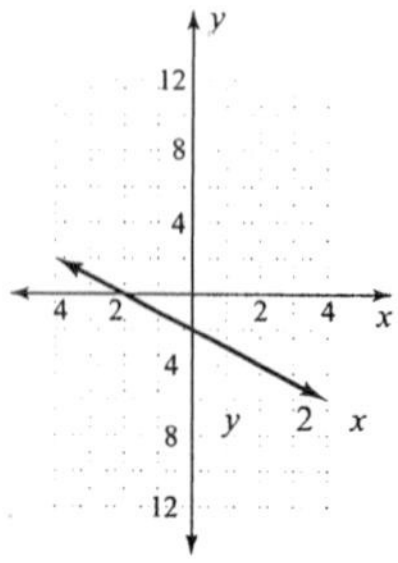

25.

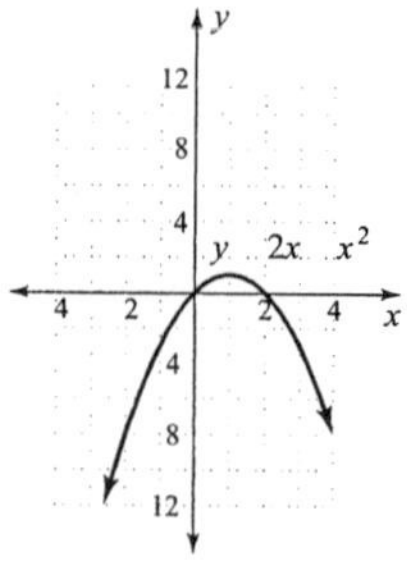

27.

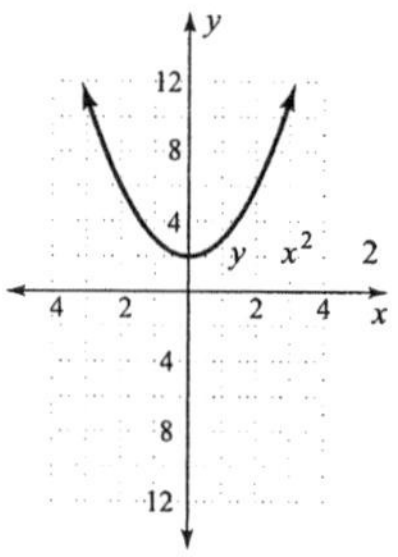

29. a. 19, 21 and 23 are straight lines **b.** 25 and 27 are parabolas

c. y = ax + b makes a straight line

31.

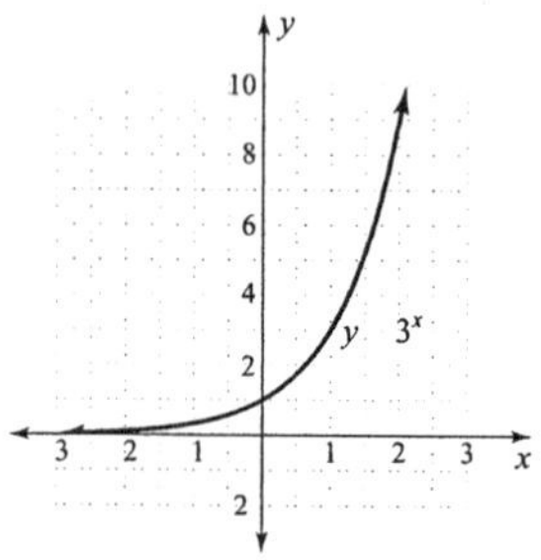

x	$y = 3^x$
−3	$3^{(-3)} = 0.037$
−2	$3^{(-2)} = 0.111$
−1	$3^{(-1)} = 0.333$
0	$3^{(-0)} = 1$
1	$3^{(1)} = 3$
2	$3^{(2)} = 9$
3	$3^{(3)} = 27$

33.

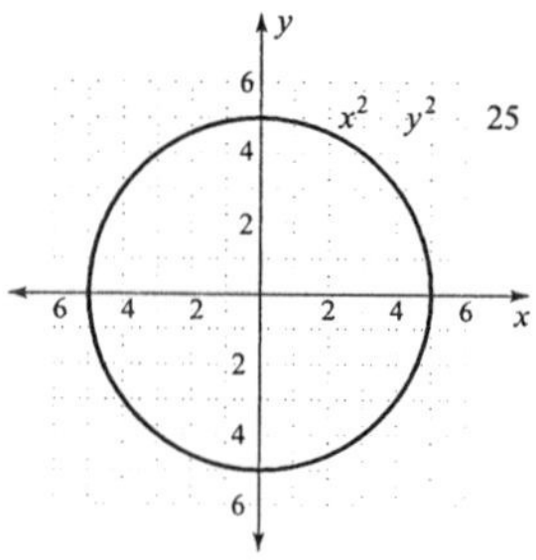

x	y	$x^2 + y^2 = 25$
±5	0	$(\pm5)^2 + (0)^2 = 25$
0	±5	$(0)^2 + (\pm5)^2 = 25$
±4	±3	$(\pm4)^2 + (\pm3)^2 = 16 + 9 = 25$
±3	±4	$(\pm3)^2 + (\pm4)^2 = 9 + 16 = 25$

35.

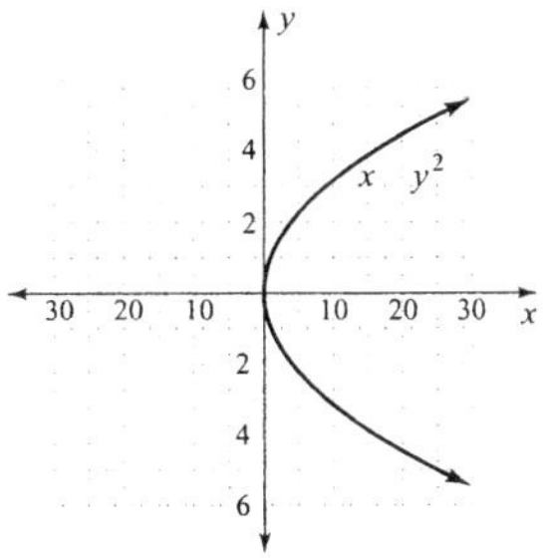

y	$x = y^2$
±2	$4 = (\pm 2)^2$
±3	$9 = (\pm 3)^2$
±4	$16 = (\pm 4)^2$
±5	$25 = (\pm 5)^2$

37.

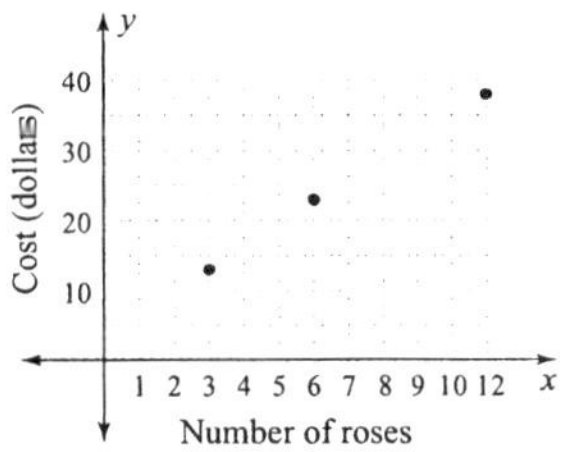

39.

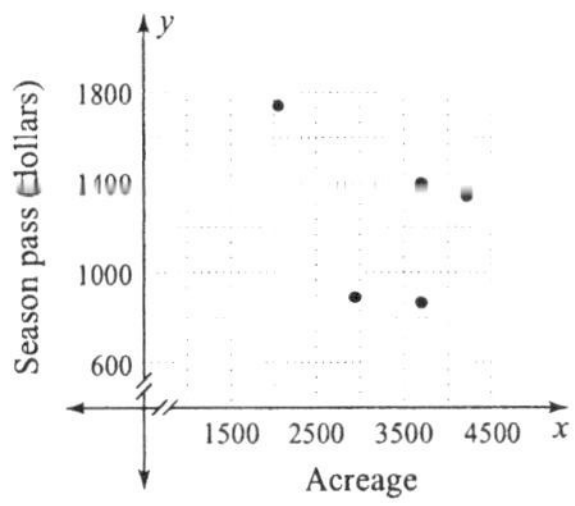

Exercises 1.5

1.

x	y = 3x –4
0	3(0) – 4 = – 4
1	3(1) – 4 = –1
2	3(2) – 4 = 2
3	3(3) – 4 = 5

x = 1 when y = –1

x = 3 when y = 5

3.

x	y = 4 – 3x
–4	4 – 3(–4) = 16
–2	4 – 3(–2) = 10
0	4 – 3(0) = 4
2	4 – 3(2) = –2
4	4 – 3(4) = –8
6	4 – 3(6) = –14

x = 6 when y = –14

x = –4 when y = 16

5.

x	y = 1 – 0.25x
a. –4	1 – 0.25(–4) = 2
0	1 – 0.25(0) = 1
b. 1	1 – 0.25(1) = 0.75
2	1 – 0.25(2) = 0.5

For exercises 7 thru 19 use calculator graphs and tables to solve for x. Answers are given for reference only.

7. a. x = 1.5 **b.** x = 3.5

9. a. x = –2 **b.** x = –3

11. a. x = 0 or x = –1 **b.** x = –3 or x = 2
c. no real number solution **d.** x = –0.5

13. a. x = –3 or x = 3 **b.** no real number solution
c. x = –3 or x = 3 **d.** x = 0

15. a. x = –1 or x = 0 **b.** x = –2 or x = 1
c. no real number solution **d.** x = –3 or x = 2

17. a. $\{0, \frac{1}{2}\}$ **b.** {1.3, –0.8}
c. {–1.5, 2} **d.** { }

19. a. $\{-1, \frac{1}{3}\}$ **b.** $\{-1\frac{2}{3}, 1\}$
c. $\{-\frac{1}{3}\}$ **d.** $\{-\frac{2}{3}, 0\}$

21.

Weight (lb)	**Index $I = W(704.5) \div H^2$**
100	$(100)(704.5) \div (62)^2 = 18.3$
110	$(110)(704.5) \div (62)^2 = 20.2$
120	$(120)(704.5) \div (62)^2 = 22.0$
130	$(130)(704.5) \div (62)^2 = 23.8$
140	$(140)(704.5) \div (62)^2 = 25.7$

23.

Weight (lb)	**Index $I = W(704.5) \div H^2$**
130	$(130)(704.5) \div (70)^2 = 18.7$
140	$(140)(704.5) \div (70)^2 = 20.1$
150	$(120)(704.5) \div (70)^2 = 21.6$
160	$(160)(704.5) \div (70)^2 = 23.0$
170	$(170)(704.5) \div (70)^2 = 24.4$

25. a. W = 130 lb **b.** W = 110 lb

27. a. W = 160 lb **b.** W = 170 lb

29. Find the output for input 2.

31. Find the x-coordinate for the point of intersection of y = 3x + 4 and y = 7.

33. Solving the equation 3x + 4 = –2 is *finding the independent variable.*

35. Locating y, given the graph and x, is *finding the dependent variable.*

37. Locating x, given the graph and y, is *finding the independent variable.*

39. Guesses will vary, equation is y = 1.07x + \$24.95. When y = \$350, x = \$303.79

41. Guesses will vary, equation is y = (\$37(3) + \$0.25x)0.95. When y = \$215.65, x = 464

43. Guesses will vary, equations are: x = first, x – \$50,000 = second, x – \$30,000 = third.
x + (x – \$50,000) + (x – \$30,000) = \$1,000,000; shares are \$360,000, \$310,000, \$330,000

For exercises 45 & 47 use calculator tables and graphs to find solution sets. Answers are given for reference only.

45. a. {–3, –1, 1} **b.** {–2.3, –2, 1.3}
c. {–3.3, 0, 0.3} **d.** {1, 4}

47. a. {–3, 0} **b.** {–2, 1}
c. {–2.7, –1, 0.7} **d.** {–2.9, –0.7, 0.5}

49.

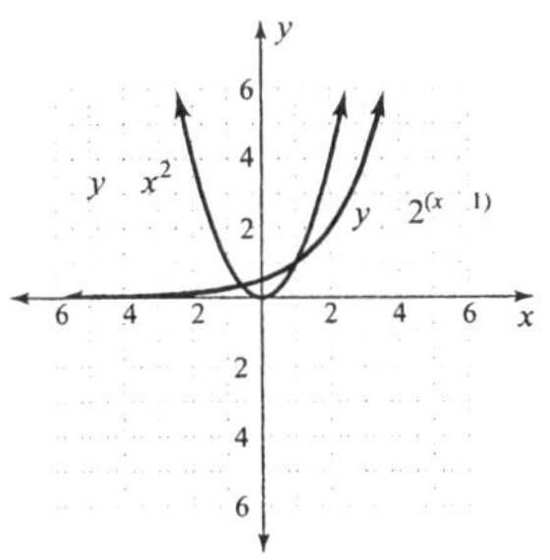

{–0.58, 1, 6.32}

51.

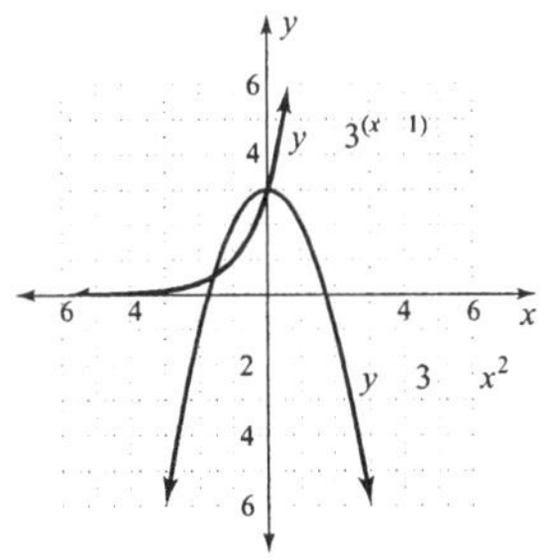

{–1.57,0}

Exercises 1.6

1. **a.** $-4-6=-4+(-6)=-10$ **b.** $-5-(-3)=-5+3=-2$

c. $-\frac{1}{2}-(-\frac{3}{2})=-\frac{1}{2}+\frac{3}{2}=1$ **d.** $100\cdot\frac{1}{4}=100\div4=25$

e. $\frac{5}{8}\div\frac{5}{8}=\frac{5}{8}\cdot\frac{8}{5}=1$ **f.** $-6\div\frac{1}{3}=-6\cdot3=-18$

3. **a.** $-3+(3)=0$ **b.** $5\cdot(\frac{1}{5})=1$ **c.** $-2+(3)=1$

d. $8\cdot(0)=0$ **e.** $\frac{7}{8}\cdot(\frac{8}{7})=1$ **f.** $\frac{3}{8}+(-\frac{3}{8})=0$

5. **a.** $x-8=-4$

$x-8+8=-4+8$

$x=4$

b. $x+3=-6$

$x+3-3=-6-3$

$x=-9$

c. $x-6=-8$

$x-6+6=-8+6$

$x=-2$

d. $\frac{x}{2}=-6$

$2\cdot\frac{x}{2}=-6\cdot2$

$x=-12$

e. $\frac{3x}{4}=21$

$(\frac{4}{3})(\frac{3x}{4})=(\frac{4}{3})(21)$

$x=28$

f. $\frac{3x}{8}=12$

$(\frac{8}{3})(\frac{3x}{8})=(\frac{8}{3})(12)$

$x=32$

g. $-\frac{2}{3}x=24$

$(-\frac{3}{2})(-\frac{2}{3}x)=(-\frac{3}{2})24$

$x=-36$

h. $-\frac{1}{4}x=8$

$(-4)(-\frac{1}{4}x)=(-4)(8)$

$x=-32$

i. $\frac{7}{8}x=35,$

$(\frac{8}{7})(\frac{7}{8}x)=(\frac{8}{7})35$

$x=40$

7. Inverse: take off jacket, take off vest, take off shirt. Dressing and undressing.

9. Inverse is not meaningful. Taking pictures.

11. Order is not important. (unless you want to dig the hole within range of the sprinkler.)

13. a. $2x - 4 = 8$
$2x - 4 + 4 = 8 + 4$
$2x = 12$
$2x \div 2 = 12 \div 2$
$x = 6$

b. $2x - 4 = 6$
$2x - 4 + 4 = 6 + 4$
$2x = 10$
$2x \div 2 = 10 \div 2$
$x = 5$

c. $2x - 4 = -6$
$2x - 4 + 4 = -6 + 4$
$2x = -2$
$2x \div 2 = -2 \div 2$
$x = -1$

d. $2x - 4 = 0$
$2x - 4 + 4 = 0 + 4$
$2x = 4$
$2x \div 2 = 4 \div 2$
$x = 2$

15. a. $\frac{5}{9}(x - 32) = -25$
$\frac{9}{5} \cdot \frac{5}{9}(x - 32) = -25 \cdot \frac{9}{5}$
$x - 32 = -45$
$x - 32 + 32 = -45 + 32$
$x = -13$

b. $\frac{5}{9}(x - 32) = 10$
$\frac{9}{5} \cdot \frac{5}{9}(x - 32) = 10 \cdot \frac{9}{5}$
$x - 32 = 18$
$x - 32 + 32 = 18 + 32$
$x = 50$

c. $\frac{5}{9}(x - 32) = 5$
$\frac{9}{5} \cdot \frac{5}{9}(x - 32) = 5 \cdot \frac{9}{5}$
$x - 32 = 9$
$x - 32 + 32 = 9 + 32$
$x = 41$

d. $\frac{5}{9}(x - 32) = -30$
$\frac{9}{5} \cdot \frac{5}{9}(x - 32) = -30 \cdot \frac{9}{5}$
$x - 32 = -54$
$x - 32 + 32 = -54 + 32$
$x = -22$

17. $3x + 3 = 0$
$3x + 3 - 3 = 0 - 3$
$3x = -3$
$3x \div 3 = -3 \div 3$
$x = -1$

19. $\frac{x}{3} - 1 = 4$
$\frac{x}{3} - 1 + 1 = 4 + 1$
$\frac{x}{3} = 5$
$3 \cdot \frac{x}{3} = 5 \cdot 3$
$x = 15$

21. $\frac{3x}{4} + 5 = 23$
$\frac{3x}{4} + 5 - 5 = 23 - 5$
$\frac{3x}{4} = 18$
$\frac{4}{3} \cdot \frac{3x}{4} = 18 \cdot \frac{4}{3}$
$x = 24$

23. $\frac{3}{8}x - 4 = 8$
$\frac{3}{8}x - 4 + 4 = 8 + 4$
$\frac{3}{8}x = 12$
$\frac{8}{3} \cdot \frac{3}{8}x = 12 \cdot \frac{8}{3}$
$x = 32$

25. a. $4 = 4$ is an Identity

b. $8 = 3$ is a Contradiction

27. a. $x - 2 = 2x - x$ is a Contradiction

b. $x^2 = x \cdot x$ is an Identity

29. a. $x(x^2) = x^3$ is an Identity

b. $\frac{x}{4} = x(\frac{1}{2} - 0.25)$ is an Identity $(\frac{1}{2} - \frac{1}{4} = \frac{1}{4})$

31.
$$x - 2(4 - x) = -17$$
$$x - 8 + 2x = -17$$
$$3x - 8 = -17$$
$$3x - 8 + 8 = -17 + 8$$
$$3x = -9$$
$$3x \div 3 = -9 \div 3$$
$$x = -3$$

33.
$$x - 3(4 + x) = 6$$
$$x - 12 - 3x = 6$$
$$-2x - 12 = 6$$
$$-2x - 12 + 12 = 6 + 12$$
$$-2x = 18$$
$$-2x \div -2 = 18 \div -2$$
$$x = -9$$

35.
$$x + 3(5 - x) = -9$$
$$x + 15 - 3x = -9$$
$$-2x + 15 = -9$$
$$-2x + 15 - 15 = -9 - 15$$
$$-2x = -24$$
$$-2x \div -2 = -24 \div -2$$
$$x = 12$$

37.
$$8 - 3(9 - x) = -14.5$$
$$8 - 27 + 3x = -14.5$$
$$3x - 19 = -14.5$$
$$3x - 19 + 19 = -14.5 + 19$$
$$3x = 4.5$$
$$3x \div 3 = 4.5 \div 3$$
$$x = 1.5$$

39.
$$14 - 9(x + 2) = 45.5$$
$$14 - 9x - 18 = 45.5$$
$$-9x - 4 = 45.5$$
$$-9x - 4 + 4 = 45.5 + 4$$
$$-9x = 49.5$$
$$-9x \div -9 = 49.5 \div -9$$
$$x = -5.5$$

41.
$$37.45 = x + 0.07x$$
$$37.45 = 1.07x$$
$$37.45 \div 1.07 = 1.07x \div 1.07$$
$$35 = x$$

43.
$$46.17 = x + 0.15x + 0.065x$$
$$46.17 = 1.215x$$
$$46.17 \div 1.215 = 1.215x \div 1.215$$
$$38 = x$$

45.
$$2(x + 3) = 1 + 4x$$
$$2x + 6 = 1 + 4x$$
$$2x + 6 - 2x - 1 = 1 + 4x - 2x - 1$$
$$5 = 2x$$
$$5 \div 2 = 2x \div 2$$
$$2.5 = x$$

47. $4 - (x - 2) = 2(x + 6)$

$4 - x + 2 = 2x + 12$

$6 - x = 2x + 12$

$6 - x + x - 12 = 2x + 12 + x - 12$

$-6 = 3x$

$-6 \div 3 = 3x \div 3$

$-2 = x$

49. $2 - 3(x + 1) = x - 3$

$2 - 3x - 3 = x - 3$

$-1 - 3x = x - 3$

$-1 - 3x + 3x + 3 - x - 3 + 3x + 3$

$2 = 4x$

$2 \div 4 = 4x \div 4$

$\frac{1}{2} = x$

51. $D = rt$

$\frac{D}{r} = \frac{rt}{r}$

$t = \frac{D}{r}$

53. $y = mx + b$

$y - mx = mx + b - mx$

$b = y - mx$

55. $V = \frac{4\pi r^3}{3}$

$\frac{3V}{4\pi} = \frac{3}{4\pi} \cdot \frac{4\pi r^3}{3}$

$r^3 = \frac{3V}{4\pi}$

56. $A = 4\pi r^2$

$\frac{A}{4\pi} = \frac{4\pi r^2}{4\pi}$

$r^2 = \frac{A}{4\pi}$

57. $pN = 1$

$\frac{pN}{N} = \frac{1}{N}$

$p = \frac{1}{N}$

59. $A = \frac{1}{2}h(a + b)$

$\frac{2A}{h} = \frac{2}{h}(\frac{1}{2}h(a + b))$

$\frac{2A}{h} - a = a + b - a$

$b = \frac{2A}{h} - a$

61. $a_n = a_1 + (n - 1)d$

$a_n - (n - 1)d = a_1 + (n - 1)d - (n - 1)d$

$a_n - (n - 1)d = a_1$

63. $A = \frac{a + b + c}{3}$

$3A = \frac{3(a + b + c)}{3}$

$3A - b - c = a + b + c - b - c$

$3A - b - c = a$

65.
$$S = \tfrac{1}{2}n(a_1 + a_n)$$
$$\frac{2S}{n} = \frac{2}{n}(\tfrac{1}{2}n(a_1 + a_n))$$
$$\frac{2S}{n} - a_n = a_1 + a_n - a_n$$
$$\frac{2S}{n} - a_n = a_1$$

67.
$$E = \frac{T_h - T_c}{T_h}$$
$$ET_h = \frac{T_h(T_h - T_c)}{T_h}$$
$$ET_h + T_c = T_h - T_c + T_c$$
$$T_c + ET_h - ET_h = T_h - ET_h$$
$$T_c = T_h - ET_h$$

69.
$$V = 344 + 0.6(T - 20)$$
$$V - 344 = 344 + 0.6(T - 20) - 344$$
$$\frac{V - 344}{0.6} = \frac{0.6(T - 20)}{0.6}$$
$$\frac{V - 344}{0.6} + 20 = T - 20 + 20$$
$$\frac{V - 344}{0.6} + 20 = T$$

71.
$$I = \frac{E}{R}$$
$$I \cdot \frac{R}{I} = \frac{E}{R} \cdot \frac{R}{I}$$
$$R = \frac{E}{I}$$

73.
$$T = \frac{0.16V}{S_e}$$
$$T\left(\frac{S_e}{T}\right) = \left(\frac{0.16V}{S_e}\right)\left(\frac{S_e}{T}\right)$$
$$S_e = \frac{0.16V}{T}$$

75.
$$S = \frac{a_1}{r - 1}$$
$$S\left(\frac{r-1}{S}\right) = \left(\frac{a_1}{r-1}\right)\left(\frac{r-1}{S}\right)$$
$$r - 1 = \frac{a_1}{S}$$
$$r - 1 + 1 = \frac{a_1}{S} + 1$$
$$r = \frac{a_1}{S} + 1 \ or \ r = \frac{a_1}{S} + \frac{S}{S} = \frac{a_1 + S}{S}$$

77. $D = \left(\frac{D}{t}\right)t$, D = D

79. $A = \frac{1}{2}\left(\frac{2A}{a+b}\right)(a+b)$, A = A

81. $E = \dfrac{\dfrac{T_c}{1-E} - T_c}{\dfrac{T_c}{1-E}}$ $E = \dfrac{\dfrac{T_c}{1-E} - \dfrac{T_c(1-E)}{1-E}}{\dfrac{T_c}{1-E}}$

$E = \dfrac{\dfrac{T_c - T_c(1-E)}{1-E}}{\dfrac{T_c}{1-E}}$ $E = \dfrac{T_c(1-(1-E)}{1-E} \cdot \dfrac{1-E}{T_c}$, E = E

Review Exercises

1. Associative properties for addition and multiplication, commutative properties for addition and multiplication, distributive property for multiplication over addition.

3. Absolute value, braces, brackets, fraction bar, parentheses, square root.

5. Independent variable, dependent variable.

7. Sum, difference, product, quotient.

9. Input-output relationships, independent variable, dependent variable.

11. Multiplicative inverses, reciprocals.

13.

	Input x	**Input** y	**Output** x + y	**Output** x − y
a.	−7	3	**−7 + 3 = −4**	**−7 − 3 = −10**
b.	3	−7	**3 + (−7) = −4**	**3 − (−7) = 10**
c.	7	**7 − y = 4** **y = 3**	**7 + 3 = 10**	4

15.

	Input x	Input y	Output $x \cdot y$	Output $x + y$
a.	3	4	12	$\mathbf{3+4=7}$
b.	-2	3	$\mathbf{(-2)(3)=-6}$	$\mathbf{-2+3=1}$
c.	-7	$\mathbf{(-7)(y)=21}$ $\mathbf{y=-3}$	21	$\mathbf{-7+(-3)=-10}$
d.	-3	5	$\mathbf{(-3)(5)=-15}$	$\mathbf{-3+5=2}$

Answers on which property was used may vary on exercises 17 - 19. One possible solution is given.

17. $\left(\frac{2}{3}+1\frac{1}{2}\right)+\frac{1}{2}=\frac{2}{3}+\left(1\frac{1}{2}+\frac{1}{2}\right)=\frac{2}{3}+2=2\frac{2}{3}$; associative property for addition

19. $25 \cdot 13 \cdot 4 = 25 \cdot 4 \cdot 13 = 100 \cdot 13 = 1300$; commutative property for multiplication

21. $-2^2+5-3(4-5)=-4+5-3(-1)=-4+5+3=4$

23. $2\pi(1.5\text{ in.})^2+2\pi(1.5\text{ in.})(4\text{ in.})=2\pi(1.5\text{ in.})(1.5\text{ in.}+4.\text{ in.})=2\pi(1.5\text{ in.})(5.5\text{ in.})$
$=16.5\pi\text{ in.}^2\approx 51.84\text{ in.}^2$

25. $\$20.00+(8-1)(-\$1.75)=\$20+7(-\$1.75)=\$20-\$12.25=\$7.75$

27. a. The difference between three and the product of two and a number.
b. The product of three and the difference between a number and 5.

29. $\frac{x}{15}+8$

31. $9-3(2x-5y)-4(5x-3y)=9-6x+15y-20x+12y=-6x-20x+15y+12y+9$
$=-26x+27y+9$

33. $(-2)^2-(-5)^2=4-25=-21$

35. $A=\pi(3.5\text{ in.})^2$
$A=12.25\pi\text{ in.}^2$

37.

x	y = −x
−2	−(−2) = 2
−1	−(−1) = 1
0	−(0) = 0
1	−(1) = −1
2	−(2) = −2

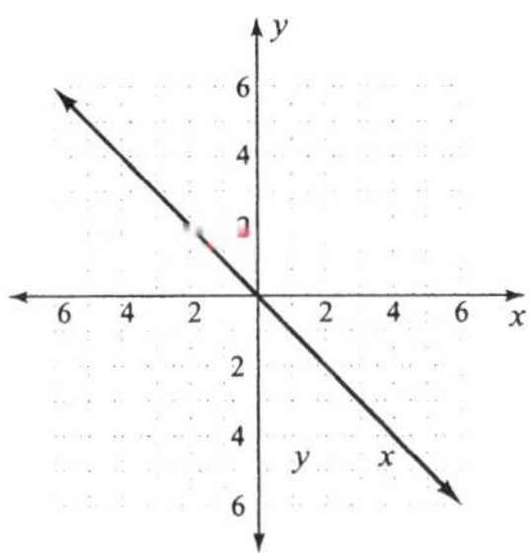

39.

x	y = 30 – 3.5x
−2	30 – 3.5(−2) = 37
−1	30 – 3.5(−1) = 33.5
0	30 – 3.5(0) = 30
1	30 – 3.5(1) = 26.5
2	30 – 3.5(2) = 23

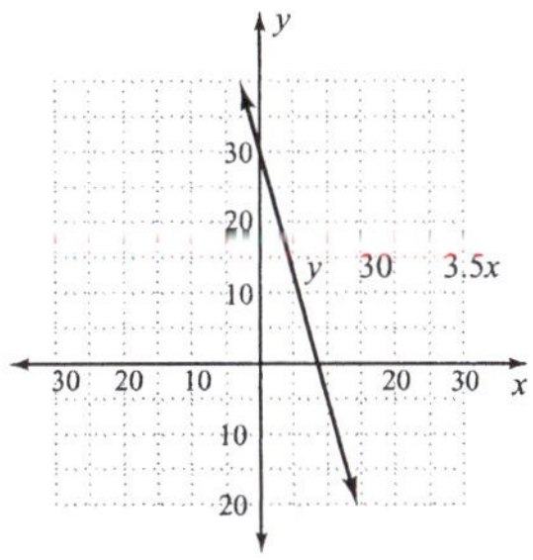

41. $2x + 2(1.5) = 11$

$2x + 3 - 3 = 11 - 3$

$2x \div 2 = 8 \div 2$

$x = 4$

43. $40 = \frac{5}{9}(F - 32)$

$\frac{9}{5}(40) = \frac{9}{5}[\frac{5}{9}(F - 32)]$

$72 + 32 = F - 32 + 32$

$104 = F$

45. $20 + (n - 1)(-1.75) = 0.75$

$20 + (n - 1)(-1.75) - 20 = 0.75 - 20$

$$\frac{(n-1)(-1.75)}{-1.75} = \frac{-19.25}{-1.75}$$

$n - 1 + 1 = 11 + 1$

$n = 12$

47. $I = Prt$

$$\frac{I}{Pt} = \frac{Ptr}{Pt}$$

$$\frac{I}{Pt} = r$$

49. $C = a + bY$

$C - a = a + by - a$

$$\frac{C-a}{y} = \frac{by}{y}$$

$$\frac{C-a}{y} = b$$

51. gcf = 3x; $6x^2 + 15x = 3x(3x) + 3x(5) = 3x(3x + 5)$

53. $3x + 1 = -2x + 1$

$3x + 1 - 1 + 2x = -2x + 1 - 1 + 2x$

$5x = 0$

$x = 0$

55. $3x + 1 = \frac{1}{2}x - \frac{3}{2}$

$3x + 1 - \frac{1}{2}x - 1 = \frac{1}{2}x - \frac{3}{2} - 1$

$2.5x = -2.5$

$x = -1$

57. a.

x	y
1	7
2	4
3	1
4	–2
5	–5

Constant difference on y is –3

When x = 0, y = 10; y = –3x + 10

b.

x	y
1	–20
2	–13
3	–6
4	1
5	8

Constant difference on y is +7

When x = 0, y = –27; y = 7x – 27

59. $x^2 = -2x + 3$ is the intersection(s) of the two graphs; $x = -3$ and $x = 1$

61. E = escrow balance, x = total payments from account, $\frac{2}{12} = 0.167$;

E = 0.167x;

For E = \$367.40, \$367.40 = 0.167x

\$367.40 ÷ 0.167 = 0.167x ÷ 0.167 x = \$2200

63.

Weight	**Index (for height 65 in.)**
100	$100(704.5) \div 65^2 = 16.7$
120	$120(704.5) \div 65^2 = 20.0$
110	$110(704.5) \div 65^2 = 18.3$
115	$115(704.5) \div 65^2 = 19.2$
114	$\mathbf{114(704.5) \div 65^2 = 19.0}$
130	$130(704.5) \div 65^2 = 21.7$
140	$140(704.5) \div 65^2 = 23.3$
145	$145(704.5) \div 65^2 = 24.2$
144	$\mathbf{144(704.5) \div 65^2 = 24.0}$

Chapter Test

1. Another name for additive inverse is *opposite.*

2. An *ordered pair* is used to locate a point on the coordinate graph.

3. An *independent* variable is the input, x, or set of numbers on the horizontal axis.

4.

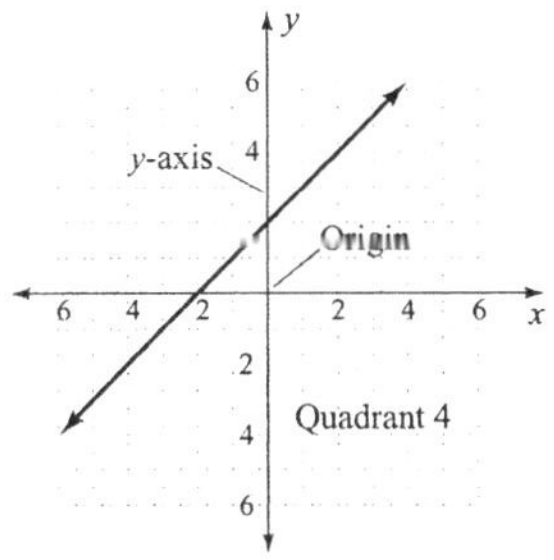

5. a. $-27 + (-3) = -30$ $\quad -27 - (-3) = -27 + 3 = -24$

$(-27)(-3) = 81$ $\quad (-27) \div (-3) = \frac{-27}{-3} = 9$

b. $-1.5 + 0.5 = -1$ $\quad -1.5 - 0.5 = -1.5 + (-0.5) = -2$

$(-1.5)(0.5) = -\frac{3}{2} \cdot \frac{1}{2} = -\frac{3}{4} = -0.75$ $\quad (-1.5) \div (0.5) = \frac{-3}{2} \div \frac{1}{2} = \frac{-3}{2} \cdot \frac{2}{1} = -3$

c. $1\frac{1}{2} + (-\frac{1}{4}) = \frac{3}{2} - \frac{1}{4} = \frac{6}{4} - \frac{1}{4} = \frac{5}{4} = 1\frac{1}{4}$ $\quad 1\frac{1}{2} - (-\frac{1}{4}) = \frac{3}{2} + \frac{1}{4} = \frac{6}{4} + \frac{1}{4} = \frac{7}{4} = 1\frac{3}{4}$

$(1\frac{1}{2})(-\frac{1}{4}) = (\frac{3}{2})(-\frac{1}{4}) = -\frac{3}{8}$ $\quad (1\frac{1}{2}) \div (-\frac{1}{4}) = (\frac{3}{2}) \div (-\frac{1}{4}) = \frac{3}{2} \cdot (-\frac{4}{1}) = \frac{-12}{2} = -6$

6. a. $(348 + 295) + 105 = 348 + (295 + 105) = 348 + 400 = 748$; associative property for addition

b. $230 + 689 + 70 = 230 + 70 + 689 = 300 + 689 = 989$; commutative property for addition

7. a. $4\{2 - 3[4 + 5(2 - 5)] + 5\} = 4\{2 - 3[4 + 5(-3)] + 5\} = 4\{2 - 3[4 + (-15)] + 5\}$

$= 4\{2 - 3[-11] + 5\} = 4\{2 + 33 + 5\} = 4\{40\} = 160$

b. $-3^2 = -(3)(3) = -9$ $\quad$ **c.** $(-3)^2 = (-3)(-3) = 9$

d. $\frac{1}{2} \cdot 25(17 + 23) = \frac{1}{2} \cdot 25(40) = \frac{1}{2} \cdot 1000 = 500$

e. $3x - (8 - 3x) = 3x - 8 + 3x = 6x - 8$

8. The product of four and the difference between a number and three.

9. $A = \dfrac{a+b+c}{3}$

$3A = a + b + c$

$3A - a - c = b$

10. $A = \dfrac{h}{2}(a+b)$

$\dfrac{2A}{h} = a+b$

$\dfrac{2A}{h} - a = b$

11. $15x^2y - 39xy^2 = 3 \cdot 5 \cdot x \cdot x \cdot y - 3 \cdot 13 \cdot x \cdot y \cdot y = 3xy(5x) - 3xy(13y)$; gcf $= 3xy$

12. a. graph crosses $y = 0$ (x-axis) at $x = -5$ **b.** graph intersects $y = -2$ at $x = 2$

c. graphs intersect at $x = -2$

13. a. $\frac{1}{3}x + \frac{5}{3} = 0$

$3(\frac{1}{3}x + \frac{5}{3}) = 3(0)$

$x + 5 = 0$

$x + 5 - 5 = 0 - 5$

$x = -5$

b. $-\frac{3}{4}x - \frac{1}{2} = -2$

$-4(-\frac{3}{4}x - \frac{1}{2}) = -4(-2)$

$3x + 2 = 8$

$3x + 2 - 2 = 8 - 2$

$3x = 6$

$\dfrac{3x}{3} = \dfrac{6}{3}$

$x = 2$

c. $\frac{1}{3}x + \frac{5}{3} = -\frac{3}{4}x - \frac{1}{2}$

$12(\frac{1}{3}x + \frac{5}{3}) = 12(-\frac{3}{4}x - \frac{1}{2})$

$4x + 20 = -9x - 6$

$4x + 20 - 20 + 9x = -9x - 6 - 20 + 9x$

$13x = -26$

$\dfrac{13x}{13} = -\dfrac{26}{13}$

$x = -2$

14. Zero is not a positive integer or a natural number.

15. $A = \dfrac{1.1}{2}(1.4 + 2.6)\ \text{m}^2$

$A = 0.55(4.0)\ \text{m}^2$

$A = 2.2\ \text{m}^2$

16.

x	y
3	1
6	2
7	**$2\frac{1}{3}$**
8	**$2\frac{2}{3}$**
9	3

Extending the table gives
$x = 7$ when $y = 2\frac{1}{3}$

17.

x	y
1	−1
2	−5
3	−9
4	−13
5	−17

Difference in y is a constant −4.
Extending the table backwards when $x = 0$, $y = 3$; equation is $y = -4x + 3$

18. 1 year = 12 months, multiply the decimal by 12 to obtain 8 months.

19. C = cost, x = months; C = \$2875x + \$800x; C = \$3675x;
\$250,000 = \$3675x
x = \$250,000 ÷ \$3675; x = 68 months or 5 years 8 months

20. a. $5 - \dfrac{\sqrt{4}}{2} - 3 = 5 - 1 - 3 = 5 - 4 = 1$ **b.** $5 - \dfrac{\sqrt{4}}{2-3} = 5 - \dfrac{2}{-1} = 5 + 2 = 7$

c. In b, the denominator set of parentheses is incomplete so the calculator assumes a parenthesis at the end of the expression, dividing $\sqrt{4}$ by $2 - 3 = -1$ instead of by 2.

21. Formula is:
$45.59(3) + 0.149x + 0.0725(45.59(3) + 0.149x) + 0.97(3) \approx 149.60 + 0.16x$

x	50	60	**59**
149.60 + 0.16x	149.60 + 0.16(50) = 157.60	149.60 + 0.16(60) = 159.20	149.60 + 0.16(59) = **159.04**

Exercises 2.1

1. a. $\frac{1}{2}+\frac{1}{2}=1;\ \frac{1}{2}\cdot\frac{1}{2}=\frac{1}{4};\ 1>\frac{1}{4}$

b. $\frac{1}{2}\div\frac{1}{2}=\frac{1}{2}\cdot\frac{2}{1}=1;\ \frac{1}{2}+\frac{1}{2}=1;\ 1=1$

c. $\frac{1}{4}+\frac{1}{4}=\frac{1}{2};\ \frac{1}{2}\cdot\frac{1}{2}=\frac{1}{4};\ \frac{1}{2}>\frac{1}{4}$

d. $\frac{1}{2}-\frac{1}{4}=\frac{1}{4};\ \frac{7}{8}-\frac{3}{8}=\frac{1}{2};\ \frac{1}{4}<\frac{1}{2}$

e. $0.3\cdot 0.5=0.15;\ 0.4\cdot 0.4=0.16;\ 0.15<0.16$

f. $1.5-3.5=-2.0;\ -2.5-(0.5)=-3.0;\ -2.0>-3.0$

	Inequality	Line Graph	Inequality in Words
3.	$x \le 2$	graph given in text	x is less than or equal to 2
5.	**$-1 < x < 5$**	graph given in text	**x is between −1 and 5**
7.	$x \ge -1$	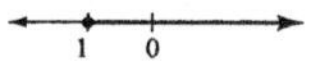	**x is greater than or equal to −1**
9.	**$-2 < x < 4$**	graph given in text	**x is between −2 and 4**
11.	**$x < 2$ or $x > 4$**	graph given in text	**x is less than 2 or x is greater than 4**
13.	**$x < -3$ or $x > 2$**	graph given in text	**x is less than −3 or x is greater than 2**
15.	**$x \le -4$ or $x \ge 1$**	graph given in text	**x is less than or equal to −4 or x is greater than or equal to 1**

17. $-3 < x < 4$ is $x > -3$ and $x < 4$

19. $x > 4$ or $x < 1$ is not appropriate for a compound inequality, (note "or")

21. $x > -1$ and $x \le 5$ is $-1 < x \le 5$

23. $-1 \le x < 1$ is $x \ge -1$ and $x < 1$

25. $3 < x$ and $4 > x$ is $3 < x < 4$

27. a. Xmin = −25, Xmax = 15 is $-25 \le x \le 15$ or x on the interval [−25, 15]

b. Ymin = −10, Ymax = 20 is $-10 \le y \le 20$ or y on the interval [−10, 20]

	Inequality	**Interval**	**Words**	**Line Graph**
29.	**$-3 < x < 5$**	**(–3, 5)**	Set of numbers greater than –3 and less than 5	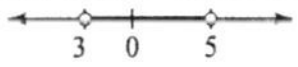
31.	**$-4 < x \le 2$**	**(–4, 2]**	Set of numbers greater than –4 and less than or equal to 2	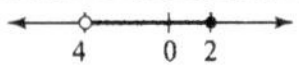
33.	**$x > 5$**	**(5, ∞)**	**Set of numbers greater than 5**	given in text
35.	$x < -2$	**(–∞, –2)**	**Set of numbers less than –2**	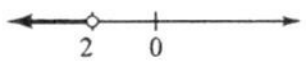
37.	**$x \le -3$**	(–∞, –3)	**Set of numbers less than or equal to –3**	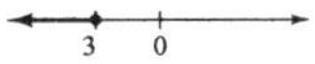
39.	**$x \ge 4$**	[4, ∞)	**Set of numbers greater than or equal to 4**	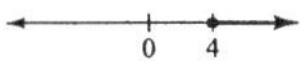

41. $y = \$4$ for $0 < x \le 2$; $y = \$4 + \$0.50(x - 2)$ for $x > 2$

43. $y = \$20$ for $0 < x \le 3$; $y = \$20 + \$5(x - 3)$ for $x > 3$, x rounded up to the next integer

45. $y = \$65$ for $0 < x \le 100$; $y = \$65 + \$0.15(x - 100)$ for $x > 100$

47. $y = \$85$ for $0 < x \le 10$; $y = \$85 + \$4.75(x - 10)$ for $x > 10$

49. $y = \$19.95$ for $0 \le x \le 100$; $y = \$19.95 + \$0.25(x - 100)$ for $x > 100$, x rounded up to next integer

51. a. is less than $<$

b. A statement that one quantity is greater than or less than another quantity is an ***inequality***.

c. is greater than or equal to $\ge$

d. Either *a* is true or *b* is true or both are true is the ***logic phrase "a or b"***.

Exercises 2.2

1. a.

x	**y = 15x – 4**
1	15(1) – 4 = 11
2	15(2) – 4 = 26
3	15(3) – 4 = 41
4	15(4) – 4 = 56

b. function; $f(x) = 15x - 4$

3. a.

y	**$x = y^2 + 2$**
–4	$(-4)^2 + 2 = 18$
–2	$(-2)^2 + 2 = 6$
0	$(0)^2 + 2 = 2$
2	$(2)^2 + 2 = 6$
4	$(4)^2 + 2 = 18$

b. one input has two outputs, not a function

5. a.

x	$y = 25 - x^2$
−5	$25 - (-5)^2 = 0$
−2	$25 - (-2)^2 = 21$
0	$25 - (0)^2 = 25$
2	$25 - (2)^2 = 21$
5	$25 - (5)^2 = 0$

b. function, $f(x) = 25 - x^2$

7. each input has one output, function

9. multiple outputs for one input, not a function

11. multiple outputs for one input, not a function

13. fails vertical-line test, all $x < 2$ have two outputs, not a function

15. each input has one output, function

17. fails vertical-line test, each input has two outputs, not a function

19. outputs are 2 more than corresponding inputs, $y = x + 2$

21. outputs are one third corresponding inputs, $y = \frac{1}{3}x$

23. a. √(100 – 36) or √(100 – 36 = 8
correct answer is **b**

b. √(100) – 36 = – 26

25. a. $(2 + 3)^2 = 25$
correct answer is **b**

b. $2 + (3^2)$ or $2 + 3^2 = 11$

27. a. abs(3) – 4 = –1
correct answer is **b**

b. abs(3 – 4) = 1

29.

x	$g(x) = 5 + 2(x - 3)$
–2	$5 + 2[(-2) - 3] = -5$
–1	$5 + 2[(-1) - 3] = -3$
0	$5 + 2[(0) - 3] = -1$
1	$5 + 2[(1) - 3] = 1$
2	$5 + 2[(2) - 3] = 3$
3	$5 + 2[(3) - 3] = 5$
4	$5 + 2[(4) - 3] = 7$

31.

x	$f(x) = x^2 - 2x - 3$
–2	$(-2)^2 - 2(-2) - 3 = 5$
–1	$(-1)^2 - 2(-1) - 3 = 0$
0	$(0)^2 - 2(0) - 3 = -3$
1	$(1)^2 - 2(1) - 3 = -4$
2	$(2)^2 - 2(2) - 3 = -3$
3	$(3)^2 - 2(3) - 3 = 0$
4	$(4)^2 - 2(4) - 3 = 5$

33.

x	$g(x) = 8 - x^2$
–2	$8 - (-2)^2 = 4$
–1	$8 - (-1)^2 = 7$
0	$8 - (0)^2 = 8$
1	$8 - (1)^2 = 7$
2	$8 - (2)^2 = 4$
3	$8 - (9)^2 = -1$
4	$8 - (4)^2 = -8$

35. a. $f(1) = 3 + 2[(1) - 1]$
$f(1) = 3 + 2(0)$
$f(1) = 3$

b. $f(5) = 3 + 2[(5) - 1]$
$f(5) = 3 + 2(4)$
$f(5) = 11$

c. $f(n) = 3 + 2[(n) - 1]$
$f(n) = 3 + 2n - 2$
$f(n) = 2n + 1$

d. $f(n + m) = 3 + 2[(n + m) - 1]$
$f(n + m) = 3 + 2n + 2m - 2$
$f(n + m) = 2n + 2m + 1$

37. a. $g(3) = (3)^2 + (3) - 2$
$g(3) = 9 + (3) - 2$
$g(3) = 10$

b. $g(1) = (1)^2 + (1) - 2$
$g(1) = 1 + (1) - 2$
$g(1) = 0$

c. $g(\square) = (\square)^2 + (\square) - 2$

d. $g(n) = (n)^2 + (n) - 2$
$g(n) = n^2 + n - 2$

e. $g(n - m) = (n - m)^2 + (n - m) - 2$
$g(n - m) = n^2 - 2nm + m^2 + n - m - 2$

39. $-5 \le x \le 5$ and $0 \le y \le 5$ is **#14**

41. $x = \Re$ and $y \ge 0$ is **#15**

43. x^2 is always positive, domain is $x = \Re$, range is $y \ge 0$

45. When $x = 0$, $y = -2$, lowest point on graph; domain is $x = \Re$, range is $y \geq -2$

47. When $x = 0$, $y = 6$, highest point on graph; domain is $x = \Re$, range is $y \leq 6$

49. a. domain **b.** negative numbers plus zero **c.** $(-\infty, 0]$

51. a. range **b.** positive numbers **c.** $(0, +\infty)$

53. a. domain **b.** positive numbers **c.** $(0, +\infty)$

55. a. range **b.** negative numbers plus zero **c.** $(-\infty, 0]$

57.

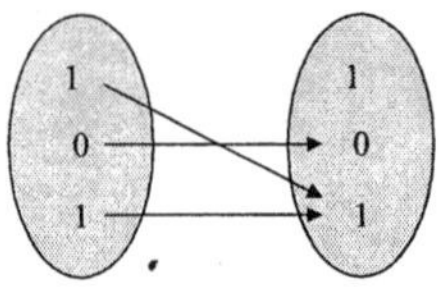

Function, each input has one output.

59.

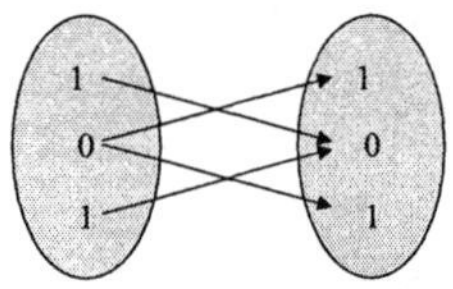

Not a function, one input has two outputs.

61. r is length, domain should be $r > 0$.

63. x is length, domain should be $x > 0$

65. input is hours, answers could vary but should be in the interval [0, 24]

67. 10 digits

69. 9 digits

71. In Exercises 1 and 18 the inputs are CD's and dress sizes. Fractional and decimal values are meaningless so the domain is positive integers (or natural numbers). In exercises 14 and 15 inputs may be fractional or decimal values so a domain of the real numbers is appropriate.

73. substitute $(a + b)$ for the input variable; $g(a + b)$

75. a.

$25 + 4(x - 10) = -15$

$25 + 4x - 40 = -15$

$4x - 15 + 15 = -15 + 15$

$4x = 0$

$x = 0$

b.

$25 + 4(x - 10) = -23$

$25 + 4x - 40 = -23$

$4x - 15 + 15 = -23 + 15$

$4x \div 4 = -8 \div 4$

$x = -2$

77. a. From the table $f(-1)$ and $f(3)$ both equal 0. The solution set is $\{-1, 3\}$.

b. $f(x) = 21$ does not appear on the table, we extend to find $f(6) = (6)^2 - 2(6) - 3 = 21$. Noting the symmetry we check $f(-4)$; $(-4)^2 - 2(-4) - 3 = 21$. The solution set is $\{-4, 6\}$.

79. a. From the table $g(-2)$ and $g(2)$ both equal 4. The solution set is $\{-2, 2\}$.

b. From the table $g(-1)$ and $g(1)$ both equal 7. The solution set is $\{-1, 1\}$.

Exercises 2.3

1.
$y = 3x - 8$
$y - y + 8 = 3x - 8 + 8 - y$
$8 = 3x - y$
$3x - y = 8$
linear in two variables

3.
$C = x + 0.15x + 0.06x$
$C - 1.21x = 1.21x - 1.21x$
$C - 1.21x = 0$
linear in two variables

5.
$x^2 = 0$
non-linear

7.
$2\pi r = 11$
$2\pi r - 11 = 11 - 11$
$2\pi r - 11 = 0$
linear in one variable

9.
$t - 6n = 4n - 2$
$t - 6n + 6n + 2 = 4n - 2 + 2 + 6n$
$t + 2 - t = 10n - t$
$2 = 10n - t$
$10n - t = 2$
linear in two variables

11.
$\frac{5}{9}(x - 32) = \frac{5}{9}x - \frac{160}{9}$
linear function

13. $2\pi x$; linear function

15. $x^2 + 2x$; non-linear

17. $gt + v_0$; linear function

19. $ax^2 + bx + c$; non-linear

21. a. x-intercept $(-1, 0)$; y-intercept $(0, 2)$ **b.** x-intercept $(3, 0)$; y-intercept $(0, -2)$

23. x-intercepts $(-1, 0)$ and $(1, 0)$; y-intercept $(0, -1)$

25. The function shown in **22b** decreases less rapidly.

27. The function shown in **21a** increases more rapidly.

29. The function shown in **24** increases for $x < 0$ and decreases for $x > 0$.

31. a. Using the intercepts (–2, 0) and (0, 2); slope is $\dfrac{2-0}{0-(-2)} = \dfrac{2}{2} = 1$

b. Using the intercepts (3, 0) and (0, –2); slope is $\dfrac{-2-0}{0-3} = \dfrac{2}{3}$

33. $\dfrac{7-5}{-2-4} = \dfrac{2}{-6} = -\dfrac{1}{3}$

35. $\dfrac{-1-(-5)}{5-(-3)} = \dfrac{4}{8} = \dfrac{1}{2}$

37. $\dfrac{-3-0}{0-5} = \dfrac{3}{5}$

39. $\dfrac{-3-0}{0-(-2)} = -\dfrac{3}{2}$

41. $\dfrac{-4.2-5.3}{2.6-1.5} = \dfrac{-9.5}{1.1} = -\dfrac{95}{11}$

43. $\dfrac{-1\frac{1}{4}-4\frac{1}{2}}{2\frac{1}{2}-(-1\frac{3}{4})} = -\dfrac{5.75}{4.25} = -\dfrac{575}{425} = -\dfrac{23}{17}$

45. $\dfrac{2\frac{1}{3}-3\frac{2}{3}}{1\frac{2}{3}-(1\frac{1}{3})} = \dfrac{-\frac{4}{3}}{3} = -\dfrac{4}{3}\cdot\dfrac{1}{3} = -\dfrac{4}{9}$

47. a. Δx = \$10 sales, Δy = \$0.60 tax; slope = $\dfrac{\$0.60}{\$10}$ = \$0.06 tax/\$ sales

b. Working backward in the table; x-intercept = (0, 0), \$0 sales means \$0 tax.

c. y-intercept is also (0, 0)

49. a. Δx = 1 trip, Δy = -\$0.75 value; slope = -\$0.75 value/trip

b. Working forward on the table; x-intercept = ($26\frac{2}{3}$, 0), maximum number of trips is 26.

c. y-intercept is in the table (0, 20); original value of mass transit ticket

51. Δx = 0.5 sec, Δy is not constant; function is not linear

53.

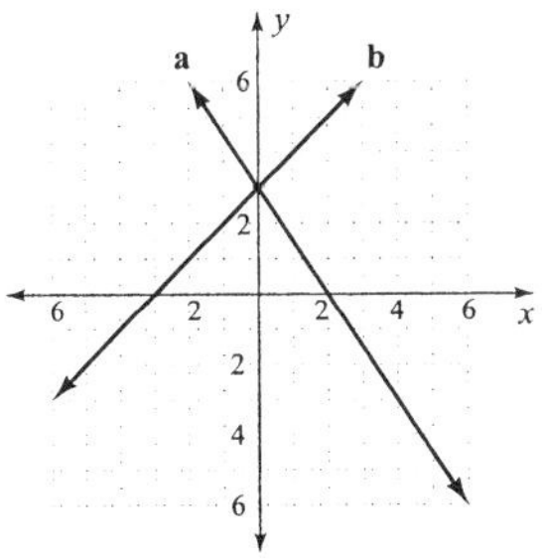

55.

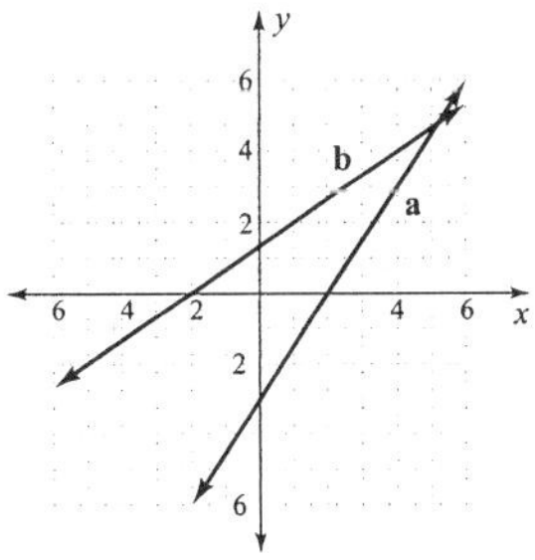

57. x-intercept $(a, 0)$, y-intercept $(0, b)$; slope is $\dfrac{b-0}{0-a} = -\dfrac{b}{a}$

59. From (4, 5) move –2 units in y and 3 units in x; $(4 + 3, 5 - 2) = (7, 3)$

61. From (–4, 1) move 3 units in y and 5 units in x; $(-4 + 5, 1 + 3) = (1, 4)$

63. a.

x	y= 10 – 0.05x
0	10 – 0.05(0) = 10
100	10 – 0.05(100) = 5
200	10 – 0.05(200) = 0

$\text{Slope} = -\dfrac{10}{200} = -0.05$

b.

x	y= 10 – 0.15x
0	10 – 0.15(0) = 10
20	10 – 0.15(20) = 7
40	10 – 0.15(40) = 4

$\text{Slope} = \dfrac{10-7}{0-20} = -\dfrac{3}{20} = -0.15$

Mid-Chapter 2 Test

1. a.

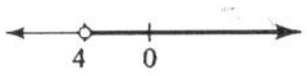

$(-4, \infty)$

b.

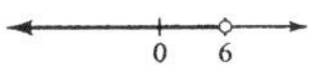

$(-\infty, 6)$

c.

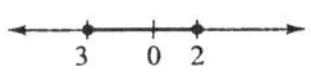

$[-3, 2]$

d.

$(3, 6]$

e.

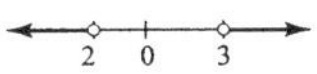

$(-\infty, -2), (3, \infty)$

f.

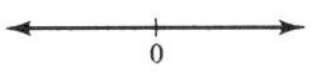

$\Re$ or $(-\infty, \infty)$

2. a. $-1 \le x \le 1$; $[-1, 1]$ **b.** $x \ge -3$; $[-3, \infty)$
c. $y \ge -2$; $[-2, \infty)$ **d.** $y > -2$ or $y < 4$; $\Re$; $(-\infty, \infty)$
e. $-2 < y \le 4$; $(-2, 4]$

3. a. The set of numbers greater than or equal to –1 and less than 3.
b. The set of inputs between –4 and –1.
c. The set of numbers less than or equal to –2.
d. The set of outputs less than or equal to –1.

4. $y = 16.45$ for $0 < x \le 30$; $y = 16.45 + 0.29(x - 30)$ for $x > 30$

5. The set of numbers $x \ge 0$ is called ***non-negative***.

6. The set of inputs in a function is called the ***domain***.

7. The ordered pair describing the intersection of a graph and the vertical axis is written ***(0, y)***.

8. a. $f(1) = 3(1) - 5$
$f(1) = 3 - 5$
$f(1) = -2$

b. $f(3) = 3(3) - 5$
$f(3) = 9 - 5$
$f(3) = 4$

c. $f(-5) = 3(-5) - 5$
$f(-5) = -15 - 5$
$f(-5) = -20$

d. $f(a) = 3(a) - 5$
$f(a) = 3a - 5$

e. $f(a + b) = 3(a + b) - 5$
$f(a + b) = 3a + 3b - 5$

9. a. $f(1) = (1)^2 - (1)$
$f(1) = 1 - 1$
$f(1) = 0$

b. $f(3) = (3)^2 - (3)$
$f(3) = 9 - 3$
$f(3) = 6$

c. $f(-5) = (-5)^2 - (-5)$
$f(-5) = 25 + 5$
$f(-5) = 30$

d. $f(a) = (a)^2 - (a)$
$f(a) = a^2 - a$

e. $f(a + b) = (a + b)^2 - (a + b)$
$f(a + b) = a^2 + 2ab + b^2 - a - b$

10. a. Domain: $\Re$, $-\infty < x < \infty$, $(-\infty, \infty)$ **b.** Range: $y \ge 0$, $[0, \infty)$
c. Graph describes a function.

11. a. Domain: $\Re$, $-\infty < x < \infty$, $(-\infty, \infty)$ **b.** Range: $\Re$, $-\infty < y < \infty$, $(-\infty, \infty)$
c. Graph describes a function.

12. a. Domain: $-5 \le x \le 1$, $[-5, 1]$ **b.** Range: $-3 \le y \le 3$, $[-3, 3]$
c. Graph does not describe a function (fails vertical line test).

13. a. $y = 5$; $a = 0, b = 1, c = 5$; $0x + 1y = 5$ **b.** $x = 4$; $a = 1, b = 0, c = 4$; $x + 0y = 4$

c. $2\pi r = 7$; $a = 2\pi, b = 0, c = 7$; $2\pi r + 0y = 7$

d. equation is not linear

14. a. To find horizontal (x) intercept let $y = 0$; $3x + 4(0) = 12$, $x = 4$, x-intercept = (4, 0)
To find vertical (y) intercept let $x = 0$; $3(0) + 4y = 12$, $y = 3$, y-intercept = (0, 3)

b. For horizontal intercept let $F = 0$; $0 = \frac{9}{5}C + 32$, $-32 = \frac{9}{5}C$, $C = -17.78$, intercept = (–17.78, 0)
For vertical intercept let $C = 0$; $F = \frac{9}{5}(0) + 32$, $F = 32$, intercept = (0, 32)

15. a. Δinput = 1, Δoutput = –3; slope = $-\frac{3}{1}$; vertical axis intercept is $(1 - 1, 2 - (-3)) = (0, 5)$

b. Δinput = 11, 6; Δoutput = 38.5, 21; $\frac{38.5}{11} = 3.5$, $\frac{21}{6} = 3.5$, slope = \$3.50/ft; vertical axis intercept is $(25 - 25, 92.5 - 25(3.50)) = (0, 5)$

16.

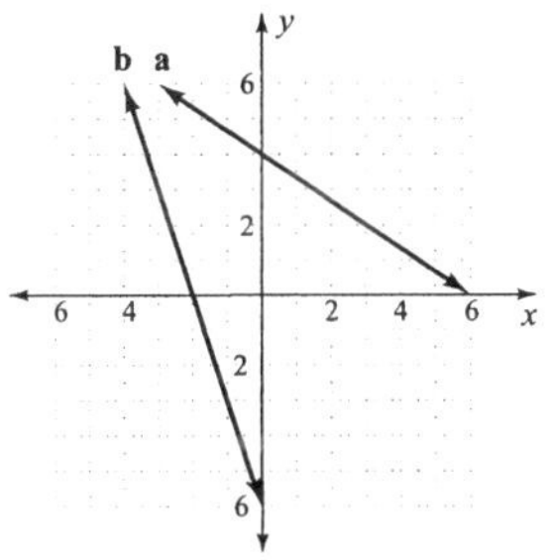

Exercises 2.4

1. $y = 0.055x$; slope = \$0.055/\$; y-intercept = 0

3. $y = 3.00x + 10$; slope = \$3.00/person; y-intercept = \$10

5. $C = 2\pi r$; slope = 2π; vertical axis intercept = 0

7. $F = \mu N$; slope = μ, vertical axis intercept = 0

9. $C = a + bY$; slope = b, vertical axis intercept = a

11. Slope = 8; y-intercept = –4: $y = 8x - 4$

13. Slope = $\frac{1}{2}$; y-intercept = –8: $y = \frac{1}{2}x - 8$

15. Slope = –2; y-intercept = 0: $y = -2x$

17. (3, 6) and (0, –2); slope $= \dfrac{-2-6}{0-3} = \dfrac{8}{3}$; $y = \frac{8}{3}x - 2$

19. (–2, 4) and (5, –3); slope $= \dfrac{-3-4}{5-(-2)} = \dfrac{-7}{7} = -1$; $b = 4 - (-1)(-2)$, $b = 2$

$y = -1x + 2$ or $y = -x + 2$

21. a. Pulse rate is a function of age. **b.** Answers will vary

c. Max. pulse rate is 220 – age. Let x = age and P = pulse rate. $P = 0.5(220 - x)$

d. $P = 0.7(220 - x)$

e. $P = 0.5(220 - 50)$ $P = 0.7(220 - 50)$

$P = 0.5(170)$ $P = 0.7(170)$

$P = 85$ $P = 119$

f. $95 = 0.5(220 - x)$ $133 = 0.7(220 - x)$

$190 = 220 - x$ $190 = 220 - x$

$x = 30$ $x = 30$

23. The fixed cost is the \$300 in fees; the variable cost per dollar is 2.5% or 0.025

Cost function is $C = 0.025x + 300$, C in \$

25. (2, 5) and (5, 4); slope $= \dfrac{4-5}{5-2} = -\dfrac{1}{3}$; $b = 5 - (-\frac{1}{3})(2)$, $b = 5\frac{2}{3}$; $y = -\frac{1}{3}x + 5\frac{2}{3}$

27. (5, 4) and (4, 1); slope $= \dfrac{1-4}{4-5} = \dfrac{-3}{-1} = 3$; $b = 4 - 3(5)$, $b = -11$; $y = 3x - 11$

29. (4, 1) and (2, 5); slope $= \dfrac{5-1}{2-4} = \dfrac{4}{-2} = -2$; $b = 1 - (-2)(4)$, $b = 9$; $y = -2x + 9$

31. slope $= \dfrac{11{,}000 - 13{,}185}{250 - 300} = 43.7$; $b = 11{,}000 - 43.7(250)$, $b = 75$;

$C = 43.70x + 75$, C in \$; fixed cost is \$75; variable cost per pair is \$43.70

33. slope $= \dfrac{212-32}{100-0} = \dfrac{180}{100} = \dfrac{9}{5}$; b = 32; F = $\frac{9}{5}$C + 32

35. Let y – cost in \$ and x – size in inches; points are (4, 0.99) and (6, 1.99);

slope $= \dfrac{1.99-0.99}{6-4} = \dfrac{1}{2}$ = 0.5; b = 0.99 – 0.5(4), b = 1.01; y = 0.5 – 1.01

37. Let y = cost in \$ and x = pounds; points are (7, 4.99) and (16, 7.99);

slope $= \dfrac{7.99-4.99}{16-7} = \dfrac{3}{9} = \dfrac{1}{3}$; b = 4.99 – $\frac{1}{3}$(7), b = 2.66; y = $\frac{1}{3}$x + 2.66

39.

Δx	x (cups)	y (\$)	Δy	slope
	1	2.79		
1			0.50	0.50
	2	3.29		
2			1.50	0.75
	4	4.79		
4			2.00	0.50
	8	6.79		

Average slope ≈ 0.60; working backward in table y-intercept = 2.19;
y = 0.60x + 2.19; LinReg gives: y = 0.578x + 2.247; r = 0.9962

41.

Δx	x (#)	y (\$)	Δy	slope
	5	3.99		
25			11.00	0.44
	30	14.99		
20			5.00	0.25
	50	19.99		

Average slope ≈ 0.35; y-intercept = 3.99 – 0.35(5) = 2.24; y = 0.35x + 2.24;
LinReg gives y = 0.359x + 2.818; r = 0.9889

43.

Δx	x (kg)	y ($)	Δy	slope
	4	4.99		
4			2.00	0.50
	8	6.99		
11.28			3.00	0.267
	19.28	9.99		

Average slope = 0.38; y-intercept = 4.99 – 0.38(4) = 3.47; y = 0.38x + 3.47;
LinReg gives y = 0.314x + 4.051; r = 0.9882

45.

Δx	x	y	Δy
	1	–8	
1			5
	2	–3	
1			5
	3	2	
1			5
	4	7	
1			5
	5	12	

slope = 5; b = –8 – 5(1) = –13
y = 5x – 13

47.

Δx	x	y	Δy
	1	9	
1			2
	2	11	
1			2
	3	13	
1			2
	4	15	
1			2
	5	17	

slope = 2; b = 9 – 2(1) = 7
y = 2x + 7

49.

Δx	x	y	Δy
	1	2	
1			6
	2	8	
1			6
	3	14	
1			6
	4	20	
1			6
	5	26	

slope = 6; b = 2 – 6(1) = –4
y = 6x – 4

Exercise 51 uses calculator tables and LinReg to fit equation - solution is given for reference only.

51. $y = 0.4x - 0.21$; $r \approx 1$

53. \$0.10 is added to the price for each quarter inch increase in diameter. Each ordered pair exactly fits the price increase rule.

Exercises 55 and 57 use calculator tables and LinReg to fit equations - solutions are given for reference only.

55. Match-up is linear, $y = -x + 17$

57.

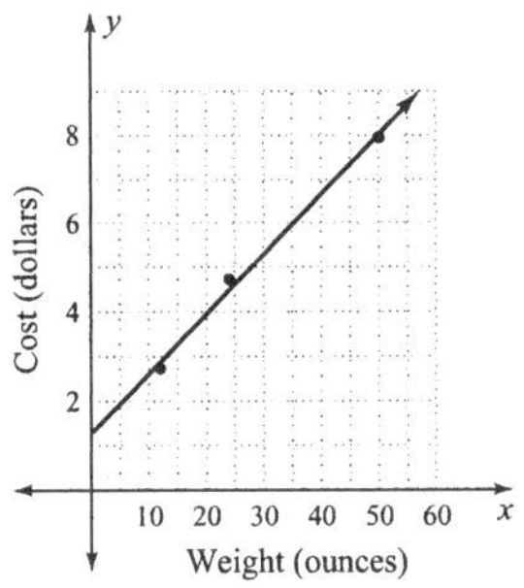

$y \approx 0.135x + 1.29$, x in oz, y in \$

59. Equation will approximate: $y \approx 1051.3x - 32.9$

61. The slope is $m = \dfrac{\Delta y}{\Delta x}$ and the output for $x = 0$ is b in $y = mx + b$.

Exercises 2.5

1. $m = \dfrac{0-(-2)}{4-4} = \dfrac{2}{0}$; undefined; $x = 4$

3. $m = \dfrac{-3-(-3)}{4-(-2)} = \dfrac{0}{6} = 0$; $y = -3$

5. $y = 0x - 3$; $y = -3$

7. $x = -1$

9. $y = 0x + 0$; $y = 0$

11. a. negative reciprocal is $-\frac{4}{3}$; $(\frac{3}{4})(-\frac{4}{3}) = -\frac{12}{12} = -1$

b. $-3\frac{1}{2} = -\frac{7}{2}$, negative reciprocal is $\frac{2}{7}$; $(-\frac{7}{2})(\frac{2}{7}) = -\frac{14}{14} = -1$

c. $4.5 = \frac{9}{2}$, negative reciprocal is $-\frac{2}{9}$; $(\frac{9}{2})(-\frac{2}{9}) = -\frac{18}{18} = -1$

d. $0.10 = \frac{1}{10}$, negative reciprocal is $-\frac{10}{1}$; $(\frac{1}{10})(-\frac{10}{1}) = -\frac{10}{10} = -1$

e. negative reciprocal is $\frac{q}{p}$; $(-\frac{p}{q})(\frac{q}{p}) = -\frac{pq}{pq} = -1$

13. a.
$2x = y - 2(3 - x)$
$2x = y - 6 + 2x$
$2x - 2x + 6 = y - 6 + 2x - 2x + 6$
$6 = y$ or $y = 6$

b.
$y - 4 = 2x + y$
$y - 4 + 4 - y = 2x + y + 4 - y$
$0 = 2x$
$x = 0$

lines are perpendicular; zero slope vs. undefined slope

15. a.
$x = -6y$
$x \div -6 = -6y \div -6$
$-\frac{x}{6} = y$ or $y = -\frac{x}{6}$

b.
$3x = y - 3x - 4$
$3x + 3x + 4 = y - 3x - 4 + 3x + 4$
$6x + 4 = y$ or $y = 6x + 4$

lines are perpendicular, slopes are negative reciprocals

17. a. $x = 4$

b.
$y = x + y - 5$
$y - y = x + y - 5 - y$
$0 = x - 5$ or $x = 5$

lines are parallel, same slope - both undefined

19. a.
$y + 2x = 2$
$y + 2x - 2x = 2 - 2x$
$y = -2x + 2$

b.
$y = 2(x - 1)$
$y = 2x - 2$

lines are neither parallel or perpendicular

21. $C = 78x$, $C = 98x$, $C = 108x$; not parallel, different slopes

23. $V = 5.00 - 0.15x$, $V = 10.00 - 0.15x$, $V = 20.00 - 0.15x$; parallel lines, same slope

25. If postage cost is \$0.37 per stamp: $C = 100(0.37)x = 37.00x$; $C = 50(0.37)x = 18.50$; $C = 20(0.37)x = 7.40x$; not parallel, different slopes

In exercises 27 to 35 change to y = mx + b form (where necessary) to find the slope of the original equation before solving the problem.

27. $2x + 3y = 6$

$3y = -2x + 6$

$y = -\frac{2}{3}x + 2$

parallel line has same slope

y-intercept is (0, 0)

$y = -\frac{2}{3}x$

29. $y = -\frac{1}{2}x + 3$

perpendicular line has negative reciprocal slope

slope = 2

y-intercept is (0, 0)

$y = 2x$

31. $y = \frac{5}{8}x - 3$

perpendicular line has negative reciprocal slope

slope $= -\frac{8}{5}$

$b = 3 - (-\frac{8}{5})(2) = 6\frac{1}{5}$

$y = -\frac{8}{5}x + 6\frac{1}{5}$

33. $4x - 3y = 12$

$4x - 12 = 3y$

$y = \frac{4}{3}x - 4$

parallel lines have same slope

$b = 1 - (\frac{4}{3})(-2) = 3\frac{2}{3}$

$y = \frac{4}{3}x + 3\frac{2}{3}$

35. $5x - 2y = 8$; $5x - 8 = 2y$; $y = \frac{5}{2}x - 4$

perpendicular lines have negative reciprocal slope

slope $= -\frac{2}{5}$; $b = -1 - (-\frac{2}{5})(2) = -\frac{1}{5}$

$y = -\frac{2}{5}x - \frac{1}{5}$

37. Starting at (1, 3) and moving clockwise around the figure:

$$\frac{4-3}{3-1} = \frac{1}{2},\ \frac{2-4}{4-3} = -2,\ \frac{1-2}{2-4} = \frac{1}{2},\ \frac{3-1}{1-2} = -2$$

opposite lines are parallel (same slope) and adjacent lines are perpendicular (negative reciprocal slopes)

39.

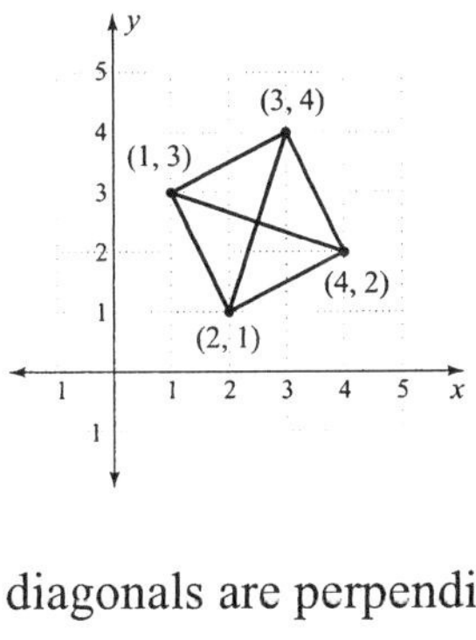

a. $\frac{4-1}{3-2} = 3$ **b.** $\frac{2-3}{4-1} = -\frac{1}{3}$

diagonals are perpendicular (negative reciprocal slopes)

41. Opposite sides should have the same slopes and adjacent sides should have negative reciprocal slopes.

$\frac{5}{2}, -\frac{5}{2}, \frac{3}{2}, -\frac{3}{2}$; not a rectangle;

$\frac{-2-(-1)}{1-(-1)} = -\frac{1}{2}, \frac{2-(-2)}{3-1} = 2, \frac{3-2}{1-3} - \frac{1}{2}, \frac{-1-3}{-1-1} = 2$; rectangle

43. Non-vertical lines are perpendicular if their slopes multiply to -1.

If non-vertical lines are perpendicular then their slopes multiply to -1.

Exercises 2.6

1. & 3.

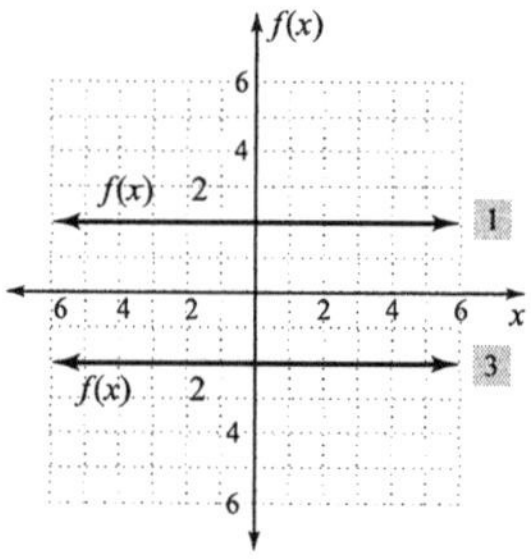

1. Domain = $\Re$, Range y = 2
3. Domain $\Re$, Range y = −2

5. Constant function; output is always \$15

7. Constant function; output is always \$100

9. Constant function; output is always 1

11. Identity function; output equals input

13. 3 = 3 is an identity

15. −a(b − c) = −ab − bc is neither

17. h(n) = n is an identity function

19. a + b + c = a + c + b is an identity

21. y = x + a − a is an identity function

23. a.

f(x)	f(x) = |x – 2|
–3	|–3 – 2| = 5
–1	|–1 – 2| = 3
0	|0 – 2| = 2
1	|1 – 2| =-1
3	|3 – 2| = 1
5	|5 – 2| = 3

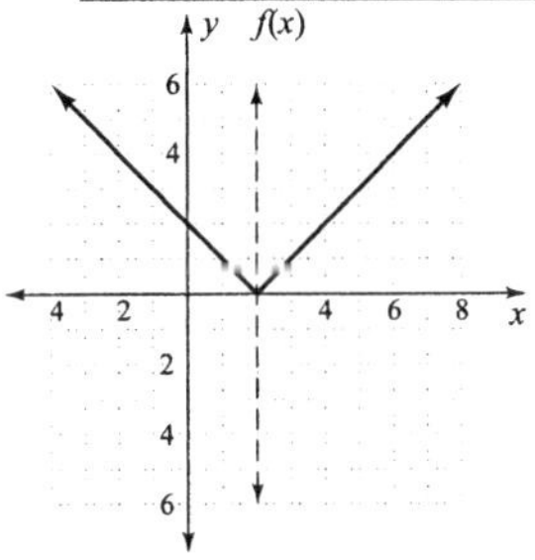

b. $f(0) = f(4) = 2$

c. x = 2

25.

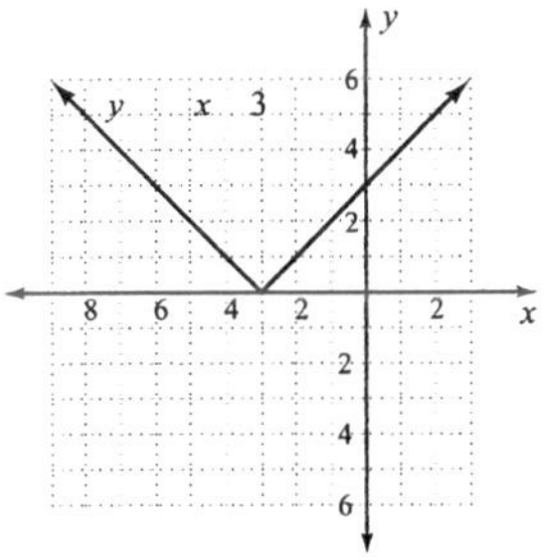

domain $\Re$, range y ≥ 0

27.

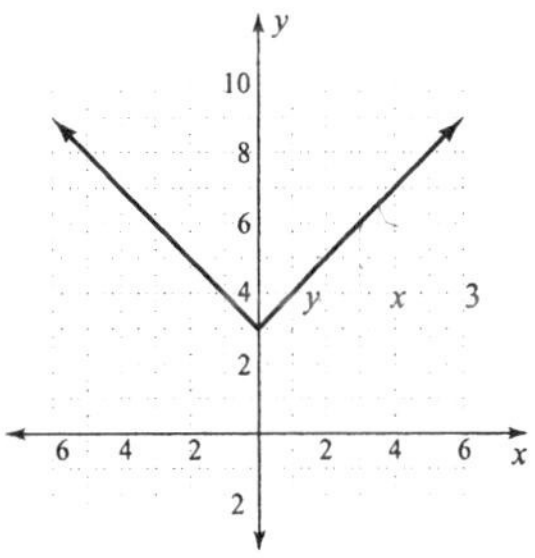

domain $\Re$, range y ≥ 3

29.

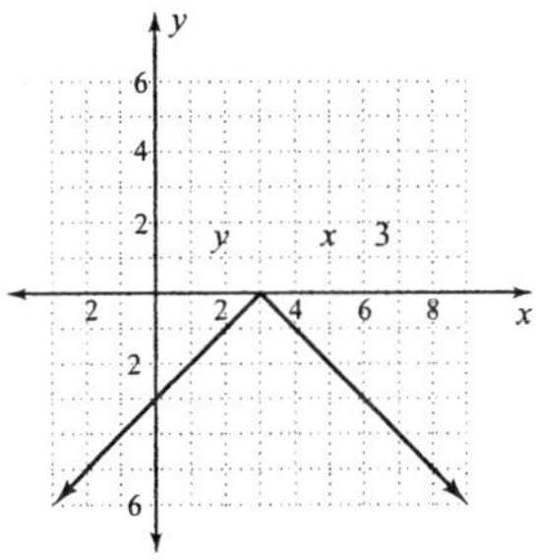

domain $\Re$, range y ≤ 0

31.

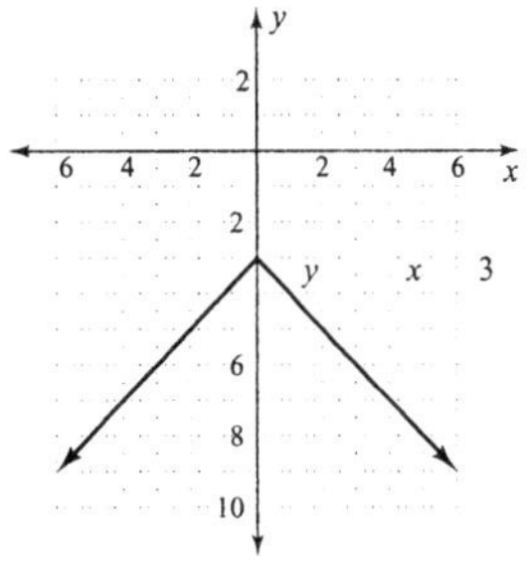

domain $\Re$, range y ≤ –3

33. a. $\{-1, 4\}$
b. $\{0, 3\}$
c. $\{\frac{3}{2}\}$
d. $\{\ \}$ or $\varnothing$

35. a. slope is -2
b. slope is 2
c. y-intercept is 3, input is 0
d. $y = -2x + 3, x \le 1.5$
e. $f(1.5) = -2(1.5) + 3 = 0$
f. $y = 2x - 3, x \ge 1.5$
g. $f(1.5) = 2(1.5) - 3 = 0$

37. a. $\{-10, 2\}$
b. $\{\ \}$ or $\varnothing$
c. $\{-6, -2\}$
d. $\{-12, 4\}$

39. $|x| = 4$
$x = 4$ or $x = -4$
$\{\pm 4\}$

41. $|x + 2| = 3$
$x + 2 = 3$ or $x + 2 = -3$
$x = 1$ or $x = -5$
$\{-5, 1\}$

43. $|x - 5| = 2$
$x - 5 = 2$ or $x - 5 = -2$
$x = 7$ or $x = 3$
$\{3, 7\}$

45. $|x - 4| = 2$
$x - 4 = 2$ or $x - 4 = -2$
$x = 6$ or $x = 2$
$\{2, 6\}$

47. a. 2 $|x + 2|$; abs(x + 2)
b. 3 $|x| + 2$; abs(x) + 2
c. 4, $\dfrac{1}{|x+2|}$; 1 ÷ (abs(x+2))
d. 1, $\dfrac{1}{|x|+2}$; 1 ÷ (abs(x) + 2)

49. a. $|153 - 423| = |-270| = 270$ mi
$D = |x_1 - x_2|$
b. $|230 - 482| = |-252| = 252$

51. a. Dot graph,
partial pages not possible

b.

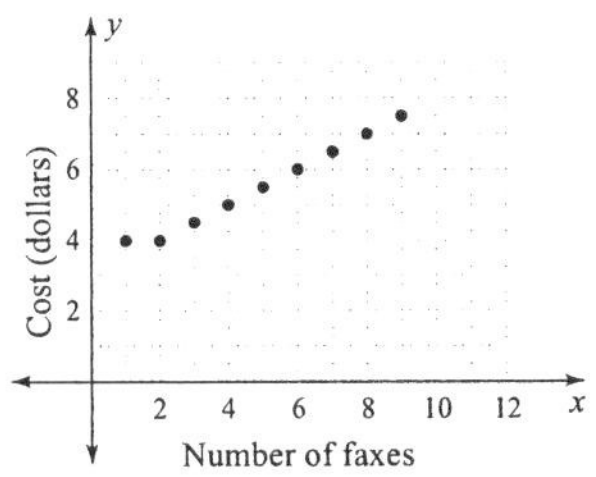

53. a. Step graph, partial hrs appropriate

b.

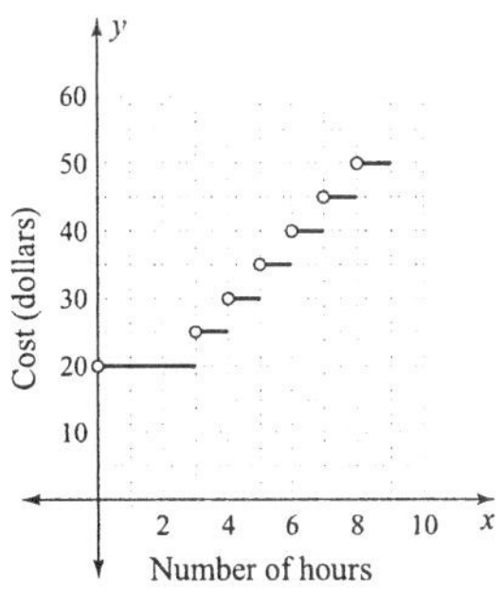

55. a. Dot graph, no partial skaters

b.

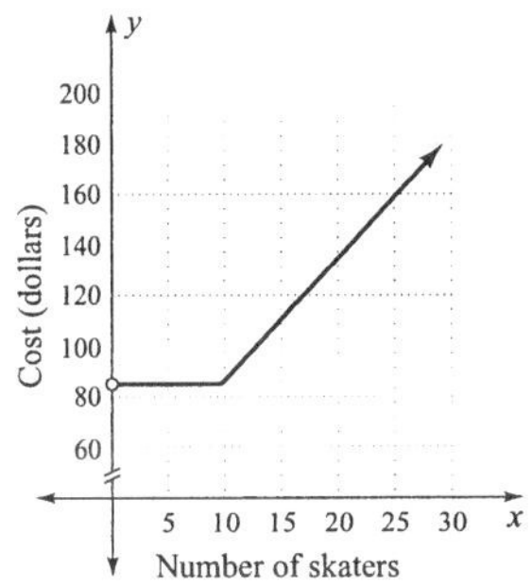

57. a. Step graph, partial min. appropriate

b.

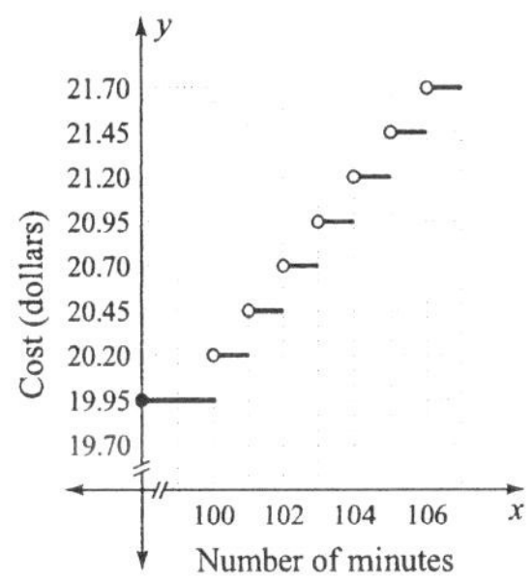

59. part of an hour, portion of a minute

Chapter 2 Review Exercises

1. The ***vertical-line test*** is used to find out if a graph is a function.

3. A ***dot graph*** has only integer inputs.

5. Limits on inputs due to an application setting is ***relevant domain.***

7. A ***linear function*** is a set of data with constant slope.

9. A function where the output exactly matches the input is an ***identity function.***

11. A function with a positive output for any real-number input is the ***absolute value function.*** (note: squaring function is not in the list)

13. Numbers that are small and placed below and to the right of letters or variables are ***subscripts.***

15. The 4 ways to find a linear equation are ***regression, point-slope, slope-intercept, sequence.***

17. a. $-8 < x \le -4$; $(-8, -4]$

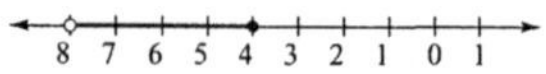

b. $-\infty < x < \infty$; $(-\infty, \infty)$

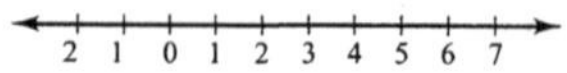

c. $-2 < x < 7$; $(-2, 7)$

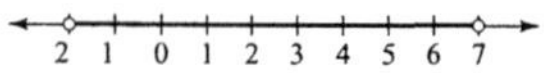

d. $x > -3$; $(-3, \infty)$

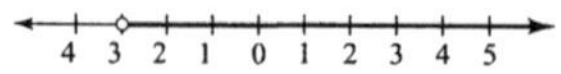

e. $x > 0$; $(0, \infty)$

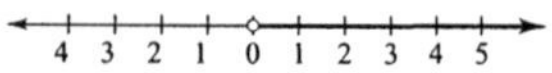

f. $x \ge 0$; $[0, \infty)$

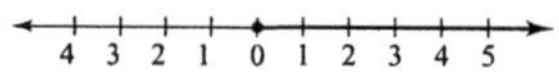

19. a. $-6.2 \le x \le 2.2$ **b.** $-3 \le y \le 3$ **c.** not a function

21. a. $\Re$ **b.** $y \ge 0$ **c.** function

23. a. $f(0) = 1$ **b.** $f(2) = 4$ **c.** $f(-1) = \frac{1}{2}$

d. $f(3) = 8$ **e.** $f(1) = 2$ **f.** $x \ge 1$

g. none **h.** $x \le 2$ **i.** 1

j. $\Re$ **k.** $y > 0$

25.
$$f(1) = 2(1)^2 - 3(1) + 1$$
$$f(1) = 2 - 3 + 1$$
$$f(1) = 0$$

27.
$$f(0.5) = 2(0.5)^2 - 3(0.5) + 1$$
$$f(0.5) = 0.5 - 1.5 + 1$$
$$f(0.5) = 0$$

29.
$$f(-2) = 2(-2)^2 - 3(-2) + 1$$
$$f(-2) = 8 + 6 + 1$$
$$f(-2) = 15$$

31.
$$f(\square) = 2(\square)^2 - 3(\square) + 1$$

33. slope is negative reciprocal, $-\frac{1}{3}$

a. y-intercept is 0; $y = -\frac{1}{3}x$

b. $b = 4 - (-\frac{1}{3})(3)$, $b = 5$; $y = -\frac{1}{3}x + 5$

	Ordered Pairs	Slope	Equation	Horiz. or Vert.	x-intercept	y-intercept
35.	(1, 3) (2, 2)	$\frac{3-2}{1-2}=-1$	b = 3 – (–1)(1), b = 4 y = –x + 4	neither	0 = –x + 4 x = 4	y = 4
37.	(–3, 3) (0, 0)	$\frac{3-0}{-3-0}=-1$	b = 0 y = –x	neither	x = 0	y = 0
39.	(–1, –1) (1, 3)	$\frac{-1-3}{-1-1}=2$	b = 3 – 2(1) b = 1 y = 2x + 1	neither	0 = 2x + 1 $x=-\frac{1}{2}$	y = 1
41.	(–3, 3) (2, 3)	$\frac{3-3}{-3-2}=0$	y = 3	horiz.	no x-intercept	y = 3

43. Ex. 35 and 37 are parallel.

45. a. y = 0.065x
b. slope = \$0.065 tax/\$1 purchased; y-intercept = 0, no tax on \$0 purchases

47. a. y = 45x + 500
b. slope = \$45/hour of repair; y-intercept = \$500, basic inspection cost

Exercise 49 uses LinReg on a graphing calculator to find equation. Solution is given for reference only.

49. $y \approx 11{,}528 - 42x$

51. Δx (ft) = 11 & 6; Δy (\$) = 16.50 & 9; slopes are 16.50 ÷ 11 = 1.50 & 9 ÷ 6 = 1.50
using the first data set: b = 45.50 – 1.50(25), b = 8; y = 1.50x + 8

53. Δx = 1, Δy = –4; working backwards when x = 0, y = 31 + 4 = 35; y = –4x + 35

55. C = \$350; constant function (monthly pass does not depend on x)

57. C = 5.95x; C in \$; increasing function (as x increases, C increases)

59. V = 350 – 5x; V in \$; decreasing function (as x increases, V decreases)

61. Let x = # of people, y = total cost; y = 85 for $0 < x \le 10$; y = 85 + 4.75(x − 10) for x > 10, inputs are positive integers only, dot graph

63. Let x = # of min.; y = cost; y = 0.26 for $0 < x \le 1$; y = 0.26 + 0.19(x − 1) for x > 1; inputs may be any non-negative number, step graph

65.

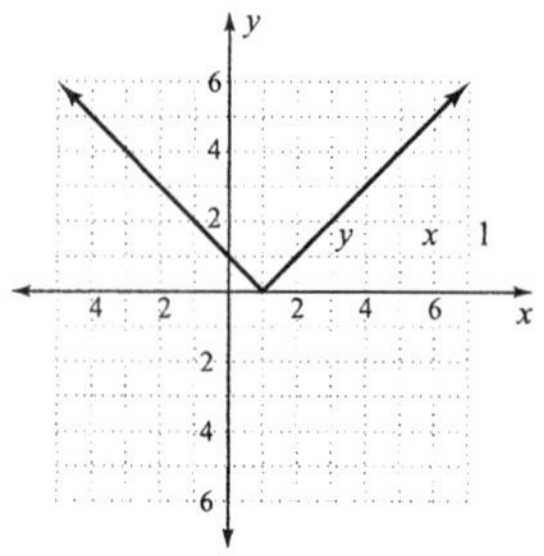

a. $|x - 1| = 4$

$x - 1 = 4$ or $x - 1 = -4$

$x = 5$ or $x = -3$

$\{-3, 5\}$

b. $|x - 1| = 2$

$x - 1 = 2$ or $x - 1 = -2$

$x = 3$ or $x = -1$

$\{-1, 3\}$

c. $|x - 1| = 0$

$x - 1 = 0$

$x = 1$

$\{1\}$

d. $|x - 1| = -2$

absolute value is always positive

{ } or ∅

67. domain = ℜ; range = 365

69. domain = ℜ; range $y \ge 0$

71. domain = ℜ; range = ℜ

73. domain = ℜ; range $y \ge 1$

75. $|0.5x - 3| = 4$

$0.5x - 3 = 4$ or $0.5x - 3 = -4$

$x = 14$ or $x = -2$

$\{-2, 14\}$

Chapter 2 Test

1. a. $x \le 5$; $(-\infty, 5]$

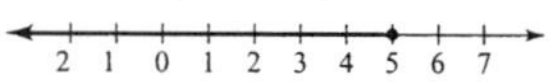

b. $-2 < x < 5$; $(-2, 5)$

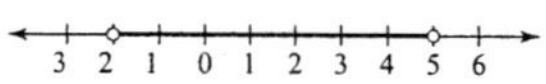

c. ℜ; $(-\infty, \infty)$

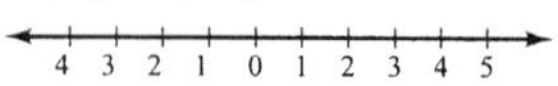

2. **a.** not a function, one input has two outputs

b. function **c.** function **d.** not a function, one input has two outputs

3. **a.** $f(-2) - 3(-2)^2 - 2(-2) - 4$

$f(-2) = 12 + 4 - 4$

$f(-2) = 12$

b. $f(0) = 3(0)^2 - 2(0) - 4$

$f(0) = 0 - 0 - 4$

$f(0) = -4$

c. $f(2) = 3(2)^2 - 2(2) - 4$

$f(2) = 12 - 4 - 4$

$f(2) = 4$

4. **a.** slope $= \dfrac{4-2}{-2-5} = -\dfrac{2}{7}$

b. $b = 4 - (-\frac{2}{7})(-2)$, $b = 3\frac{3}{7}$; $y = -\frac{2}{7}x + 3\frac{3}{7}$

c. parallel line = same slope: $-\frac{2}{7}$

d. perpendicular line = negative reciprocal slope; $\frac{7}{2}$; $b = -1 - (\frac{7}{2})2$, $b = -8$; $y = \frac{7}{2}x - 8$

5. **a.** The slope of a horizontal line is ***zero.***

b. A line that falls from left to right has a ***negative*** slope and is said to be a ***decreasing*** function.

c. If the slope of a graph between all pairs of points is constant, the graph is a ***linear*** function.

d. A horizontal linear graph is also called a ***constant*** function.

e. Linear equations have a ***constant*** slope.

f. The set of inputs to a number pattern is the ***positive integers or natural numbers.***

6. **a.** $y = 7x + 2.50$, y in \$ **b.** slope is \$7 per mile

7. **a.** reasonable inputs and output would be non-negative numbers; x = time in minutes, y = cost in \$

b. (1, 0.13), (19, 2.11)

c. slope $= \dfrac{2.11 - 0.13}{19 - 1} = 0.11$, $b = 0.13 - 0.11(1)$, $b = 0.02$; $y = 0.11x + 0.2$

8. From LinReg on graphing calculator: $y \approx 10.1x - 13.8$

9. **a.** $\Delta y = 8$, next number is $42 + 8 = 50$; when $x = 0$, $y = 10 - 8 = 2$; $y = 8x + 2$

b. $\Delta y = 7$, next number is $12 + 7 = 19$; when $x = 0$, $y = -16 - 7 = -23$; $y = 7x - 23$

10.

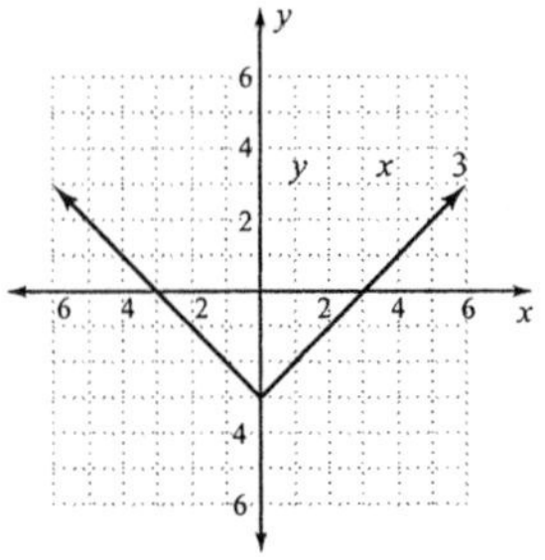

11.

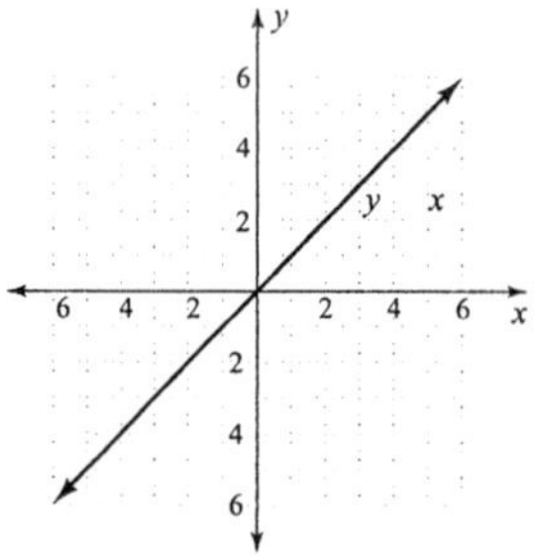

12.

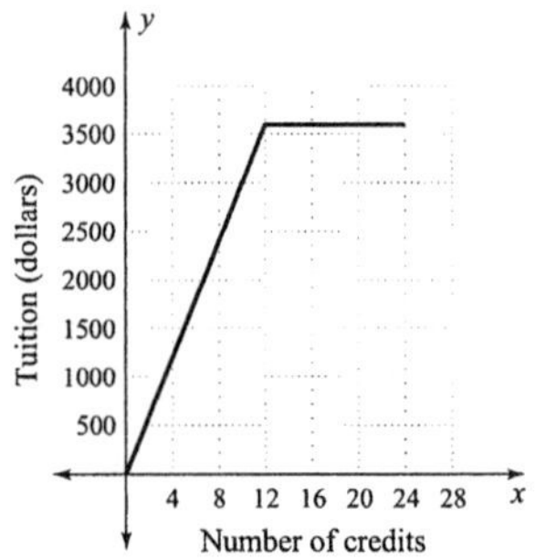

13.

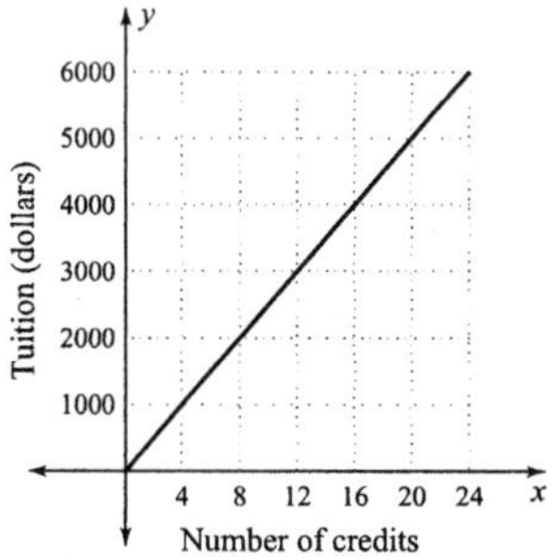

14.

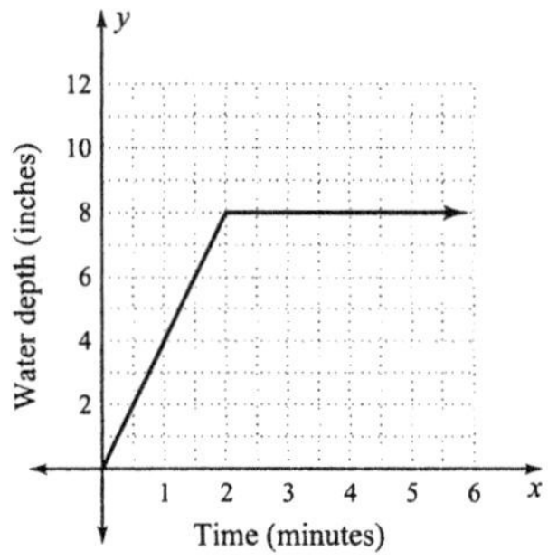

15.

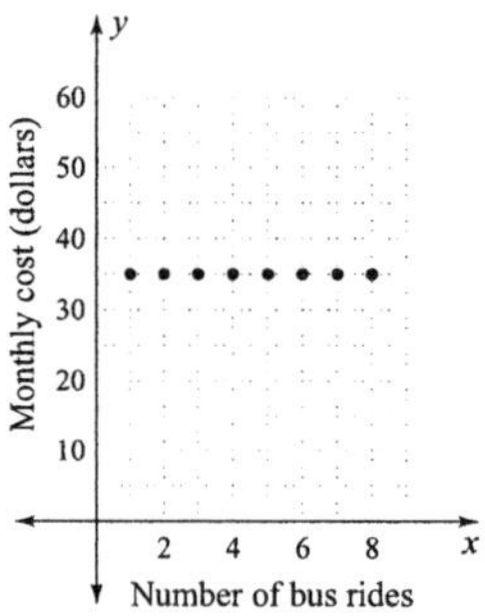

16.

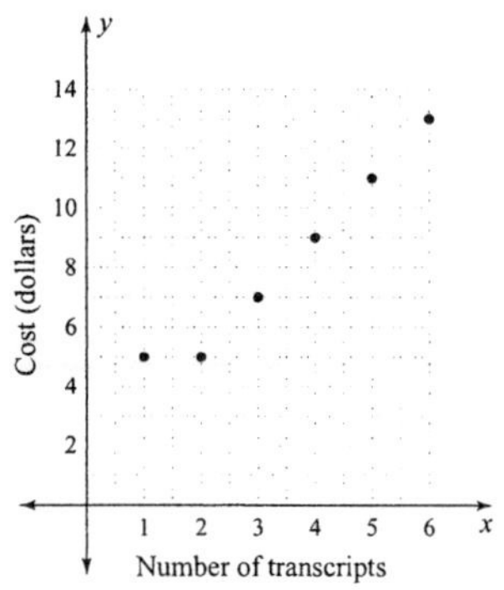

Transcript Copies	1	2	3	4	5	6
Cost $	5	5	7	9	11	13

Points should not be connected, only whole copies are reasonable.

17. $|2x + 2.5| = 5$

$2x + 2.5 = 5$ or $2x + 2.5 = -5$

$2x = 2.5$ or $2x = -7.5$

$x = 1.25$ or $x = -3.75$

$\{-3.75, 1.25\}$

Cumulative Review Chapters 1 and 2

1.

Input x	Input y	Output xy	Output x + y	Output x − y
2	4	**(−2)(4) = −8**	**−2 + 4 = 2**	**−2 − 4 = −6**
−3	7	**(−3)(7) = −21**	**−3 + 7 = 4**	**−3 − 7 = −10**
2	**−6 ÷ 2 = −3**	−6	**2 + (−3) = −1**	**2 − (−3) = 5**
−3	**6 ÷ (−3) = −2**	6	**−3 + (−2) = −5**	**−3 − (−2) = −1**
−1	**−7 − (−1) = −6**	**(−1)(−6) = 6**	−7	**−1 − (−6) = 5**
−7 − (−2) = −5	−2	**(−5)(−2) = 10**	−7	**−5 − (−2) = −3**
1 − (−2) = 3	−2	**(3)(−2) = −6**	1	**3 − (−2) = 5**
2	**−7 − 2 = −9**	**(2)(−9) = −18**	−7	**2 − (−9) = 11**

3. **a.** Two numbers, n and $-n$, that add to zero are ***opposites.***

b. Two numbers or expressions, a and b, that are multiplied to obtain the product ab are ***factors.***

c. Two numbers, n and $\frac{1}{n}$, that multiply to 1 are ***reciprocals.***

d. Removing a common factor from two or more terms is ***factoring.***

e. Collections of objects or numbers are ***sets.***

5. Factoring ab + ac changes ***a sum to a product.***

7. To divide real numbers, we may change ***division to multiplication by the reciprocal.***

9. $a(b + c) - b(a + c) + c(a - b) = ab + ac - ab - bc + ac - bc = 2ac - 2bc$

11. $\frac{1}{4}\pi(15 \text{ ft})^2 = 56.25\pi \text{ ft}^2$

13. $\dfrac{16 + 21x}{6}$, does not simplify

15. $15 - 4x = 5(6 - x)$

$15 - 4x = 30 - 5x$

$-4x + 5x = 30 - 15$

$x = 15$

17. $3x = x + 15$

$3x - x = 15$

$2x = 15$

$x = 7.5$

19.

x	$f(x) = x + 2$
−1	$(-1) + 2 = 1$
0	$0 + 2 = 2$
1	$1 + 2 = 3$
2	$2 + 2 = 4$

21.

x	$f(x) = x^2$
−1	$(-1)^2 = 1$
0	$(0)^2 = 0$
1	$(1)^2 = 1$
2	$(2)^2 = 4$

23. slope $= \dfrac{1-(-3)}{-6-4} = -\dfrac{2}{5}$; $b = 1 - (-\frac{2}{5})(-6) = -\frac{7}{5}$; $y = -\frac{2}{5}x - \frac{7}{5}$

25. slope is negative reciprocal or $-\frac{1}{3}$; $b = 0$; $y = -\frac{1}{3}x$

27. next pair is (4, 4); $f(x) = x$

29. next pair is (4, 4); $f(x) = 4$

31. Let x = number of workers and y = cost in \$; $y = 65x + 500$

33. a. {−1} **b.** {−4, 2} **c.** {−5, 3}

d. {−3, 1} **e.** { }

Exercises 3.1

1.

$3x + y = 4$

$3x + y - 3x = -3x + 4$

$y = -3x + 4$

3.

$x - 3y = 7$

$x - 3y + 3y = 3y + 7$

$x = 3y + 7$

5.

$4y - 2x = 5$

$4y - 2x + 2x = 5 + 2x$

$4y - 5 = 5 + 2x - 5$

$2x = 4y - 5$

$x = \dfrac{4y-5}{2}$

$x = 2y - \frac{5}{2}$

7.

$4x - 2y = -3$

$4x - 2y + 2y = 2y - 3$

$4x + 3 = 2y - 3 + 3$

$2y = 4x + 3$

$y = \dfrac{4x+3}{2}$

$y = 2x + \frac{3}{2}$

9.

$y = 3x - 4$

$3y - 2x = 9$

$3(3x - 4) - 2x = 9$

$9x - 12 - 2x = 9$

$7x - 12 = 9$

$7x = 21$

$\mathbf{x = 3}$

$y = 3(3) - 4$

$\mathbf{y = 5}$

$3(5) - 2(3) ?= 9$

$15 - 6 = 9$ ✓

11.

$0.6x + y = 3$

$1.4x + y + 3 = 0$

$y = 3 - 0.6x$

$y = -1.4x - 3$

$3 - 0.6x = -1.4x - 3$

$-0.6x + 1.4x = -3 - 3$

$0.8x = -6$

$\mathbf{x = -7.5}$

$y = 3 - 0.6(-7.5)$

$\mathbf{y = 7.5}$

$1.4(-7.5) + 7.5 + 3 ?= 0$

$-10 + 10 = 0$ ✓

Solution check will be left to the student for the remainder of this section.

13.

$x + 2y = 9.4$
$3x - 5y = 4$
$x = 9.4 - 2y$
$3(9.4 - 2y) - 5y = 4$
$28.2 - 6y - 5y = 4$
$28.2 - 11y = 4$
$28.2 - 4 = 11y$
$24.2 = 11y$
$\mathbf{y = 2.2}$
$x = 9.4 - 2(2.2)$
$\mathbf{x = 5}$

15.

$y = 1.8x - 6.6$
$y + 1.4x = 3$
$y = 3 - 1.4x$
$1.8x - 6.6 = 3 - 1.4x$
$1.8x + 1.4x = 3 + 6.6$
$3.2x = 9.6$
$\mathbf{x = 3}$
$y = 1.8(3) - 6.6$
$\mathbf{y = -1.2}$

17.

$3y + 2x = -2$ (1)
$3y - 5x = 26$ (2)
$-1(3y - 5x) = -1(26)$ (2)
$-3y + 5x = -26$ (2)
$3y + 2x = -2$ (1) *add*
$0y + 7x = -28$
$7x = -28$
$\mathbf{x = -4}$
$3y + 2(-4) = -2$
$3y - 8 = -2$
$3y = 6$
$\mathbf{y = 2}$

19.

$5x = 14.7 + 6y$ (1)
$2x + 6y = -4.2$ (2)
$5x - 6y = 14.7$ (1) *add*
$7x + 0y = 10.5$
$7x = 10.5$
$\mathbf{x = 1.5}$
$5(1.5) = 14.7 + 6y$
$7.5 = 14.7 + 6y$
$7.5 - 14.7 = 6y$
$-7.2 = 6y$
$\mathbf{y = -1.2}$

21.

$2x + 4y = 0$ (1)
$3x - 2y = 24$ (2)
$2(3x - 2y) = 2(24)$ (2)
$6x - 4y = 48$ (2)
$2x + 4y = 0$ (1) *add*
$8x + 0y = 48$
$8x = 48$
$\mathbf{x = 6}$
$2(6) + 4y = 0$
$12 + 4y = 0$
$4y = -12$
$\mathbf{y = -3}$

23.

$5x - 2y = -9$ (1)
$6x + 7y = 8$ (2)
$7(5x - 2y) = 7(-9)$ (1)
$2(6x + 7y) = 2(8)$ (2)
$35x - 14y = -63$ (3)
$12x + 14y = 16$ (4) *add*
$47x + 0y = -47$
$\mathbf{x = -1}$
$6(-1) + 7y = 8$
$-6 + 7y = 8$
$7y = 14$
$\mathbf{y = 2}$

25.
$A + B = 18$
$A - B = -58$ *add*
$2A = -40$
$\mathbf{A = -20}$
$-20 + B = 18$
$\mathbf{B = 38}$

27.
$C + D = 2$
$2C + D = -11$ *subtract*
$-C = 13$
$\mathbf{C = -13}$
$-13 + D = 2$
$\mathbf{D = 15}$

29.
$4E = F$
$3E + 4F = 34.2$ *substitute*
$3E + 4(4E) = 34.2$
$3E + 16E = 34.2$
$19E = 34.2$
$\mathbf{E = 1.8}$
$4(1.8) = F$
$\mathbf{F = 7.2}$

31.
$3G + 2H = 0$
$H + G = 0.8$
$H = 0.8 - G$ *substitute*
$3G + 2(0.8 - G) = 0$
$3G + 1.6 - 2G = 0$
$\mathbf{G = -1.6}$
$H + (-1.6) = 0.8$
$\mathbf{H = 2.4}$

33. We might ask: What is the total value of the money? The reader will need to supply the following facts: dimes are \$0.10 and quarters are \$0.25

35. We might ask: What is measure of each angle? The reader will need to supply the following facts: sum of three interior angles in a triangle is 180°

37. We might ask: What is the perimeter? What are the measures of the other two sides? The reader will need to supply the following facts: opposite sides of a rectangle are equal

39.
Let x = sales \$.
Package 3 = $550 + 0.04x$
Package 1 = $250 + 0.10x$
$550 + 0.04x = 250 + 0.10x$
$550 - 250 = 0.10x - 0.4x$
$300 = 0.06x$
x = \$5000
$550 + 0.04(5000) =$ **\$750**

41.
Let x = sales \$
Package 4 = $750 + 0.04x$
Package 2 = $350 + 0.08x$
$750 + 0.04x = 350 + 0.08x$
$750 - 350 = 0.08x - 0.04x$
$400 = 0.04x$
x = \$10,000
$750 + 0.04(10,000) =$ **\$1150**

43. bottle + cork = 1.10
bottle = cork + 1
(cork + 1) + cork = 1.10
2cork + 1 = 1.10
2cork = 0.10
cork = \$0.05
bottle = 0.05 + 1
bottle = \$1.05

45. length + width = 20
length = width + 4
(width + 4) + width = 20
2width + 4 = 20
2width = 16
width = 8"
length = 8 + 4
length = 12"

For exercise 47 , remember perimeter = 2l + 2w

47. $2l + 2w = 380$
$l = w + 30$
$2(w + 30) + 2w = 380$
$2w + 60 + 2w = 380$
$4w = 320$
$w = 80$ yds
$l = 80 + 30$
$l = 110$ yds

For exercises 49 to 53, remember supplementary angles add to 180° and complementary angles add to 90°

49. A and B = 2 complementary angles
A + B = 90
A − B = 25
2A = 115
A = 57.5°
57.5 + B = 90
B = 32.5°

51. D + E = 180
$\frac{1}{3}$D + E = 90
$\frac{2}{3}$D + 0E = 90
D = 135°
135 + E = 180
E = 45°

53. A + B = 90
2A + B = 180
−A + 0B = −90
A = 90°
90 + B = 90
B = 0°
not possible with non-zero angles

55. $acx + bcy = ce$

$acx + ady = af$ *subtract*

$bcy - ady = ce - af$

$y(bc - ad) = ce - af$

$$y = \frac{ce - af}{bc - ad}$$

57. Substitution is more convenient when one variable has coefficient 1 and is easy to solve for.

Exercises 3.2

1.

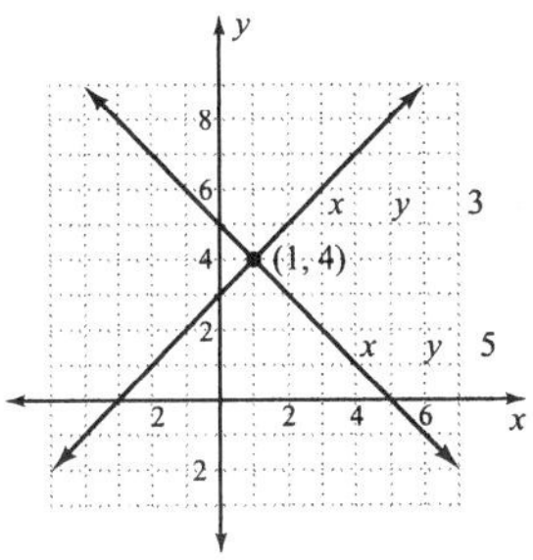

(1, 4)

3.

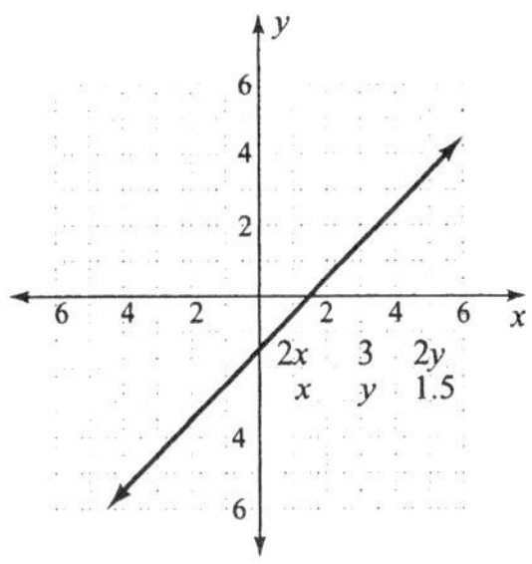

Coincident lines

x = 1.5 + y; 2(1.5 + y) – 3 = 2y

3 + 2y – 3 = 2y

2y = 2y; identity

5.

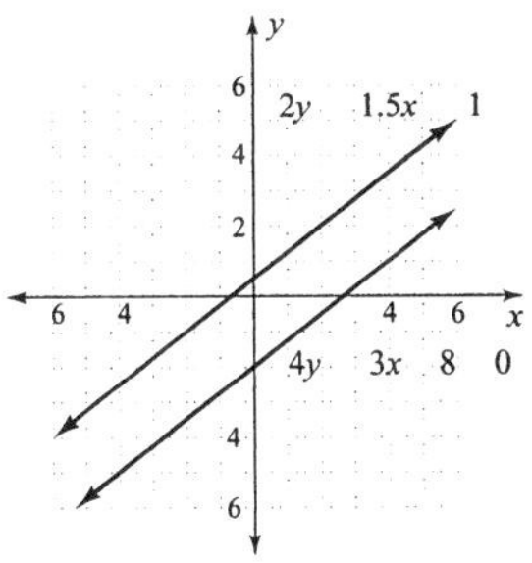

Parallel lines

2(2y) = 2(1.5x + 1)

4y = 3x + 2

4y = 3x – 8

3x + 2 = 3x – 8

2 = –8; contradiction

7.

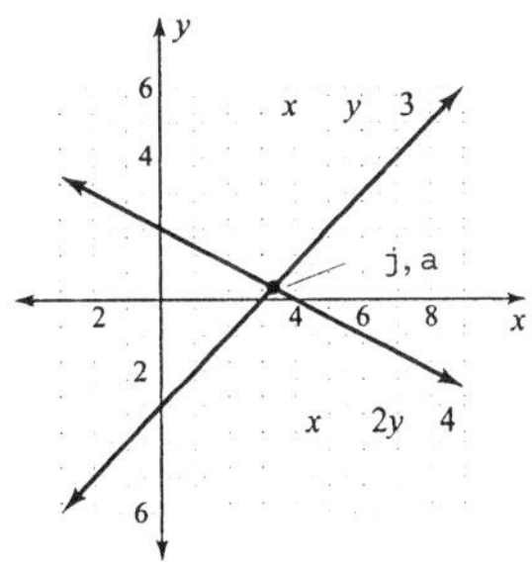

$\left(\frac{10}{3}, \frac{1}{3}\right)$

9.

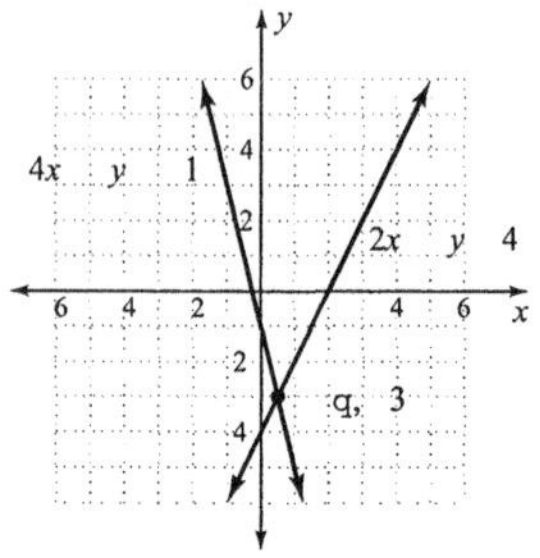

$\left(\frac{1}{2}, -3\right)$

11.

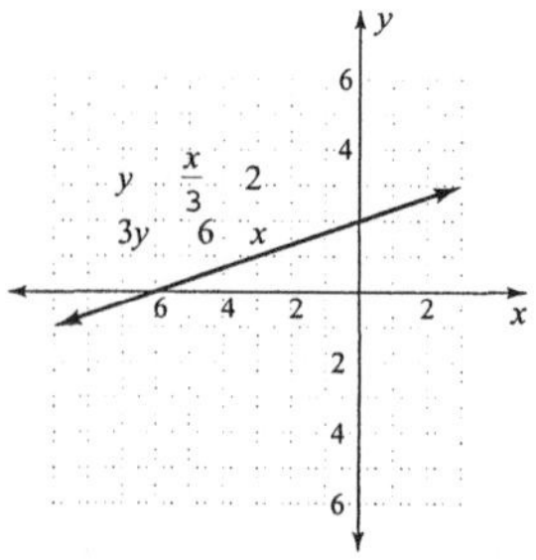

Coincident lines

$$3\left(\frac{x}{3}+2\right)-6=x$$

$x+6-6=x$

$x=x$; identity

13.

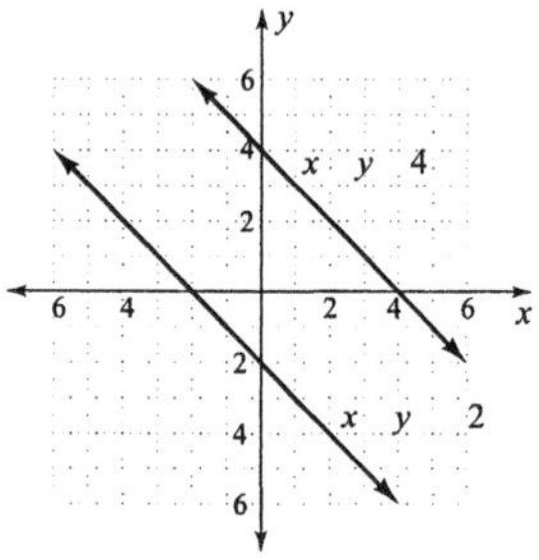

Parallel lines

y = 4 – x

x + 2 = –(4 – x)

x + 2 = –4 + x

2 = –4; contradiction

15.

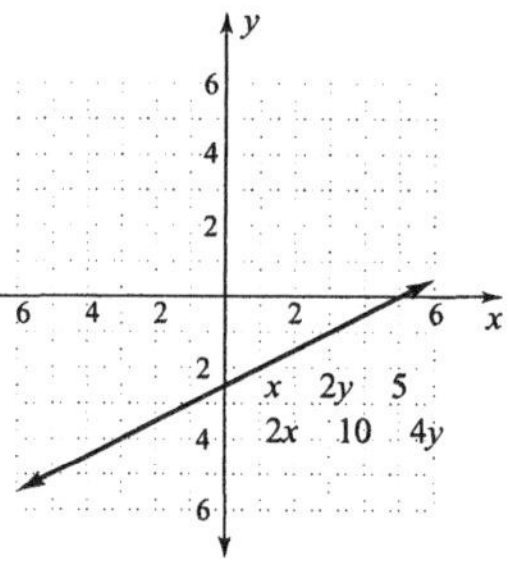

Coincident lines

x = 2y + 5

2(2y + 5) – 10 = 4y

4y + 10 – 10 = 4y

4y = 4y; identity

17. Package 1 = 250 + 0.10x

Package 2 = 350 + 0.08x

250 + 0.10x = 350 + 0.08x

0.02x = 100

x = \$5000

250 + 0.10(5000) = \$750

19. Package 3 = 550 + 0.04x

Package 4 = 750 + 0.04x

550 + 0.04x = 750 + 0.04x

0 = 200; contradiction

parallel graphs, pkg 4 is always better (note y-intercept value)

21. Increasing the percent of sales is the same as increasing the slope, the graph becomes steeper.

23. Packages 3 and 5 have the same fixed amount of \$550, therefore they have the same vertical axis intercept.

25. Parallel lines have no intercept, write a pair of equations for parallel lines.

27. Yes, all straight line equations can be written in the standard form of $ax + by = c$.

29. $0x + y = 3$; $a = 0$, $b = 1$ and $c = 3$

31. a. $n = 0$ is the additive identity **b.** $n = 1$ is the multiplicative identity

c. The product of a number and its multiplicative inverse is ***the identity, 1.***

d. The linear function $y = x$ *or* $f(x) = x$ is the identity function.

Exercises 3.3

1. Quantity: \$1600, \$12,000; Value: 5% and 8% annual interest

3. Quantity: 20 hr, 15 hr; Value: \$6.50 and \$7.25 per hour

5. Quantity: 100 mL, 1000 mL; Value: 8% and 0% solution

For GPA calculations: A = 4 pts, B = 3 pts, C = 2 pts, D = 1 pt and F = 0

7.

Credit hours	Grade point	Total points
4	4	(4)(4) = 16
3	3	(3)(3) = 9
3	2	(3)(2) = 6
1	3	(1)(3) = 3
Total = 11	**GPA = 34 ÷ 11 = 3.09**	Total = 34

$4A + 3B + 3C + 1B = \text{total points}$

$4(4) + 3(3) + 3(2) + 1(3) = 34$

$4 + 3 + 3 + 1 = 11$

9.

Credit hours	Grade points	Total points
75	3.08	75(3.08) = 231
x	4.0	4x
75 + x	3.25	3.25(75 + x) = 231 + 4x

Solving for x: $3.25(75 + x) = 231 + 4x$

$243.75 + 3.25x = 231 + 4x$

$12.75 = 0.75x$

$x = 17$ credit hours of A

11.

Kg	Cost/kg	Total
H (Honduran)	18.70	18.70H
I (Indonesian)	23.65	23.65I
H + I = 200	19.80	200(19.80) = 18.70H + 23.65I

$H + I = 200 \rightarrow H = 200 - I$

$18.70H + 23.65I = 3960$

$18.70(200 - I) + 23.65I = 3960$

$3740 - 18.70I + 23.65I = 3960$

$4.95I = 220$

$\mathbf{I = 44.44 \approx 44}$ **kg**

$H = 200 - 44$

$\mathbf{H \approx 156}$ **kg**

13. $9d + 13q = 3.5$ (a)

$12d + 9q = 3.0$ (b)

$4(9d + 13q) = 4(3.5)$

$-3(12d + 9q) = -3(3.0)$

$36d + 52q = 14$ (c)

$-36d - 27q = -9$ (d) {add (c) and (d)}

$0d + 25q = 5$

$\mathbf{q = 0.2}$ **oz**

$9d + 13(0.2) = 3.5$

$9d = 0.9$

$\mathbf{d = 0.1}$ **oz**

15.

$	Rate	Total
x	0.06	0.06x
75,000 – x	0.09	0.09(75,000 – x)
75,000	0.08	0.08(75,000)

$0.06x + 0.09(75{,}000 - x) = 6{,}000$

$0.06x + 6{,}750 - 0.09x = 6{,}000$

$-0.03x = -750$

$x = 25{,}000$

$25,000 should be invested at 6% and $50,000 at 9%

17.

$	Rate	Total
c	0.21	0.21c
b	0.11	0.11b
10,000	0.13	0.13(10,000)

$b + c = 10{,}000 \rightarrow b = 10{,}000 - c$

$0.11b + 0.21c = 1{,}00$

$0.11(10{,}000 - c) + 0.21c = 1{,}300$

$1{,}100 - 0.11c + 0.21c = 1{,}300$

$0.10c = 200$

c = $2,000

$b = 10{,}000 - 2{,}000$

b = $8,000

19.

$	Rate	Total
x	0.045	0.045x
23,600 – x	0.085	0.085(23,600–x)
23,600		1,782

$0.045x + 0.085(23{,}600 - x) = 1{,}782$

$0.045x + 2006 - 0.085x = 1782$

$224 = 0.04x$

$x = 5{,}600$

$5,600 invested at 4.5% and $18,000 at 8.5%

21.

Liters	%	Total
x	0.50	0.50x
y	0.00	0y
0.5	0.05	0.05(0.5)

$x + y = 0.5$

$0.50x = 0.025$

$\mathbf{x = 0.05\ L}$

$0.05 + y = 0.5$

$\mathbf{y = 0.45\ L}$

23.

Liters	%	Total
x	0.03	0.03x
y	0.20	0.20y
1,000	0.064	0.064(1,000)

$x + y = 1{,}000 \rightarrow y = 1{,}000 - x$

$0.03x + 0.20y = 64$

$0.03x + 0.20(1{,}000 - x) = 64$

$0.03x + 200 - 0.20x = 64$

$136 = 0.17x$; $\mathbf{x = 800\ L}$

$y = 1{,}000 - 800$; $\mathbf{y = 200\ L}$

25. $x + y = 1{,}000$

$0x + 0.20y = 0.03(1{,}000)$

y = 150 gal hydrogen peroxide

$x + 150 = 1{,}000$

x = 850 gal water

27. $A + B = 1{,}000$

$0.15A + 0.10B = 0.12(1{,}000)$

$0.15A + 0.10(1{,}000 - A) = 120$

$0.15A + 100 - 0.10A = 120$

$0.05A = 20$; $\mathbf{A = 400\ lb}$

$400 + B = 1{,}000$; $\mathbf{B = 600\ lb}$

<u>Mid-Chapter 3 Test</u>

1. $5x - 2y = 4$

$5x - 2y + 2y - 4 = 4 - 4 + 2y$

$5x - 4 = 2y$

$y = \frac{5}{2}x - 2$

2. $2x - \frac{1}{2}y = 5$

$2x - \frac{1}{2}y + \frac{1}{2}y - 5 = 5 - 5 + \frac{1}{2}y$

$2x - 5 = \frac{1}{2}y$

$y = 4x - 10$

3. $2y - 3x = 8$

$y = 1.5x - 3$

$2(1.5x - 3) - 3x = 8$

$3x - 6 - 3x = 8$

$-6 = 8$; contradiction

parallel lines

4. $3y = 2x - 6$

$y = -\frac{5}{6}x + 1$

$3(-\frac{5}{6}x + 1) = 2x - 6$

$-\frac{5}{2}x + 3 = 2x - 6$

$-\frac{5}{2}x + 3 + \frac{5}{2}x + 6 = 2x - 6 + 6 + \frac{5}{2}x$

$9 = 4.5x$

$x = 2$

$y = -\frac{5}{6}(2) + 1$

$y = -\frac{2}{3}$; $(2, -\frac{2}{3})$

5.
$y = -0.8x - 1.8$
$y = -1.2x + 3$
$-0.8x - 1.8 = -1.2x + 3$
$0.4x = 4.8$
$x = 12$
$y = -0.8(12) - 1.8$
$y = -11.4$
$(12, -11.4)$

6.
$y = 2.5x + 4$
$2y - 5x = 8$
$2(2.5x + 4) - 5x = 8$
$5x + 8 - 5x = 8$
$8 = 8$; identity
coincident lines

7.
$x + y = \frac{1}{2}$
$x - y = \frac{3}{4}$ *add*
$2x + 0y = \frac{5}{4}$
$x = \frac{5}{8}$

8.
$2x - 12 = -3y$ (a); $2x + 3y = 12$ (a std form)
$6 - 4y = 3x$ (b); $3x + 4y = 6$ (b std form)
$3(2x + 3y) = 3(12)$ (c); $6x + 9y = 36$ (c)
$-2(3x + 4y) = -2(6)$ (d); $-6x - 8y = -12$ (d)
(c) + (d): $\mathbf{y = 24}$
$2x - 12 = -3(24)$
$2x = -60$; $\mathbf{x = -30}$

9.
Graphs represent:
(a) $y = -x + 2$ and (b) $y = \frac{2}{3}x + 2$
intersection is (0, 2)

10.
$A + B = 180$
$A - B = 90$
$2A = 90$
$\mathbf{A = 45°}$
$B = 180 - 45$, $\mathbf{B = 135°}$

11.
$C + B = 54{,}568$
$C = B + 8270$
$B + 8270 + B = 54{,}568$
$2B = 46{,}298$
$B = 23{,}149$
$C = 23{,}149 + 8270$
$C = 31{,}419$
Chamberlain scored 31,419 pts
Baylor scored 23,149 pts

12.

Hours	Rate	Total
20	6.00	120
x	12.00	12x
20 + x	8.00	160 + 8x

$160 + 8x = 120 + 12x$
$40 = 4x$
$x = 10$ hours

13.
$x + y = 1$ pt
$0.03x + 0y = 0.02(1)$
$x = \frac{2}{3}$ pt 3% hydrogen peroxide
$y = 1 - \frac{2}{3}$; $y = \frac{1}{3}$ pt water

Exercises 3.4

1.

Class c	Study s	Work w	Total
2	2(3) = 6	2(2) = 4	12
5	5(3) = 15	5(2) = 10	30
c	3c	2c	c + s + w = 60

s = 3c; w = 2c; c + s + w = 60

c + 3c + 2c = 60

6c = 60

c = 10 hr

s = 3(10); s = 30 hr

w = 2(10); w = 20 hr

3.

Angle x	Angle y	Angle z	Total 180°
30	30	30 + 21 = 51	111
40	40	40 + 21 = 61	141
x	y = x	x + 21	x + y + z

x = y; z = x + 21 (or x = z – 21); x + y + z = 180

x + x + x + 21 = 180

3x = 159

x = 53°; y = 53°

z = 53 + 21; z = 74°

5.

	Protein	Carbs	Fat	Total
	5	10	13	28
grams	p	c = 2p	f	28
	20	40	117	177
calories	4p	4c	9f	222

p + c + f = 28; c = 2p; 4p + 4c + 9f = 222

p + 2p + f = 28; 3p + f = 28; f = 28 – 3p

4p + 4(2p) + 9f = 222; 12p + 9f = 222

12p + 9(28 – 3p) = 222; 12p + 252 – 27p = 222; 252 – 15p = 222; 30 = 15p

p = 2 g; c = 2(2); **c = 4 g**; f = 28 – 3(2); **f = 22 g**

7.

$2a + 3b + 2c = 1$ (1)

$3a + 2c = 10$ (2)

$a + b = 1$ (3)

solve (3) for b: $b = 1 - a$

solve (2) for c: $c = 5 - 1.5a$

substitute into (1):

$2a + 3(1 - a) + 2(5 - 1.5a) = 1$

solve for a:

$2a + 3 - 3a + 10 - 3a = 1$

$13 - 4a = 1$

$4a = 12$

$\mathbf{a = 3}$

$b = 1 - 3$

$\mathbf{b = -2}$

$c = 5 - 1.5(3)$

$\mathbf{c = 0.5}$

9.

$2a + b - 4c = 17$ (1)

$5a - 2b - 27$ (2)

$b + 3c + 7 = 0$ (3)

solve (2) for a: $a = 0.4b + 5.4$

solve (3) for c: $c = -\frac{1}{3}b - \frac{7}{3}$

substitute into (1):

$2(0.4b + 5.4) + b - 4(-\frac{1}{3}b - \frac{7}{3}) = 17$

$0.8b + 10.8 + b + 1.3b + 9.3 = 17$

$3.1b + 20.1 = 17$

$3.1b = -3.1$

$\mathbf{b = -1}$

$a = 0.4(-1) + 5.4$

$\mathbf{a = 5}$

$c = -\frac{1}{3}(-1) - \frac{7}{3}$

$\mathbf{c = -2}$

11.

$x + y + z = 2$

$x + 3z = 0 \rightarrow x = -3z$

$2z = 7 + x + y$

$-3z + y + z = 2$

$y - 2z = 2$ (a)

$2z = 7 + (-3z) + y$

$-y + 5z = 7$ (b)

add (a) and (b):

$3z = 9$

$\mathbf{z = 3}$

$x = -3(3)$

$\mathbf{x = -9}$

$-9 + y + 3 = 2$

$\mathbf{y = 8}$

13.

$2x + 3y - z = 11$

$3x + y + 2z = 13$

$2y = z + 7$

$z = 2y - 7$

$2x + 3y - (2y - 7) = 11$

$2x + y = 4$ (a)

$3x + y + 2(2y - 7) = 13$

$3x + 5y - 14 = 13$

$3x + 5y = 27$ (b)

$-5(2x + y) = -5(4)$

$-10x - 5y = -20$ (c)

add (b) and (c):

$-7x = 7$

$\mathbf{x = -1}$

$2(-1) + y = 4$

$\mathbf{y = 6}$

$z = 2(6) - 7$

$\mathbf{z = 5}$

15.
$x + y - z = 2$ (1)
$x - y + z = 3$ (2)
$-x + y + z = 4$ (3)
add (1) and (2)
$2x = 5$
x = 2.5
add (2) and (3)
$2z = 7$
z = 3.5
substitute into (1)
$2.5 + y - 3.5 = 2$
y = 3

17.
$2a + 3b + c = -4$ (1)
$3a + 2b + 4c = 19$ (2)
$a - b - 3c = -19$ (3)
$-3(a - b - 3c) = -3(-19)$
$-3a + 3b + 9c = 57$ (4)
add (2) and (4)
$5b + 13c = 76$ (a)
$-2(a - b - 3c) = -2(-19)$
$-2a + 2b + 6c = 38$ (5)
add (1) and (5)
$5b + 7c = 34$ (b)
subtract (a) and (b)
$6c = 42$; **c = 7**
$5b + 7(7) = 34$
$5b = -15$; **b = –3**
$a - (-3) - 3(7) = -19$
$a - 18 = -19$; **a = –1**

19.
$a + b + c = 10$ (1)
$-a + b = 4$ (2)
$2a - b + c = 0$ (3)
subtract (1) from (3)
$a - 2b = -10$ (4)
add (2) and (4)
$-b = -6$; **b = 6**
$-a + 6 = 4$; **a = 2**
$2 + 6 + c = 10$; **c = 2**

21.
$x + y - z = 2$ (1)
$x - z = 4$ (2)
$x + y = 3$ (3)
subtract (1) and (3)
$-z = -1$; **z = 1**
$x - 1 = 4$
x = 5
$5 + y = 3$
y = –2

23.
$a + 2b + 3c = 2$ (1)
$4a + 5b + 6c = 3$ (2)
$a + b + c = 1$ (3)
subtract (3) from (1): $b + 2c = 1$ (4)
$-4(a + b + c) = -4(1)$
$-4a - 4b - 4c = -4$ (5)
add (2) and (5): $b + 2c = -1$ (6)
from (4) and (6): $1 = -1$
contradiction; inconsistent

25.
$2a + b + c = 3$ (1)
$a + 2b - 2c = 2$ (2)
$a - b + 3c = 1$ (3)
subtract (2) and (3): $3b - 5c = 1$ (4)
$-2(a - b + 3c) = -2(1)$
$-2a + 2b - 6c = -2$ (5)
add (1) and (5) $3b - 5c = 1$ (6)
from (4) and (6): $1 = 1$
identity; dependent

27. $2x - y + z = 1$ (1)
$x + z = -2$ (2)
$x - y = 3$ (3)
add (2) and (3)
$2x - y + z = 1$ (4)
from (1) and (4); $1 = 1$
identity; dependent

29. $2x + y - z = 3$ (1)
$5x - 4z = 0$ (2)
$5y + 3z = 0$ (3)
$5x = 4z$
$x = 0.8z$
$5y = -3z$
$y = -0.6z$
$2(0.8z) + (-0.6z) - z = 3$
$1.6z - 0.6z - z = 3$
$0 = 3$; contradiction; inconsistent

31. a. Intersection is a plane; dependent
b. intersection is a plane; dependent
c. no common intersection; inconsistent
d. no common intersection; inconsistent

33. $p + d + q = 61$
$0.01p + 0.10d + 0.25q = 4.69$
$3q = d$
$p + 3q + q = 61$
$p + 4q = 61$ (a)
$0.01p + 0.10(3q) + 0.25q = 4.69$
$0.01p + 0.55q = 4.69$
$100(0.01p + 0.55q) = 100(4.69)$
$p + 55q = 469$ (b)
subtract (a) and (b)
$51q = 408$; $\mathbf{q = 8}$
$3(8) = d$; $\mathbf{d = 24}$
$p + 24 + 8 = 61$; $\mathbf{p = 29}$

35. Let x = Okamoto, y = Strange; z = Pavin:
$x + y = z + 412{,}545$
$y = x + 459{,}907$
$x + z = 1{,}445{,}464$
$z = 1{,}445{,}464 - x$
$x + x + 459{,}907 = 1{,}445{,}464 - x + 412{,}545$
$2x + 459{,}907 = 1{,}858{,}009 - x$
$3x = 1{,}398{,}102$
x = $466,034 (Okamoto)
$y = 466{,}034 + 459{,}907$
y = $925,941 (Strange)
$z = 1{,}445{,}464 - 466{,}034$
z = $979,430 (Pavin)

37. $c + f + p = 29$
$p = f$
$4c + 9f + 4p = 121$
$c + p + p = 29$
$c + 2p = 29 \rightarrow c = 29 - 2p$
$4(29 - 2p) + 9p + 4p = 121$
$116 - 8p + 13p = 121$
$5p = 5$; **p = 1 g; f = 1 g;**
$c + 1 + 1 = 29$; **c = 27 g**

39. $c + f + p = 32$
$c = f + 12$
$4c + 9f + 4p = 148$
$(f + 12) + f + p = 32$
$2f + p = 20 \rightarrow p = 20 - 2f$
$4(f + 12) + 9f + 4(20 - 2f) = 148$
$4f + 48 + 9f + 80 - 8f = 148$
$5f = 20$; **f = 4 g**; $c = 4 + 12$; **c = 16 g**
$16 + 4 + p = 32$; **p = 12 g**

41. $n + d + q = 44$

$0.05n + 0.10d + 0.25q = 5.80$

$q = d + 3$

$n + d + (d + 3) = 44; \quad n + 2d = 41 \rightarrow n = 41 - 2d$

$0.05(41 - 2d) + 0.10d + 0.25(d + 3) = 5.80$

$2.05 - 0.10d + 0.10d + 0.25d + 0.75 = 5.80$

$0.25d + 2.80 = 5.80$

$0.25d = 3.00 \rightarrow \mathbf{d = 12}$

$q = 12 + 3; \quad \mathbf{q = 15}$

$n + 12 + 15 = 44; \quad \mathbf{n = 17}$

43. An inconsistent system has no solutions, a dependent system has an infinite number of solutions; it is not possible to have both at the same time.

<u>**Exercises 3.5**</u>

1. The graph of $y = -2$ is below the graph of $y = -2x + 3$ for $x < 2.5$.

$-2 < -2x + 3$

$-5 < -2x$

$2.5 > x$ or $x < 2.5$

3. The graph of $y = -2x + 3$ is above the graph of $y = 0$ for $x < 1.5$.

$-2x + 3 > 0$

$-2x > -3$

$x < 1.5$

5. The graph of $y = x + 4$ is above the graph of $y = -\frac{1}{2}x + 1$ for $x > -2$.

$x + 4 > -\frac{1}{2}x + 1$

$1.5x > -3$

$x > -2$

7. The graph of $y = 0$ is below the graph of $y = x + 4$ for $x > -4$.

$0 < x + 4$

$-4 < x$ or $x > -4$

9. The graph of $-\frac{1}{2}x + 1$ is below or the same as the graph of $y = 0$ for $x \geq 2$.

$-\frac{1}{2}x + 1 \leq 0$

$1 \leq \frac{1}{2}x$

$2 \leq x$ or $x \geq 2$

11. The graph of $y = -2x + 3$ is above or the same as the graph of $y = 3$ for $x \leq 0$.

$-2x + 3 \geq 3$

$-2x \geq 0$

$x \leq 0$

13. The graph of $y = x + 4$ is below or the same as the graph of $y = 4$ for $x \le 0$.

$x + 4 \le 4$

$x \le 0$

15. 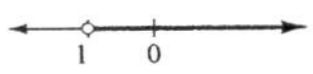

$x > -1$

17. 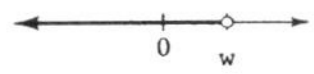

$x < \frac{3}{2}$

19. 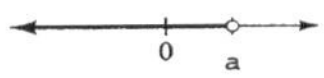

$x < \frac{1}{3}$

21. 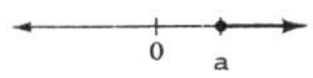

$x \ge \frac{1}{3}$

23. 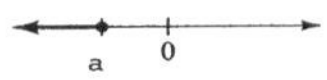

$x \le -\frac{1}{3}$

25. $y = x - 2$ is below $y = x$ for all $\Re$.

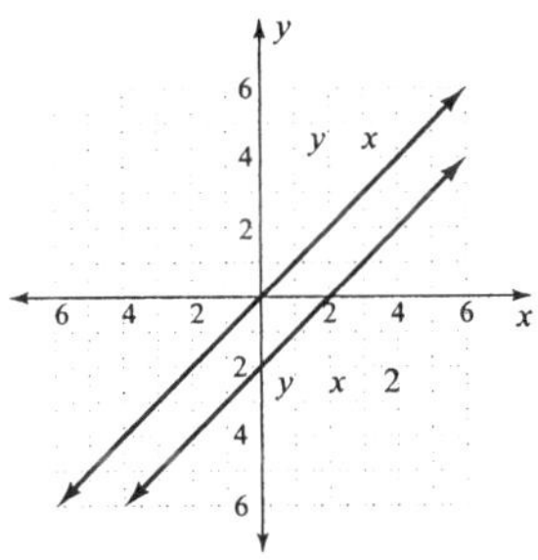

27. one example is $x - 2 > x$

29. $1 - 2x < x$

$1 < 3x$

$\frac{1}{3} < x$ or $x > \frac{1}{3}$

31. $-\frac{1}{2}x + 1 < 4 - x$

$\frac{1}{2}x < 3$

$x < 6$

33. $0.4x - 6 \le 4$

$0.4x \le 10$

$x \le 25$

35. $8 - 0.75x > 11$

$-0.75x > 3$

$x < -4$

37. $3(1 - x) - 1 < 2.5x - 9$

$3 - 3x - 1 < 2.5x - 9$

$11 < 5.5x$

$2 < x$ or $x > 2$

39. $5 - 3.5(x + 1) \ge 4 - 1.5x$

$5 - 3.5x - 3.5 \ge 4 - 1.5x$

$-2.5 \ge 2x$

$-1.25 \ge x$ or $x \le -1.25$

41. $10 < 5x < 20$
$10 \div 5 < 5x \div 5 < 20 \div 5$
$2 < x < 4$

43. $10 < -5x < 20$
$10 \div -5 > -5x \div -5 > 20 \div -5$
$-2 > x > -4$

45. $-3 < 2 - x < 4$
$-3 - 2 < 2 - x - 2 < 4 - 2$
$-5 < -x < 2$
$-5 \div -1 > -x \div -1 > 2 \div -1$
$5 > x > -2$

47. $-16 \le 4 - 5x \le 29$
$-16 - 4 \le 4 - 5x - 4 \le 29 - 4$
$-20 \le -5x \le 25$
$-20 \div -5 \ge -5x \div -5 \ge 25 \div -5$
$4 \ge x \ge -5$

49. a. $x = -2$ **b.** $\Re$ **c.** $-6 < x < 2$
d. $\Re$ **e.** no solution **f.** $x = 1$ or $x = -5$
g. $x < -5$ or $x > 1$

51. $|2x - 3| > 7$
$2x - 3 > 7$ or $-(2x - 3) > 7$
$2x > 10$ or $-2x + 3 > 7$
$x > 5$ or $-2x > 4$
$x < -2$
$x < -2$ or $x > 5$

53. $|5 - 4x| < 11$
$5 - 4x < 11$ or $-(5 - 4x) < 11$
$-4x < 6$ or $-5 + 4x < 11$
$x > -1.5$ or $4x < 16$
$x < 4$
$-1.5 < x < 4$

55. $|3 - \frac{1}{2}x| > 4.5$
$3 - \frac{1}{2}x > 4.5$ or $-(3 - \frac{1}{2}x) > 4.5$
$-\frac{1}{2}x > 1.5$ or $-3 + \frac{1}{2}x > 4.5$
$x < -3$ or $\frac{1}{2}x > 7.5$
$x > 15$
$x < -3$ or $x > 15$

57. $|0.5 - 0.25x| \le 2$
$0.5 - 0.25x \le 2$ or $-(0.5 - 0.25x) \le 2$
$-0.25x \le 1.5$ or $-0.5 + 0.25x \le 2$
$x \ge -6$ or $0.25x \le 2.5$
$x \le 10$
$-6 \le x \le 10$

59. $0.025x \le 5$
$x \le \$200$

61. $5 \le 0.02x \le 25$
$\$250 \le x \le \$1250$

63. $3.75 + 0.50x < 9.00 + 0.25x$
$0.50x < 5.25 + 0.25x$
$0.25x < 5.25$
$x < 21$ photos

65. $30 + 6x < 50 + 5x$
$6x < 20 + 5x$
$x < 20$ guests

67. There is a \$20 difference in the basic cost. The per guest charge differs by \$1. It will take \$20 ÷ \$1/guest = 20 guests for the cost to be equal.

69. To check the solution to a compound inequality, choose a number that satisfies the answer and see if it also satisfies the inequality.

71. The negative changes the relative value of the sides: for example,
$20 > 10 \Rightarrow (-1)(20) > (-1)(10) \Rightarrow -20 < -10$

73. A dot at the endpoint indicates that value is part of the solution set; a small circle indicates that value is excluded from the solution set.

Chapter 3 Review Exercises

1.
$x + 2y = -4$
$3x - 4y = 4$
$x = -4 - 2y$
$3(-4 - 2y) - 4y = 4$
$-12 - 6y - 4y = 4$
$-10y = 16$
$\mathbf{y = -1.6}$
$x + 2(-1.6) = -4$
$\mathbf{x = -0.8}$

3.
$y = 2x - 1$
$\frac{1}{2}(y + 1) = x$
$\frac{1}{2}(y + 1)(2) = 2x$
$y + 1 = 2x$
$y = 2x - 1$ same as first equation
coincident lines

5.
$y = 3x - 1$
$y - 2 = 3x$
$y = 3x + 2$; same slope as first eqn.
parallel lines

7.
$5x - 102 = 9y$
$3.5x + 1.5y = 9$
$5x - 9y = 102$ (a)
$6(3.5x + 1.5y) = 6(9)$
$21x + 9y = 54$ (b)
add (a) and (b); $26x = 156$; $\mathbf{x = 6}$
$5(6) - 102 = 9y$; $-72 = 9y$; $\mathbf{y = -8}$

9.
$y = -\frac{1}{3}x - \frac{1}{3}$
$2.5x - 4y = 20.5$
$2.5x - 4(-\frac{1}{3}x - \frac{1}{3}) = 20.5$
$2.5x + \frac{4}{3}x + \frac{4}{3} = 20.5$
$3\frac{5}{6}x = 19\frac{1}{6}$
$\mathbf{x = 5}$
$y = -\frac{1}{3}(5) - \frac{1}{3}$
$\mathbf{y = -2}$

11.
$0.3x - 0.4y = 1.4$
$0.2x - 0.5y = 0.7$
$-1.5(0.2x - 0.5y) = -1.5(0.7)$
$-0.3x + 0.75y = -1.05$
add last eqn. to first eqn.;
$0.35y = 0.35$; $\mathbf{y = 1}$
$0.3x - 0.4(1) = 1.4$
$0.3x = 1.8$; $\mathbf{x = 6}$

13.
$\frac{1}{2}x + \frac{3}{4}y = 2$
$\frac{1}{3}x - \frac{1}{2}y = -1$
$2(\frac{1}{2}x + \frac{3}{4}y) = 2(2)$
$-3(\frac{1}{3}x - \frac{1}{2}y) = -3(-1)$
$x + 1.5y = 4$
$-x + 1.5y = 3$ *add*
$3y = 7$
$\mathbf{y = 2\frac{1}{3}}$
$\frac{1}{2}x + \frac{3}{4}(\frac{7}{3}) = 2$
$0.5x + 1.75 = 2$
$0.5x = 0.25$
$\mathbf{x = 0.5}$

15.
$a + b + c = 0$ (1)
$a - b + c = 6$ (2)
$a - 2b - c = 7$ (3)
subtract (2) from (1)
$2b = -6$; $\mathbf{b = -3}$
substitute into (2) and (3)
$a + 3 + c = 6$; $a + c = 3$ (4)
$a - 2(-3) - c = 7$; $a - c = 1$ (5)
add (4) and (5)
$2a = 4$; $\mathbf{a = 2}$
$2 + (-3) + c = 0$
$\mathbf{c = 1}$

17.
$3x - y + z = 4$
$-x + z = 2$; $x = z - 2$
$-y + 4z = 10$; $y = 4z - 10$
$3(z - 2) - (4z - 10) + z = 4$
$3z - 6 - 4z + 10 + z = 4$
$4 = 4$; identity
dependent equations
infinite number of solutions

19.
$2x + y - z = 3$ (1)
$x + y + 2z = 4$ (2)
$x - y - 3z = 6$ (3)
subtract (1) and (2)
$x - 3z = -1$ $x = 3z - 1$ (4)
subtract (4) from (3); $-y = 7$; $\mathbf{y = -7}$
substitute y and (4) into (2)
$(3z - 1) + (-7) + 2z = 4$
$5z - 8 = 4$; $\mathbf{z = 2.4}$
$x - (-7) - 3(2.4) = 6$; $\mathbf{x = 6.2}$

21. Lines represent $y = 2x - 3$ and $y = -x - 6$; solution is the intersection $(-1, -5)$.

23. Lines represent $y = \frac{1}{2}x + 3$ and $y = -\frac{1}{4}x$; solution is the intersection $(-4, 1)$.

25.

Cr. Hrs.	Grade Pt.	Total
20	2.80	20(2.80) = 56.0
x	3.25	3.25x
20 + x	3.00	3.00(20 + x)

$3.00(20 + x) = 56 + 3.25x$

$60 + 3x = 56 + 3.25x$

$4 = 0.25x$; $x = 16$ credit hours

27

Pounds	Protein	Total
C	0.13	0.13C
D	0.08	0.08D
1000	0.09	90

$C + D = 1000$; $C = 1000 - D$

$0.13C + 0.08D = 90$

$0.13(1000 - D) + 0.08D = 90$

$130 - 0.13D + 0.08D = 90$

$40 = 0.05D$

D = 800 lb

$C = 1000 - 800$; **C = 200 lb**

29. $L + W + H = 12$

$L = H + 2$

$W = \frac{1}{2}(L + H)$; $2W = L + H$

$(H + 2) + W + H = 12$

$2H + W = 10$; $W = 10 - 2H$

$2W = (H + 2) + H$

$2W = 2H + 2$

$2(10 - 2H) = 2H + 2$

$20 - 4H = 2H + 2$

$18 = 6H$; **H = 3 in.**

$W = 10 - 3(3)$; **W = 4 in.**

$L = 3 + 2$; **L = 5 in.**

31. $L = W + H$; $L - W - H = 0$

$W = H + 2$

$L + W + H = 20$

add first and third eqns

$2L = 20$; **L = 10 in.**

$10 = (H + 2) + H$

$8 = 2H$; **H = 4 in.**

$W = 4 + 2$; **W = 6 in.**

33. Let x = # of Jeanne's perfect games and y = # of Tish's perfect games

$x = y + 2$

$x + y = 40$

$(y + 2) + y = 40$

$2y = 38$; $\mathbf{y = 19}$

$x = 19 + 2$; $\mathbf{x = 21}$

35. $A + B = 180$

$A = 5B$

$5B + B = 180$

$6B = 180$

$\mathbf{B = 30°}$

$A = 5(30)$

$\mathbf{A = 150°}$

37. $A + B = 90$

$2A + 2B = 180$

$2(A + B) = 180$

$A + B = 90$; identity

dependent equations

infinite number of solutions

39. $p + f + c = 49$

$4p + 9f + 4c = 201$

$c = p + 10$

$p + f + (p + 10) = 49$

$2p + f = 39$

$4p + 9f + 4(p + 10) = 201$

$8p + 9f = 161$ (1)

$-4(2p + f) = -4(39)$

$-8p - 4f = -156$ (2)

add (1) and (2)

$5f = 5$; **f = 1 g**

$2p + 1 = 39$; **p = 19 g**

$c = 19 + 10$; **c = 29 g**

41. $n + d + q = 74$

$0.05n + 0.10d + 0.25q = 12.40$

$q = 2n - 5$

$n + d + (2n - 5) = 74$

$3n + d = 79 \quad d = 79 - 3n$

$0.05n + 0.10d + 0.25(2n - 5) = 12.40$

$0.55n + 0.10d = 13.65$

$0.55n + 0.10(79 - 3n) = 13.65$

$0.55n + 7.9 - 0.30n = 13.65$

$0.25n = 5.75$

$\mathbf{n = 23}$

$q = 2(23) - 5$; $\mathbf{q = 41}$

$23 + d + 41 = 74$; $\mathbf{d = 10}$

43. $2t + 1p + 0.5n = 17$

$26t + 3.50p + 1.50n = 152.50$

$n = t + p$

$2t + p + 0.5(t + p) = 17 \quad 2.5t + 1.5p = 17$

$26t + 3.50p + 1.50(t + p) = 152.50 \quad 27.5t + 5p = 152.50$ (1)

$-11(2.5t + 1.5p) = -11(17) \quad -27.5t - 16.5p = -187$ *add to (1)*

$-11.5p = -34.5$; $\mathbf{p = 3} \quad 2.5t + 1.5(3) = 17$; $\mathbf{t = 5} \quad n = 5 + 3$; $\mathbf{n = 8}$

45.a. 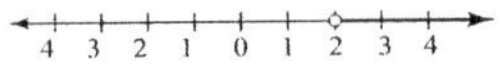

$-\frac{3}{4}x - \frac{1}{2} < -2$

$4(-\frac{3}{4}x - \frac{1}{2}) < 4(-2)$

$-3x - 2 < -8$

$-3x < -6$

$x > 2$

b. 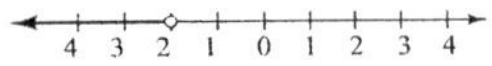

$\frac{1}{3}x + \frac{5}{3} < -\frac{3}{4}x - \frac{1}{2}$

$12(\frac{1}{3}x + \frac{5}{3}) < 12(-\frac{3}{4}x - \frac{1}{2})$

$4x + 20 < -9x - 6$

$13x < -26$

$x < -2$

c. 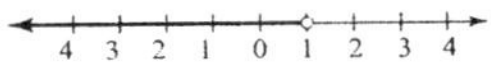

$\frac{1}{3}x + \frac{5}{3} < 2$

$3(\frac{1}{3}x + \frac{5}{3}) < 3(2)$

$x + 5 < 6$

$x < 1$

47. $-2x + 1 \leq 3$

$-2x \leq 2$

$x \geq -1$

49. $\frac{1}{2}x - \frac{3}{2} < -1$

$2(\frac{1}{2}x - \frac{3}{2}) < 2(-1)$

$x - 3 < -2$

$x < 1$

51. $\frac{1}{2}x - \frac{3}{2} \leq -2x + 1$

$2(\frac{1}{2}x - \frac{3}{2}) \leq 2(-2x + 1)$

$x - 3 \leq -4x + 2$

$5x \leq 5;\; x \leq 1$

53. a. $19 < x\left(\frac{704.5}{74^2}\right) < 24$

$19(74^2) < 704.5x < 24(74^2)$

$147.7 < x < 186.5$

b. $19 < x\left(\frac{704.5}{62^2}\right) < 24$

$19(62^2) < 704.5x < 24(62^2)$

$103.7 < x < 131.0$

55. $8 < 0.05x < 9;\; 160 < x < 180$

57. $|4x - 3| < 6$

$4x - 3 < 6$ or $-(4x - 3) < 6$

$4x < 9$ or $-4x + 3 < 6$

$x < 2.25$ or $-4x < 3$

$x > -0.75$

$-0.75 < x < 2.25$

59. $|0.75x - 5| > 1$

$0.75x - 5 > 1$ or $-(0.75x - 5) > 1$

$0.75x > 6$ or $-0.75x + 5 > 1$

$x > 8$ or $-0.75 > -4$

$x < 5\frac{1}{3}$

$x < 5\frac{1}{3}$ or $x > 8$

Chapter 3 Test

1.

$3x - 4y = 12$

$x - \frac{3}{2}y = -9$

$x = 1.5y - 9$ *substitute into 1st eqn.*

$3(1.5y - 9) - 4y = 12$

$4.5y - 27 - 4y = 12$

$0.5y = 39;$ $\mathbf{y = 78}$

$x = 1.5(78) - 9;$ $\mathbf{x = 108}$

2.

$5x - 2y = 5$

$5x = 2(2 + y)$

$5x - 2y = 4$ *subtract from 1st eqn.*

$0 = 1$; contradiction

no solution; parallel lines

3.

$2x - 3y = 6$

$4x + 5y = -32$

$-2(2x - 3y) = -2(6)$

$-4x + 6y = -12$ *add to 2nd eqn.*

$11y = -44;$ $\mathbf{y = -4}$

$2x - 3(-4) = 6$

$2x = -6;$ $\mathbf{x = -3}$

4.

$x + 2y = -8;$ $x = -2y - 8$

$0.5x + y = -4$

$0.5(-2y - 8) + y = -4$

$-y - 4 + y = -4$

$-4 = -4$; identity

infinite number of solutions

coincident lines

5.

$2x + y + 2z = 5$ (1)

$x + 3z = 4;$ $x = 4 - 3z$ (2)

$4x + 3y = 0$ (3)

$-3(2x + y + 2z) = -3(5)$

$-6x - 3y - 6z = -15$ (4)

add (3) and (4)

$-2x - 6z = -15$ (5)

substitute (2) into (5)

$-2(4 - 3z) - 6z = -15$

$-8 + 6z - 6z = -15$

$-8 = -15$; contradiction

no solution; inconsistent

6.

$3x + y + z = 8$ (1)

$2x - y - 2z = 17$ (2)

$-x - y + 3z = -9$ (3)

add (1) and (2), then add (1) and (3)

$5x - z = 25$ $z = 5x - 25$ (4)

$2x + 4z = -1$ (5)

substitute (4) into (5)

$2x + 4(5x - 25) = -1$

$2x + 20x - 100 = -1$

$22x = 99;$ $\mathbf{x = 4.5}$

$z = 5(4.5) - 25;$ $\mathbf{z = -2.5}$

$3(4.5) + y + (-2.5) = 8$

$\mathbf{y = -3}$

7. $L = W + H;\; L - W - H = 0$
$W = H + 8$
$L + W + H = 24$
add 1st and 3rd equations
$2L = 24;$ **$L = 12$ in.**
$12 + (H + 8) + H = 24$
$2H + 20 = 24;$ **$H = 2$ in.**
$W = 2 + 8;$ **$W = 10$ in.**

8. $C + P = 3922$
$C + 10 = P$
$C + (C + 10) = 3922$
$2C + 10 = 3922$
$2C = 3912$
$C = 1956$
$1956 + 10 = P;$ **$P = 1966$**

9. $0.16x = 0.014(400)$
$x + w = 400$
$0.16x = 5.6$
$x = 35$ gal
$35 + w = 400$
$w = 365$ gal

10. $p + n + q = 59$
$q = n + 10$
$0.01p + 0.05n + 0.25q = 6.35$
$p + n + (n + 10) = 59$
$p + 2n = 49;\; p = 49 - 2n$
$0.01p + 0.05n + 0.25(n + 10) = 6.35$
$0.01p + 0.30n = 3.85$
$0.01(49 - 2n) + 0.30n = 3.85$
$0.49 - 0.02n + 0.30n = 3.85$
$0.28n = 3.36;$ **$n = 12$**
$q = 12 + 10;$ **$q = 22$**
$p + 12 + 22 = 59;$ **$p = 25$**

11. $j + a + c = 11$ (1)
$0.89j + 0.59a + 3.98c = 15.07$ (2)
$17j + 8a + 16c = 158$ (3)
$-0.89(j + a + c) = -0.89(11)$
$-0.89j - 0.89a - 0.89c = -9.79$ (4) add (2) and (4)
$-0.30a + 3.09c = 5.28$ (5)
$-17(j + a + c) = -17(11)$
$-17j - 17a - 17c = -187$ (6) add (3) and (6)
$-9a - c = -29;\; c = -9a + 29$ substitute into (5)
$-0.30a + 3.09(-9a + 29) = 5.28$
$-0.30a - 27.81a + 89.61 = 5.28$
$-28.11a = -84.33;$ **$a = 3$**
$c = -9(3) + 29;$ **$c = 2$**
$j + 3 + 2 = 11;$ **$j = 6$**

12. a. The graphs intersect in a plane or line - **true**
b. The system is dependent - **true**
c. The system contains parallel planes - **false**
d. The algebra yielded a contradiction - **false**

13. Lines represent $y = x - 1$ and $y = -\frac{1}{2}x - 4$. Solution is intersection $(-2, -3)$

14. $-0.5 < 0.5 - 0.25x < 2.5$

$-1 < -0.25x < 2$

$4 > x > -8$ or $-8 < x < 4$

15. $|2 - 0.5x| < 4$

$2 - 0.5x < 4$ or $-(2 - 0.5x) < 4$

$-0.5x < 2$ or $-2 + 0.5x < 4$

$x > -4$ or $0.5x < 6$

$x < 12$

$-4 < x < 12$

16. $19 \le \frac{704.5}{68^2}x \le 24$

$\frac{68^2 \cdot 19}{704.5} \le x \le \frac{68^2 \cdot 24}{704.5}$

$124.7 \text{ lb} \le x \le 157.5 \text{ lb}$

Exercises 4.1

1. a.

Outputs	Differences 1st	2nd
1		
	2	
3		4
	6	
9		12
	18	
27		36
	54	
81		108
	162	
243		

each number is 3 times the previous number; next number is 3(243) = 729
sequence is neither linear or quadratic

b.

Outputs	Differences 1st
11	
	7
18	
	7
25	
	7
32	
	7
39	

each number is 7 more than the previous number; next number is 39 + 7 = 46
sequence is linear

c.

Outputs	Differences 1st
56	
	−9
47	
	−9
38	
	−9
29	
	−9
20	

each number is 9 less than the previous number; next number is 20 – 9 = 11; sequence is linear

d.

Outputs	Differences 1st	2nd
−1		
	2	
1		4
	6	
7		4
	10	
17		4
	14	
31		

using 2nd difference, next 1st difference is 18, next number is 49 sequence is quadratic

3. a. equation has degree 3, neither linear or quadratic
b. equation has degree 5, neither linear or quadratic
c. equation has degree 2, quadratic
d. equation has degree 1, linear

5. **a.** In $f(r) = \pi r^2 + 2\pi r(3 \text{ meters})$, the variable is **r**; $a = \pi$, $b = 2\pi(3)$, $c = 0$

b. In $f(T) = \dfrac{g}{4\pi^2}T^2$, the variable is **T**; $a = \dfrac{g}{4\pi^2}$, $b = 0$, $c = 0$

c. $y = \frac{1}{2}x(x + 1) = \frac{1}{2}x^2 + \frac{1}{2}x$, the variable is **x**, $a = \frac{1}{2}$, $b = \frac{1}{2}$, $c = 0$

7. **a.** equation has degree 3, not quadratic

b. note right side will have degree 2; quadratic

$x - 4 = x(x + 4)$

$x - 4 = x^2 + 4x$

$0 = x^2 + 3x + 4$; $a = 1, b = 3, c = 4$

c. note right side will have degree 2; quadratic

$x - 2(x - 4) = x(x - 4)$

$x - 2x + 8 = x^2 - 4x$

$-x + 8 = x^2 - 4x$

$0 = x^2 - 3x - 8$; $a = 1, b = -3, c = -8$

d. equation has degree 1, not quadratic

9.

x	$y = 2x^2 + x + 1$
−3	$2(-3)^2 + (-3) + 1 = 16$
−2	$2(-2)^2 + (-2) + 1 = 7$
−1	$2(-1)^2 + (-1) + 1 = 2$
0	$2(0)^2 + (0) + 1 = 1$
1	$2(1)^2 + (1) + 1 = 4$
2	$2(2)^2 + (2) + 1 = 11$
3	$2(3)^2 + (3) + 1 = 22$

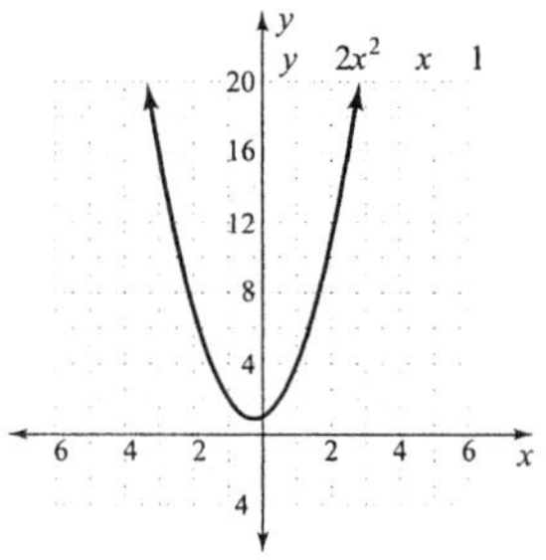

11.

x	$y = -x^2 - 2$
−3	$-(-3)^2 - 2 = -11$
−2	$-(-2)^2 - 2 = -6$
−1	$-(-1)^2 - 2 = -3$
0	$-(0)^2 - 2 = -2$
1	$-(1)^2 - 2 = -3$
2	$-(2)^2 - 2 = -6$
3	$-(3)^2 - 2 = -11$

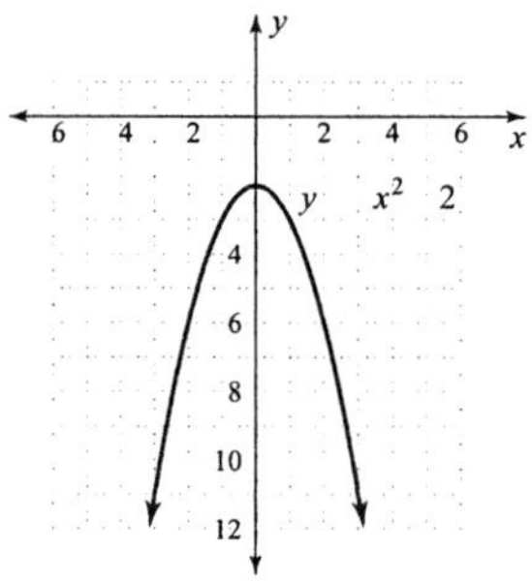

13. The x-intercepts are where $f(x) = 0$; (1, 0) and (3, 0). The y-intercept is where $x = 0$; (0, 3); noting symmetry in table, the vertex is (2, −1)

15.

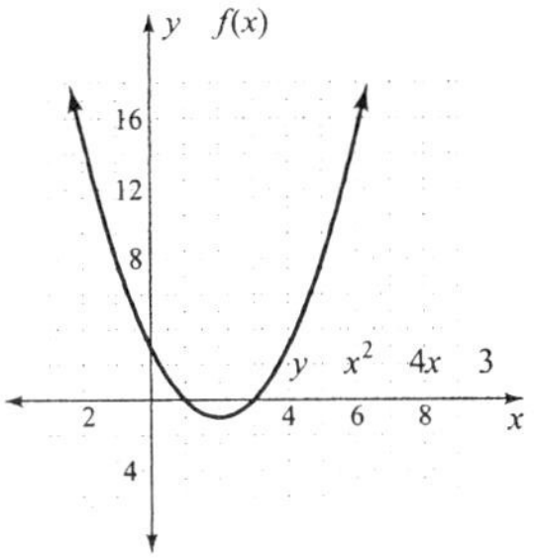

17. $f(6) = 6^2 - 4(6) + 3$

$f(6) = 36 - 24 + 3$

$f(6) = 15$

19. a. from the table $\{-1, 5\}$

b. from the table $\{1, 3\}$

c. from the table and ex. 17 $\{-2, 6\}$

from the table $\{\ \}$ or $\varnothing$

21. a. from the graph x-intercepts are at $(-2, 0)$ and $(4, 0)$

b. from the graph y-intercept is $(0, -8)$

c. axis of symmetry goes through the vertex at $x = 1$

d. vertex is at $(1, -9)$

e. from the graph: $x^2 - 2x - 8 = -9$ $\{1\}$

f. from the graph: $x^2 - 2x - 8 = -5$ $\{-1, 3\}$

g. from the graph: $x^2 - 2x - 8 = 0$ $\{-2, 4\}$

h. range of $f(x)$ is $y \geq -9$

23.

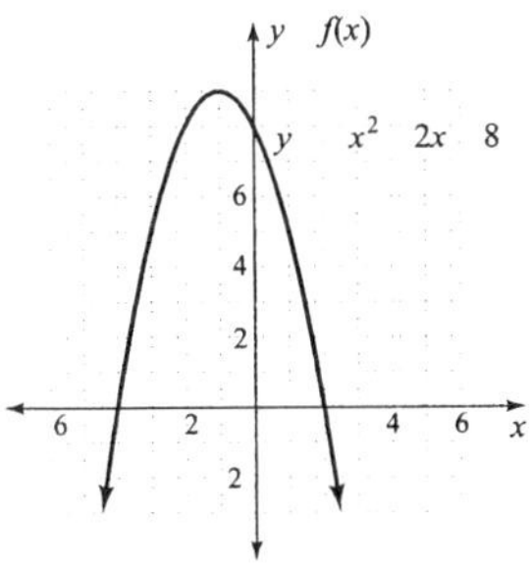

x	$f(x) = -x^2 - 2x + 8$
−4	$-(-4)^2 - 2(-4) + 8 = 0$
−3	$-(-3)^2 - 2(-3) + 8 = 5$
−2	$-(-2)^2 - 2(-2) + 8 = 8$
−1	$-(-1)^2 - 2(-1) + 8 = 9$
0	$-(0)^2 - 2(0) + 8 = 8$
1	$-(1)^2 - 2(1) + 8 = 5$
2	$-(2)^2 - 2(2) + 8 = 0$

a. x-intercepts are $(-4, 0)$ and $(2, 0)$

b. y-intercept is $(0, 8)$

c. axis of symmetry $x = -1$

d. vertex $(-1, 9)$

e. $-x^2 - 2x + 8 = 0$; $\{-4, 2\}$

f. $-x^2 - 2x + 8 = 8$; $\{-2, 0\}$

g. $-x^2 - 2x + 8 = 10$; $\{\ \}$ or $\varnothing$

h. $-x^2 - 2x + 8 = 5$; $\{-3, 1\}$

i. range is $y \leq 9$

25.

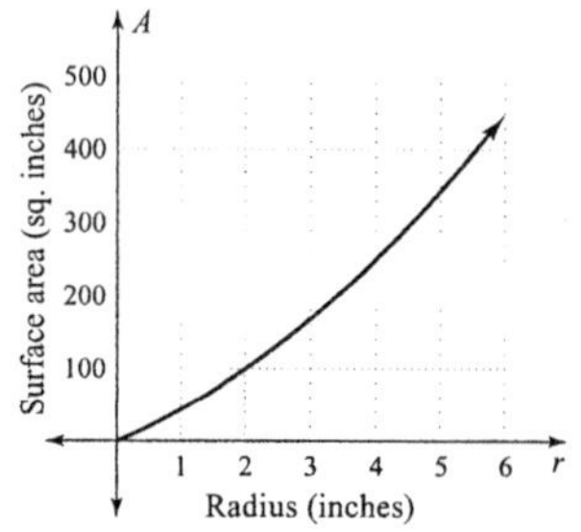

r	$A = 2\pi r^2 + 2\pi r(6)$
0	$2\pi(0)^2 + 12\pi(0) = 0$
1	$2\pi(1)^2 + 12\pi(1) \approx 44.0$
2	$2\pi(2)^2 + 12\pi(2) \approx 100.5$
3	$2\pi(3)^2 + 12\pi(3) \approx 169.6$
4	$2\pi(4)^2 + 12\pi(4) \approx 251.3$
5	$2\pi(5)^2 + 12\pi(5) \approx 345.6$
6	$2\pi(6)^2 + 12\pi(6) \approx 452.4$

a. from the table & graph ≈ 3.4 in **b.** from the table & graph ≈ 5.5 in.

c. The top and bottom surface areas vary with the square of the radius.

27. If $a = 0$ then $y = ax^2 + bx + c$ becomes $y = bx + c$, a linear equation. The statement that $a \neq 0$ in a quadratic equation is to distinguish between a quadratic and linear equation.

29. The range for the function is bounded by the y-coordinate of the vertex.

31. The other x-intercept will be an equal distance from the axis of symmetry, in this case 4 units to the left of the y-axis ($x = 0$) at $x = -4$.

Exercises 4.2

1.

Input	Output	Differences 1st	2nd
1	7		
		9	
2	16		2
		11	
3	27		2
		13	
4	40		2
		15	
5	55		

quadratic $2a = 2$; $a = 1$;
working backward, when $x = 0$, $y = 0$;
$c = 0$; $7 = 1(1)^2 + b(1) + 0$; $b = 6$
$y = x^2 + 6x$

3.

Input	Output	Differences 1st	2nd
1	4		
		4	
2	8		0
		4	
3	12		0
		4	
4	16		0
		4	
5	20		

linear; $m = 4$; when $x = 0$, $y = 0$
$b = 0$; $y = 4x$

5.

Input	Output	Differences	
		1st	**2nd**
1	8		
		19	
2	27		4
		23	
3	50		4
		27	
4	77		4
		31	
5	108		

quadratic; $2a = 4$, $a = 2$; when $x = 0$
$y = -7$, $c = -7$; $8 = 2(1)^2 + b(1) - 7$;
$b = 13$; $y = 2x^2 + 13x - 7$

7.

Input	Output	Differences	
		1st	**2nd**
1	5		
		7	
2	12		2
		9	
3	21		2
		11	
4	32		2
		13	
5	45		

quadratic; $2a = 2$, $a = 1$; when $x = 0$
$y = 0$; $c = 0$; $5 = 1(1)^2 + b(1) + 0$
$b = 4$; $y = x^2 + 4x$

9.

Input	Output	Differences	
		1st	**2nd**
1	13		
		12	
2	25		0
		12	
3	37		0
		12	
4	49		0
		12	
5	61		

linear; $m = 12$; when $x = 0$, $y = 1$; $b = 1$
$y = 12x + 1$

11.

Input	Output	Differences	
		1st	**2nd**
1	1		
		0	
2	1		1
		1	
3	2		0
		1	
4	3		1
		2	
5	5		

neither linear or quadratic

13.

Input	Output	Differences	
1	9	**1st**	**2nd**
		7	
2	16		−2
		5	
3	21		−2
		3	
4	24		−2
		1	
5	25		

quadratic; $2a = -2$, $a = -1$, when $x = 0$
$y = 0$, $c = 0$; $9 = -1(1)^2 + b(1) + 0$
$b = 10$; $y = -x^2 + 10x$

15.

Input	Output	Differences	
1	2	**1st**	**2nd**
		1	
2	3		2
		3	
3	6		2
		5	
4	11		2
		7	
5	18		2
		9	
6	27		

$2a = 2$, $a = 1$; when $x = 0$, $y = 3$, $c = 3$
$2 = 1(1)^2 + b(1) + 3$; $b = -2$
$y = x^2 - 2x + 3$
$(7)^2 - 2(7) + 3 = 38$
$(8)^2 - 2(8) + 3 = 51$
$(9)^2 - 2(9) + 3 = 66$

17.

$2a = 8$, $a = 4$; when $x = 0$, $y = -1$
$c = -1$; $3 = 4(1)^2 + b(1) - 1$;
$b = 0$; $y = 4x^2 - 1$
$4(6)^2 - 1 = 143$
$4(7)^2 - 1 = 195$
$4(8)^2 - 1 = 255$

Input	Output	Differences	
1	3	**1st**	**2nd**
		12	
2	15		8
		20	
3	35		8
		28	
4	63		8
		36	
5	99		

For exercises 19 to 27 equations are found on a graphing calculator. Results are given for reference only.

19. a. Unless there is a reason to price larger amounts relatively higher, the linear equation is reasonable.

b. $y = 0.245x + 0.891$, $r = 0.9897$ **c.** $y = 0.0140x^2 + 0.0233x + 1.424$

d. $4.81 (linear); $5.38 (quadratic)

21. a. Quadratic, a linear equation would suggest that posts of 2 or 4 inches are free.

b. $y = 0.5x - 2.16$, $r = 0.9911$ **c.** $0.0375x^2 - 0.175x + 0.26$

d. \$5.84 (linear); \$7.49 (quadratic)

23. a. From calculator regression: $y = -80x + 656$; $y = -16x^2 + 576$

b. The slope suggest a drop of 80 feet per second from a starting height of 656 feet.

c. Initial speed is zero feet per second. It cannot travel as far in the first second as in later seconds. The graph is not linear.

d. The constant term is the starting height, 576 feet.

e. Let $y = 0$; $0 = -16x^2 + 576$; $16x^2 = 576$; $x^2 = 36$; $x = 6$ sec (discard -6 as extraneous)

25. a. B(−50, 0), C(50, 0), D(0, 30); $y = -0.12x^2 + 30$

b. for B(0, 0): C (100, 0), D(50, 30); $y = -0.012x^2 + 1.2x$

for C(0, 0): B(−100, 0), D(−50, 30); $y = -0.012 - 1.2x$

for D(0, 0): B(−50, −30), C(50, −30), $y = -0.012x^2$

27. for A(0,0): B(200, 150), C(400,0) $y = -0.00375x^2 + 1.5x$

for B(0, 0): A(−200, −150), C(200, −150), $y = -0.00375x^2$

for C(0, 0): A(−400, 0), B(−200, 150), $y = -0.00375x^2 - 1.5x$

for D(0, 0): A(−200, 0), B(0, 150), C(200, 0), $Y = -0.00375x^2 + 150$

29. a.

x	$f(x) = ax^2 + bx + c$	1st diff	2nd diff
1	a + b + c		
		3a + b	
2	4a + 2b + c		2a
		5a + b	
3	9a + 3b + c		2a
		7a + b	
4	16a + 4b + c		2a
		9a + b	
5	25a + 5b + c		

b. The second differences are a constant: 2a

c. The lead term of first differences is 3a + b; could be used to solve for b.

29. d. & e.

x	f(x) = ax + b	1st diff.
1	a + b	
		a
2	2a + b	
		a
3	3a + b	
		a
4	4a + b'	
		a
5	5a + b	

f. first difference = a = m (slope)

Exercises 4.3

1. a. 1 · 48, 2 · 24, 3 · 16, 4 · 12, 6 · 8 **b.** 1 · 36, 2 · 18, 3 · 12, 4 · 9, 6 · 6

c. 1 · 72, 2 · 36, 3 · 24, 4 · 18, 6 · 12, 8 · 9

3. a. $\sqrt{y} + x\sqrt{y}$ is not a polynomial **b.** $x^2 + y^2$ is a binomial

5. a. 2πr is a monomial **b.** $\frac{1}{2}gt^2 + vt + h$ is a trinomial

7.

m	n	m + n	m · n
−3	−5	**−3 + (−5) = −8**	**(−3)(−5) = 15**
3	**4**	7	12
2	6	**2 + 6 = 8**	**2 · 6 = 12**
3	**5**	8	15
−4	**−6**	−10	24
−2	**−12**	−14	24
−3	−8	**−3 + (−8) = −11**	**(−3)(−8) = 24**
2	**−6**	−4	−12

9. a. $(x^3 + 2x^2 + 4x) - (2x^2 + 4x + 8) = x^3 + 2x^2 + 4x - 2x^2 - 4x - 8 = x^3 - 8$

b. $(x^3 - 6x^2 + 9x) - (3x^2 - 18x + 27) = x^3 - 6x^2 + 9x - 3x^2 + 18x - 27 = x^3 - 9x^2 + 27x - 27$

11. a. $(x - 6) + (2x + 5) = x - 6 + 2x + 5 = 3x - 1$

b. $(3x - 1) + (4x - 5) = 3x - 1 + 4x - 5 = 7x - 6$

13. a. $x(x - 2) + 2(x - 2) = x^2 - 2x + 2x - 4 = x^2 - 4$

b. $4(3 - x) + x(3 - x) = 12 - 4x + 3x - x^2 = 12 - x - x^2$

15. a. $(x^3 + 2x^2 + x) + (x^2 + 2x + 1) = x^3 + 2x^2 + x + x^2 + 2x + 1 = x^3 + 3x^2 + 3x + 1$

b. $(a^3 - a^2b + ab^2) + (a^2b - ab^2 + b^3) = a^3 - a^2b + ab^2 + a^2b - ab^2 + b^3 = a^3 + b^3$

17. $x(x^2 - 2xy + y^2) - y(x^2 - 2xy + y^2) = x^3 - 2x^2y + xy^2 - x^2y + 2xy^2 - y^3 =$
$x^3 - 3x^2y + 3xy^2 - y^3$

19.

Multiply	$2x$	$+3$
$3x$	$(3x)(2x) = 6x^2$	$(3)(3x) = 9x$
-1	$(-1)(2x) = -2x$	$(-1)(3) = -3$

$(2x + 3)(3x - 1) = 6x^2 + 7x - 3$

21.

Multiply	$3x$	$+1$
$3x$	$(3x)(3x) = 9x^2$	$(1)(3x) = 3x$
$+1$	$(1)(3x) = 3x$	$(1)(1) = 1$

$(3x + 1)(3x + 1) = 9x^2 + 6x + 1$

23.

Multiply	x^2	$-2x$	$+4$
x	$(x)(x^2) = x^3$	$(x)(-2x) = -2x^2$	$(4)(x) = 4x$
$+2$	$(2)(x^2) = 2x^2$	$(2)(-2x) = -4x$	$(2)(4) = 8$

$(x + 2)(x^2 - 2x + 4) = x^3 + 8$

25. $(x + 6)(x - 3) =$
$x^2 - 3x + 6x - 18 =$
$x^2 + 3x - 18$

27. $(x - 6)(x + 3) =$
$x^2 + 3x - 6x - 18 =$
$x^2 - 3x - 18$

29. $(x - 9)(x - 2) =$
$x^2 - 2x - 9x + 18 =$
$x^2 - 11x + 18$

31. $(x + 4)(x - 4) =$
$x^2 - 4x + 4x - 16 =$
$x^2 - 16$

33. $(x - 5)(x - 5) =$
$x^2 - 5x - 5x + 25 =$
$x^2 - 10x + 25$

35. $(2x - 3)(x + 4) =$
$2x^2 + 8x - 3x - 12 =$
$2x^2 + 5x - 12$

37. $(2x + 3)(x - 4) =$
$2x^2 - 8x + 3x - 12 =$
$2x^2 - 5x - 12$

39. $(3x - 1)(3x + 2) =$
$9x^2 + 6x - 3x - 2 =$
$9x^2 + 3x - 2$

41. $(x + 2)(x + 2) =$
$x^2 + 2x + 2x + 4 =$
$x^2 + 4x + 4$

43. $(2x - 1)(2x + 1) =$
$4x^2 + 2x - 2x - 1 =$
$4x^2 - 1$

45. $(x - 2)(x^2 + 2x + 4) =$
$x^3 + 2x^2 + 4x - 2x^2 - 4x - 8 =$
$x^3 - 8$

47.

Multiply	x^2	$-3x$	-2
x^2	$\mathbf{(x^2)(x^2) = x^4}$	$\mathbf{(x^2)(-3x) = -3x^3}$	$\mathbf{(x^2)(-2) = -2x^2}$
$+3x$	$\mathbf{(3x)(x^2) = 3x^3}$	$\mathbf{(3x)(-3x) = -9x^2}$	$\mathbf{(3x)(-2) = -6x}$
-2	$\mathbf{(-2)(x^2) = -2x^2}$	$\mathbf{(-2)(-3x) = 6x}$	$\mathbf{(-2)(-2) = 4}$

$(x^2 - 3x - 2)(x^2 + 3x - 2) = x^4 - 13x^2 + 4$

49.

Factor	**6x**	**+1**
2x	$12x^2$	$+2x$
−3	$-18x$	-3

$12x^2 - 16x - 3 = (6x + 1)(2x - 3)$

51.

Factor	**x**	**−3**
3x	$3x^2$	$-9x$
−4	$-4x$	$+12$

$3x^2 - 13x + 12 = (x - 3)(3x - 4)$

53. $x^2 + 7x + 12 =$
$x^2 + 4x + 3x + 12 =$
$x(x + 4) + 3(x + 4) =$
$(x + 4)(x + 3)$

55. $x^2 + x - 12 =$
$x^2 + 4x - 3x - 12 =$
$x(x + 4) - 3(x + 4) =$
$(x + 4)(x - 3)$

57. $x^2 - 7x + 12 =$
$x^2 - 3x - 4x + 12 =$
$x(x - 3) - 4(x - 3) =$
$(x - 3)(x - 4)$

59. $x^2 - 11x - 12 =$
$x^2 + x - 12x - 12 =$
$x(x + 1) - 12(x + 1) =$
$(x + 1)(x - 12)$

61. $x^2 + 3x - 28 =$
$x^2 + 7x - 4x - 28 =$
$x(x + 7) - 4(x + 7) =$
$(x + 7)(x - 4)$

63.

Factor	**2x**	**+5**
3x	$6x^2$	+15x
+2	+4x	+10

diagonal sum = 19x;
diagonal product = $60x^2$;
factors are 4x and 15x
$(2x + 5)(3x + 2)$

65.

Factor	**2x**	**+5**
3x	$6x^2$	+15x
–2	–4x	–10

diagonal sum = 11x;
diagonal product = $-60x^2$;
factors are –4x and 15x
$(2x + 5)(3x - 2)$

67.

Factor	**6x**	**–5**
x	$6x^2$	–5x
–2	–12x	+10

diagonal sum = –17x;
diagonal product = $60x^2$;
factors are –5x and –12x
$(6x - 5)(x - 2)$

69.

Factor	**3x**	**–2**
5x	$15x^2$	–10x
+3	+9x	–6

diagonal sum = –1x;
diagonal product = $-90x^2$
factors are –10x and 9x
$(3x - 2)(5x + 3)$

71.

Factor	**3x**	**+5**
2x	$6x^2$	+10x
+1	+3x	+5

diagonal sum = 13x;
diagonal product = $30x^2$;
factors 10x and 3x
$(3x + 5)(2x + 1)$

73.

Factor	**x**	**+6**
10x	$10x^2$	+60x
+1	+x	+6

diagonal sum = 61x;
diagonal product = $60x^2$;
factors are 60x and x
$(x + 6)(10x + 1)$

75. $6x^2 - 32x + 10 = 2(3x^2 - 16x + 5)$

Factor	**x**	**−5**
3x	$3x^2$	−15x
−1	−x	+5

diagonal sum = −16x;
diagonal product = 15x;
factors are −x and −15x
$(x - 5)(3x - 1)$

77. $x^3 + 2x^2 + 4x = x(x^2 + 2x + 4)$

79. $x^3 - 6x^2 + 9x = x(x^2 - 6x + 9)$
$x^2 - 6x + 9 =$
$x^2 - 3x - 3x + 9$
$x(x - 3) - 3(x - 3) =$
$(x - 3)(x - 3)$
$x^3 - 6x^2 + 9x = x(x - 3)^2$

81. $a^3 - a^2b + ab^2 = a(a^2 - ab + b^2)$

83. $a^3 + a^2b + ab^2 = a(a^2 + ab + b^2)$

85. All are in the form $(a \pm b)^2$
Exercises 32 & 33 are similar.

87. $3x^2 - 48 = 3(x^2 - 16)$
$x^2 - 16 = x^2 + 4x - 4x - 16 =$
$x(x + 4) - 4(x + 4) =$
$(x + 4)(x - 4)$
$3x^2 - 48 = 3(x + 4)(x - 4)$

89. $12x^2 - 27 = 3(4x^2 - 9)$

Factor	**2x**	**−3**
2x	$4x^2$	$-6x$
+3	$6x$	-9

diagonal sum is 0;
diagonal product is $-36x^2$
factors are $6x$ and $-6x$
$12x^2 - 27 = 3(2x - 3)(2x + 3)$

91. $20x^2 - 45 = 5(4x^2 - 9) =$
$5(2x - 3)(2x + 3)$ *see ex. 89*

93. $28 - 63x^2 = 7(4 - 9x^2) =$
$7(2 + 3x)(2 - 3x)$

95. a. Two or more numbers or expressions being multiplied are ***factors.***
b. A one-term polynomial is a ***monomial.***
c. A two-term polynomial is a ***binomial.***
d. A three-term polynomial is a ***trinomial.***
e. The result of multiplying a number times itself is a ***square.***
f. The result of multiplying three identical factors is a ***perfect cube.***

97. a.

Multiply	x^2	$+x$	$+1$
x^2	$(x^2)(x^2) = x^4$	$(x^2)(x) = x^3$	$(x^2)(1) = x^2$
$-x$	$(-x)(x^2) = -x^3$	$(-x)(x) = -x^2$	$(-x)(1) = -x$
-1	$(-1)(x^2) = -x^2$	$(-1)(x) = -x$	$(-1)(1) = -1$

$x^4 - x^2 - 2x - 1$

b.

Multiply	x^2	$-x$	$+1$
x^2	$(x^2)(x^2) = x^4$	$(x^2)(-x) = -x^3$	$(x^2)(1) = x^2$
$+x$	$(x)(x^2) = x^3$	$(x)(-x) = -x^2$	$(x)(1) = x$
-1	$(-1)(x^2) = -x^2$	$(-1)(-x) = x$	$(-1)(1) = -1$

$x^4 - x^2 + 2x - 1$

97. c.

Multiply	x^2	$+x$	$+1$
x^2	$(x^2)(x^2) = x^4$	$(x^2)(x) = x^3$	$(x^2)(1) = x^2$
$+x$	$(x)(x^2) = x^3$	$(x)(x) = x^2$	$(x)(1) = x$
-1	$(-1)(x^2) = -x^2$	$(-1)(x) = -x$	$(-1)(1) = -1$

$x^4 + 2x^3 + x^2 - 1$

d.

Multiply	x^2	$-x$	$+1$
x^2	$(x^2)(x^2) = x^4$	$(x^2)(-x) = -x^3$	$(x^2)(1) = x^2$
$-x$	$(-x)(x^2) = -x^3$	$(-x)(-x) = x^2$	$(-x)(1) = -x$
-1	$(-1)(x^2) = -x^2$	$(-1)(-x) = x$	$(-1)(1) = -1$

$x^4 - 2x^3 + x^2 - 1$

e.

Multiply	x^2	$+x$	$+1$
x^2	$(x^2)(x^2) = x^4$	$(x^2)(x) = x^3$	$(x^2)(1) = x^2$
$+x$	$(x)(x^2) = x^3$	$(x)(x) = x^2$	$(x)(1) = x$
$+1$	$(1)(x^2) = x^2$	$(1)(x) = x$	$(1)(1) = 1$

$x^4 + 2x^3 + 3x^2 + 2x + 1$

Mid-Chapter 4 Test

1. **a.** The highest or lowest point on a parabola is the ***vertex.***
b. The line of symmetry for $y = x^2$ is ***the y-axis or x = 0.***
c. A polynomial with 2 terms is a ***binomial.***
d. The standard form for a quadratic equation is $y = ax^2 + bx + c$.
e. The constant term is a quadratic equation is the place where the graph crosses the *y*-axis.

2.

x	$y = x^2 - 4x + 3$
a. −1	$(-1)^2 - 4(-1) + 3 = 8$
0	$(0)^2 - 4(0) + 3 = 3$
b. 1	$(1)^2 - 4(1) + 3 = 0$
c. 2	$(2)^2 - 4(2) + 3 = -1$
b. 3	$(3)^2 - 4(3) + 3 = 0$
4	$(4)^2 - 4(4) + 3 = 3$
a. 5	$(5)^2 - 4(5) + 3 = 8$

3.

$y = \frac{1}{2}n(n + 1) + 1$

$y = \frac{1}{2}n^2 + \frac{1}{2}n + 1$

$a = \frac{1}{2}, b = \frac{1}{2}\ c = 1$

4.

$y = 1000(1 + x)(1 + x)$

$y = 1000(1 + x + x + x^2)$

$y = 1000(x^2 + 2x + 1)$

$y = 1000x^2 + 2000x + 1000$

$a = 1000, b = 2000, c = 1000$

5. a.

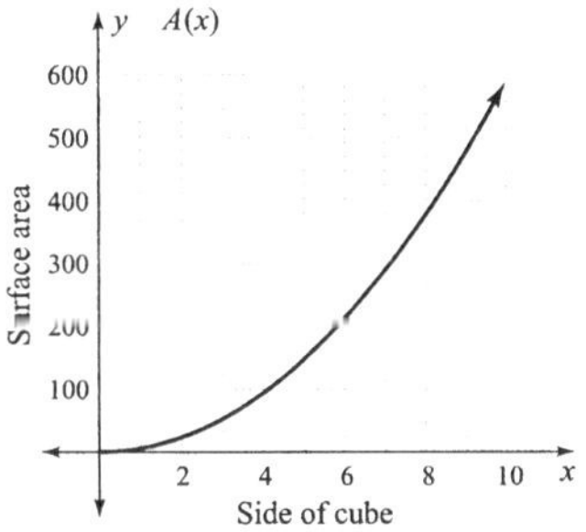

b. The surface area is 4 times greater.

x	$A = 6x^2$
0	$6(0)^2 = 0$
2	$6(2)^2 = 24$
4	$6(4)^2 = 96$
5	$6(5)^2 = 150$
6	$6(6)^2 = 216$
8	$6(8)^2 = 384$
10	$6(10)^2 = 600$

6. a.

x	y	1st diff.	2nd diff.
1	15		
		–7	
2	8		–2
		–5	
3	3		–2
		–3	
4	0		–2
		–1	
5	–1		

quadratic; $2a = -2$; $3a + b = -7$,

$a = -1$; $3(-1) + b = -7$, $b = -4$

when $x = 0$, $y = 24$, $y = -x^2 - 7x + 24$

b.

x	y	1st diff.	2nd diff.
1	4		
		5	
2	9		0
		5	
3	14		0
		5	
4	19		0
		5	
5	24		

linear; $m = 5$, when $x = 0$, $y = -1$

$y = 5x - 1$

6. c.

x	y	1st diff.	2nd diff.
1	−3		
		7	
2	4		0
		7	
3	11		0
		7	
4	18		0
		7	
5	25		

linear; m = 7, when x = 0, y = −10

d.

x	y	1st diff.	2nd diff.
1	1		
		−2	
2	−1		4
		2	
3	1		−4
		−2	
4	−1		4
		2	
5	1		

neither; $(-1)^2$ or y = 1 for odd x and y = −1 for even x

7. $f(x) \approx 4.9x^2$; quadratic regression was selected because second differences are a constant 9.81.

8. a. $x^3 - 6x^2 + 9x - x^2 + 18x - 27$
$x^3 - 7x^2 + 27x - 27$

b. $x^3 + 3x^2 + 9x + (-3x^2 - 9x - 27)$
$x^3 + 3x^2 + 9x - 3x^2 - 9x - 27$
$x^3 - 27$

c. $16a + 4b + c - (9a + 3b + c)$
$16a + 4b + c - 9a - 3b - c$
$7a + b$

d. $9a + 3b + c - (4a + 2b + c)$
$9a + 3b + c - 4a - 2b - c$
$5a + b$

9. a. $(x - 5)(x + 5)$
$x^2 + 5x - 5x - 25$
$x^2 - 25$

b. $(1 - x)(1 - x)$
$1 - x - x + x^2$
$x^2 - 2x + 1$

c. $(1 - x)(x + 3)$
$x + 3 - x^2 - 3x$
$-x^2 - 2x + 3$

d. $(2x + 3)(3 - 2x)$
$6x - 4x^2 + 9 - 6x$
$-4x^2 + 9$

10. a.

Factor	3x	+4
x	$3x^2$	4x
−3	−9x	−12

diagonal sum = +5x; diagonal product = $-36x^2$
factors are 4x and −9x; $3x^2 + 5x - 12 = (3x + 4)(x - 3)$

b. $x^2 - x = x(x - 1)$

10. c.

Factor	3x	+4
2x	$6x^2$	8x
−1	−3x	−4

diagonal sum = 5x
diagonal product = $-24x^2$
factors are 8x and −3x
$6x^2 + 5x - 4 = (3x + 4)(2x - 1)$

d. $3x^3 + 6x^2 + 3x = 3x(x^2 + 2x + 1)$

Factor	x	+1
x	x^2	x
+1	x	1

diagonal sum = 2x
diagonal product = x^2
factors are x & x
$3x^3 + 6x^2 + 3x = 3x(x + 1)^2$

11. a. $h_0 = 0$; $v_0 = 60$; $h = -\frac{1}{2}(32)t^2 + 60t + 0$

$h = -16t^2 + 60t$

b.

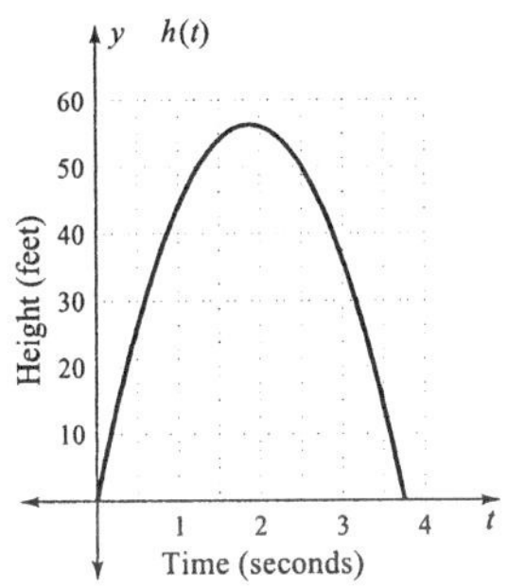

t	$h = -16t^2 + 60t$
0	$-16(0)^2 + 60(0) = 0$
1	$-16(1)^2 + 60(1) = 44$
2	$-16(2)^2 + 60(2) = 56$
3	$-16(3)^2 + 60(3) = 36$
4	$-16(4)^2 + 60(4) = -16$

c. x-intercepts from table and graph are: (0, 0) and (3.75, 0); the first is when the ball is thrown, the second when it hits the ground.

d. $56 = -16t^2 + 60t$; from the table x = 2; from the graph x = 1.75 {1.75, 2}, the ball passes through 56 ft on the way up and again on the way down.

e. from the graph, the ball reaches 56.25 ft.

Exercises 4.4

1. $(x - 3)(x + 4) =$
$x^2 + 4x - 3x - 12 =$
$x^2 + x - 12$

3. $(2x + 3)(x - 4) =$
$2x^2 - 8x + 3x - 12 =$
$2x^2 - 5x - 12$

5. $(x + 3)(x + 3) =$
$x^2 + 2(3x) + 9 =$
$x^2 + 6x + 9$ (perfect sq. trinomial)

7. $(x + 6)(x - 6) =$
$x^2 - 6x + 6x - 36 =$
$x^2 - 36$ (difference of squares)

9. $(2x - 5)^2 =$
$(2x)^2 + 2(2x)(-5) + (-5)^2 =$
$4x^2 - 20x + 25$ (perfect sq. trinomial)

11. $x^2 + \underline{2(4)}x + 16 = (x + 4)^2$
$x^2 + \underline{\mathbf{8}}x + 16 = (x + 4)^2$

13. $x^2 - \underline{2\left(\sqrt{49}\right)}x + 49 = (x - \underline{\sqrt{49}}\,)^2$
$x^2 - \underline{\mathbf{14}}x + 49 = (x - \underline{\mathbf{7}})^2$

15. $x^2 - 5x + \underline{(-2.5)^2} = (x - 2.5)^2$
$x^2 - 5x + \underline{\mathbf{6.25}} = (x - 2.5)^2$

17. $x^2 + 7x + \underline{\left(\frac{7}{2}\right)^2} = (x + \frac{7}{2})^2$
$x^2 + 7x + \underline{\mathbf{12.25}} = (x + \underline{\mathbf{3.5}})^2$

19. $x^2 - 24x + \underline{(\frac{24}{2})^2} = (x - \underline{\frac{24}{2}})^2$
$x^2 - 24 + \underline{\mathbf{144}} = (x - \underline{\mathbf{12}})^2$

21. $x^2 + 4x + 4 =$
$x^2 + 2(2x) + 2^2 =$
$(x + 2)^2$

23. $n^2 + 2n + 1 =$
$n^2 + 2(n) + 1^2 =$
$(n + 1)^2$

25. $49x^2 + 14x + 1 =$
$(7x)^2 + 2(7x)(1) + 1^2$
$(7x + 1)^2$

27. $9x^2 - 6xy + y^2 =$
$(3x)^2 + 2(3x)(-y) + (-y)^2 =$
$(3x - y)^2$

29. $x^2 + 5x + 6\frac{1}{4} =$
$x^2 + 2(x)(\frac{5}{2}) + (\frac{5}{2})^2 =$
$(x + 2\frac{1}{2})^2$

31. $x^2 - 4 =$
$x^2 - 2^2 =$
$(x - 2)(x + 2)$

33. $n^2 + 16$
cannot be factored

35. $49x^2 - 1 =$
$(7x)^2 - 1^2 =$
$(7x - 1)(7x + 1)$

37. $y^2 - 36z^2 =$
$y^2 - (6z)^2 =$
$(y + 6z)(y - 6z)$

39. $9x^2 - y^2 =$
$(3x)^2 - y^2 =$
$(3x + y)(3x - y)$

41. $x^2 - 144 =$
$x^2 - (12)^2 =$
$(x - 12)(x + 12)$

43. $0.25x^2 - 0.01 =$
$(0.5x)^2 - 0.1^2 =$
$(0.5x + 0.1)(0.5x - 0.1)$

45.

Multiply	**n**	
n	n^2	**–4n**
	+4n	+16

Diagonals must add to zero and multiply to $+16n^2$. Both results are not possible.

47. $y = (x + 5)^2$ is $y = x^2$ shifted to the left 5 units. x-intercept = (–5, 0)

49. To find $y = (2x - 5)^2$ from $y = (2x)^2$, shift right 5 units.

51. a. vertex of $y = (3x)^2$ is (0, 0); vertex of $y = (3x + 6)^2$ is (–2, 0); vertex of $y = (sx - t)^2$ would be on x-axis at $\frac{t}{s}$.

b. vertex will be at $(-\frac{1}{2}, 0)$

c. vertex will be at $(\frac{4}{3}, 0)$

53. a. shift is right $\left|\frac{t}{s}\right|$ when $t > 0$, left $\left|\frac{t}{s}\right|$ when $t < 0$

b. $t = 1$, $s = 2$; shift is right $\frac{1}{2}$ unit

c. $(2x + 3)^2 = (2x - (-3))^2$; $s = 2$, $t = -3$; shift is left $\frac{3}{2}$ units

55. $y = (x - 8)(x + 8)$ is the same as $y = x^2 - 64$; shift $y = x^2$ down 64 units

57. $(x + 1)(x^2 - x + 1) =$
$x^3 + 1$
sum of cubes

59. $(x - 3)(x^2 - 6x + 9) =$
$x^3 - 6x^2 + 9x - 3x^2 + 18x - 27$
$x^3 - 9x^2 + 27x - 27$

61. $(x + 1)(x^2 - 2x + 1) =$
$x^3 - 2x^2 + x + x^2 - 2x + 1$
$x^3 - x^2 - x + 1$

63.

Factor	x^2	**–4x**	**+16**
x	x^3	$-4x^2$	**+16x**
+4	$+4x^2$	**–16x**	+64

$x^3 + 64 = (x + 4)(x^2 - 4x + 16)$

65. a. $x^3 + 27 = x^3 + 3^3 =$

$(x + 3)(x^2 - 3x + 3^2) =$

$(x + 3)(x^2 - 3x + 9)$

b. $x^3 + 8 = x^3 + 2^3 =$

$(x + 2)(x^2 - 2x + 2^2) =$

$(x + 2)(x^2 - 2x + 4)$

c. $x^3 - 64 = x^3 - 4^3 =$

$(x - 4)(x^2 + 4x + 4^2) =$

$(x - 4)(x^2 + 4x + 16)$

67. a. $x^3 - 0.001 = x^3 - 0.1^3 =$

$(x - 0.1)(x^2 + 0.1x + 0.1^2) =$

$(x - 0.1)(x^2 + 0.1x + 0.01)$

b. $x^3 - \frac{1}{27} = x^3 - (\frac{1}{3})^3 =$

$(x - \frac{1}{3})(x^2 + \frac{1}{3}x + (\frac{1}{3})^2) =$

$(x - \frac{1}{3})(x^2 + \frac{1}{3}x + \frac{1}{9})$

69. They are not the same. $y = (x + 2)^3$ is $y = x^3$ shifted to the left 2 units; $y = x^3 + 8$ is $y = x^3$ moved up 8 units.

71. a. $A = \pi(R^2 - r^2)$

$A = \pi(R + r)(R - r)$

b. $R - r$ is the distance between the 2 radii.

73. x can be any real number

75. a. $(19)(21) = 20^2 - 1 = 400 - 1 = 399$

b. $(39)(41) = 40^2 - 1 = 1600 - 1 = 1599$

77. a. $(8)(12) = (10 - 2)(10 + 2) = 10^2 - 4$

b. $(18)(22) = (20 - 2)(20 + 2) = 20^2 - 4$

c. $(98)(102) = (100 - 2)(100 + 2) = 100^2 - 4$

All use the pattern $x^2 - 4 = (x - 2)(x + 2)$

Exercises 4.5

For exercises 1 through 23 results of factoring are given but details are not. Please refer to sections 4.3 and 4.4 for factoring details.

1. $x^2 - x - 6 = 0$

$(x + 2)(x - 3) = 0$

$x + 2 = 0$ or $x - 3 = 0$

$x = -2$ or $x = 3$

$\{-2, 3\}$

3. $2x^2 + x - 1 = 0$

$(2x - 1)(x + 1) = 0$

$2x - 1 = 0$ or $x + 1 = 0$

$x = \frac{1}{2}$ or $x = -1$

$\{-1, \frac{1}{2}\}$

5. $x^2 - 121 = 0$
$(x + 11)(x - 11) = 0$
$x + 11 = 0$ or $x - 11 = 0$
$x = -11$ or $x = 11$
$\{\pm 11\}$ (ds)

7. $3x^2 - 48 = 0$
$3(x + 4)(x - 4) = 0$
$x + 4 = 0$ or $x - 4 = 0$
$x = -4$ or $x = 4$
$\{\pm 4\}$ (ds)

9. $4x^2 - 2x = 0$
$2x(2x - 1) = 0$
$2x = 0$ or $2x - 1 = 0$
$x = 0$ or $x = \frac{1}{2}$
$\{0, \frac{1}{2}\}$

11. $2x^2 = 6x$
$2x^2 - 6x = 0$
$2x(x - 3) = 0$
$2x = 0$ or $x - 3 = 0$
$x = 0$ or $x = 3$
$\{0, 3\}$

13. $2x^2 - 6x + 4 = 0$
$2(x^2 - 3x + 2) = 0$
$2(x - 2)(x - 1) = 0$
$x - 2 = 0$ or $x - 1 = 0$
$x = 2$ or $x = 1$
$\{1, 2\}$

15. $2x^2 - 11x + 12 = x - 6$
$2x^2 - 12x + 18 = 0$
$2(x^2 - 6x + 9) = 0$
$2(x - 3)^2 = 0$
$x - 3 = 0;\ x = 3$
$\{3\}$ (pst) double root

17. $\frac{1}{2}x^2 - \frac{3}{2}x = 5$
$x^2 - 3x = 10$
$x^2 - 3x - 10 = 0$
$(x - 2)(x + 5) = 0$
$x - 2 = 0$ or $x + 5 = 0$
$x = 2$ or $x = -5$
$\{-5, 2\}$

19. $0.25x^2 + 0.50x - 6 = 0$
$x^2 + 2x - 24 = 0$
$(x - 4)(x + 6) = 0$
$x - 4 = 0$ or $x + 6 = 0$
$x = 4$ or $x = -6$
$\{-6, 4\}$

21. $0.4x^2 + 0.2x = 0.2$
$4x^2 + 2x = 2$
$4x^2 + 2x - 2 = 0$
$2(2x + x - 1) = 0$
$2(2x - 1)(x + 1) = 0$
$2x - 1 = 0$ or $x + 1 = 0$
$x = 0.5$ or $x = -1$
$\{-1, 0.5\}$

23. $x^2 - \frac{5}{3}x = \frac{2}{3}$
$3x^2 - 5x = 2$
$3x^2 - 5x - 2 = 0$
$(3x + 1)(x - 2) = 0$
$3x + 1 = 0$ or $x - 2 = 0$
$x = -\frac{1}{3}$ or $x = 2$
$\{-\frac{1}{3}, 2\}$

25. $x - 4 = 0$ or $x + 3 = 0$
$x = 4$ or $x = -3$
x-intercepts $-3, 4$

27. $2x - 5 = 0$ or $x + 3 = 0$
$x = 2.5$ or $x = -3$
x-intercepts $-3, 2.5$

29. $3x - 4 = 0$ or $2x + 1 = 0$
$x = 1\frac{1}{3}$ or $x = -0.5$
x-intercepts $-0.5, 1\frac{1}{3}$

31. $3 = a(-4 + 1)(-4 + 5)$
$3 = a(-3)(1)$
$3 = -3a; \; a = -1$

33. $7.5 = a(0 + 1)(0 + 5)$
$7.5 = a(1)(5)$
$7.5 = 5a$
$a = 1.5$

35. $y = a(x - 3)(x + 2)$
$12 = a(-3 - 3)(-3 + 2)$
$12 = a(-6)(-1)$
$12 = 6a$
$a = 2$
$y = 2(x - 3)(x + 2)$
$y = 2(x^2 - x - 6)$
$y = 2x^2 - 2x - 12$

37. $y = a(x - 3)(x + 2)$
$12 = a(0 - 3)(0 + 2)$
$12 = a(-3)(2)$
$12 = -6a$
$a = -2$
$y = -2(x - 3)(x + 2)$
$y = -2(x^2 - x - 6)$
$y = -2x^2 + 2x + 12$

39. $y = a(x + 3)(x - 5)$
$6 = a(3 + 3)(3 - 5)$
$6 = a(6)(-2)$
$6 = -12a$
$a = -0.5$
$y = -0.5(x + 3)(x - 5)$
$y = -0.5(x^2 - 2x - 15)$
$y = -0.5x^2 + x + 7.5$

41. $y = a(x + 2)(x - 5)$
$-24 = a(2 + 2)(2 - 5)$
$-24 = a(4)(-3)$
$-24 = -12a$
$a = 2$
$y = 2(x + 2)(x - 5)$
$y = 2(x^2 - 3x - 10)$
$y = 2x^2 - 6x - 20$

43. $y = a(x - 10)(x - 30)$
$12 = a(20 - 10)(20 - 30)$ $\quad 12 = a(10)(-10)$ $\quad 12 = -100a$ $\quad a = -0.12$
$y = -0.12(x - 10)(x - 30)$ $\quad y = -0.12(x^2 - 40x + 300)$ $\quad y = -0.12x^2 + 4.8x - 36$

45. $y = -\frac{3}{40}(x - 10)(x - 40)$; to find vertex, average x-intercept values, substitute into original equation to find y.

47. To find x-intercepts use the zero product rule and solve the factor equations.

49. $x = -2$ or $x = -1$

$x + 2 = 0$ or $x + 1 = 0$

$0 = a(x + 2)(x + 1)$

a is any real number

51. $x = -\frac{3}{2}$ or $x = \frac{1}{2}$

$2x = -3$ or $2x = 1$

$2x + 3 = 0$ or $3x - 1 = 0$

$0 = a(2x + 3)(2x - 1)$

a is any real number

53. a. $h_0 = 320$ ft, $v_0 = 16$ ft/sec, to find elapsed time to ground: $0 = -16t^2 + 16t + 320$; divide both sides by -16: $0 = t^2 - t - 20$; factor: $0 = (t - 5)(t + 4)$; solve: $t - 5 = 0$, $t = 5$ sec or $t + 4 = 0$, $t = -4$ sec (discard negative)

Time to reach the ground = 5 sec.

To find speed when it reaches the ground use $v = -32t + 16$ and substitute $t = 5$ sec:

$v = -32(5) + 16$; $\mathbf{v = -144}$ **ft/sec** or ≈ 98 mph.

The worker's friend should have similar protect as a major league baseball catcher.

b. $v_0 = -16$ ft/sec: $0 = -16t^2 - 16t + 320$; $0 = t^2 + t - 20$; $0 = (t + 5)(t - 4)$;

$t + 5 = 0$, $t = -5$ (discard negative) or $t - 4 = 0$, $t = 4$

Time to reach the ground = 4 sec.

$v = -32(4) - 16$; $\mathbf{v = 144}$ **ft/sec**

Speed is the same, same precautions are recommended.

55. a. $(8.5 - 2x)(11 - 2x)$

b. $8.5 - 2x = 0$ or $11 - 2x = 0$

$x = 4.25$ or $x = 5.5$

c. $x = 4.25$ uses the entire width of the paper, it is impossible to make squares with $x = 5.5$

d. $(8.5 - 2x)(11 - 2x) = 51$

$93.5 - 39x + 4x^2 = 51$

$4x^2 - 39x + 42.5 = 0$

$(4x - 5)(x - 8.5) = 0$

$4x - 5 = 0$ or $x - 8.5 = 0$

$x = 1.25$ or $x = 8.5$

$x = 1.25$ is a reasonable solution

$x = 8.5$ is impossible in this problem setting.

57. $-0.12x^2 + 2.4x = 11.25$

$-0.12x^2 + 2.4x - 11.25$

clear decimals: multiply by 100

$-12x^2 + 240x - 1125 = 0$

remove common factor

$-3(4x^2 - 80x + 375) = 0$

divide both sides by -3

$4x^2 - 80x + 375$

Factor	2x	–25
2x	$4x^2$	–50x
–15	–30x	375

diagonal product is $1500x^2$;

diagonal sum is $-80x$

factors are $-50x$ and $-30x$

$(2x - 15)(2x - 25) = 0$

$2x - 15 = 0$ or $2x - 25 = 0$

$x = 7.5$ or $x = 12.5$

$\{7.5, 12.5\}$

Chapter 4 Review

1. $y = 3(x - 3)^2 - 4$

$y = 3(x^2 + 2(x)(-3) + (-3)^2) - 4$

$y = 3(x^2 - 6x + 9) - 4$

$y = 3x^2 - 18x + 27 - 4$

$y = 3x^2 - 18x + 23$

$a = 3, b = -18, c = 23$

3. $x(x - 3) = x + 7$

$x^2 - 3x = x + 7$

$x^2 - 4x - 7 = 0$

$a = 1, b = -4, c = -7$

5. The vertex of the graph of a quadratic function is the location of the maximum or minimum y-value. ***true***

7. The vertex of the parabola lies on the axis of symmetry of the graph. ***true***

9. a.

x	$f(x) = x^2 - x - 6$
–2	$(-2)^2 - (-2) - 6 = 0$
–1	$(-1)^2 - (-1) - 6 = -4$
0	$(0)^2 - (0) - 6 = -6$
1	$(1)^2 - (1) - 6 = -6$
2	$(2)^2 - (2) - 6 = -4$

b. extend table or use symmetry $(-2, 0)$, $(3, 0)$

c. Average x-values of intercepts: $(-2 + 3) \div 2 = 0.5$; $x = 0.5$

d. $(0.5)^2 - (0.5) - 6 = -6.25$ vertex is at $(0.5, -6.25)$

9. e.

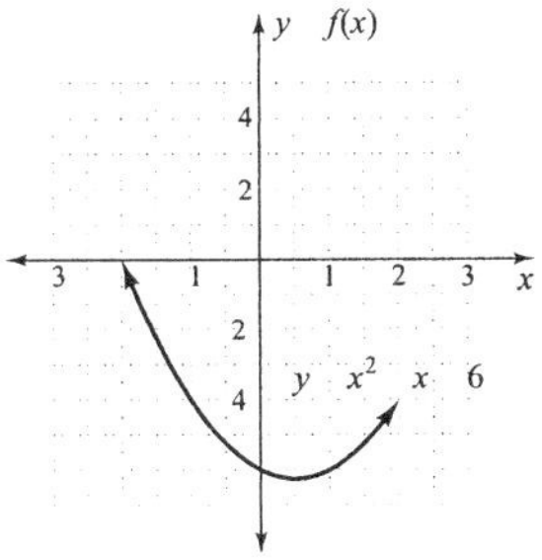

f. $x^2 - x - 6 = 6$

$x^2 - x - 12 = 0$

$(x - 4)(x + 3) = 0$

$x - 4 = 0$ or $x + 3 = 0$

$x = 4$ or $x = -3$

$\{-3, 4\}$

g. $x^2 - x - 6 = 0$

note: x-intercepts

$\{-2, 3\}$

h. $x^2 - x - 6 = -7$

$x^2 - x + 1 = 0$

can not be factored $\{\ \}$ or $\varnothing$

i. range is $y \geq -6.25$

11.

r	$A = 4\pi r^2$
0	$4\pi(0)^2 = 0$
1	$4\pi(1)^2 = 4\pi$
2	$4\pi(2)^2 = 16\pi$
3	$4\pi(3)^2 = 36\pi$
4	$4\pi(4)^2 = 64\pi$
5	$4\pi(5)^2 = 100\pi$
6	$4\pi(6)^2 = 144\pi$
7	$4\pi(7)^2 = 196\pi$
8	$4\pi(8)^2 = 256\pi$
9	$4\pi(9)^2 = 320\pi$
10	$4\pi(10)^2 = 400\pi$

13. a. from the graph: $2x^2 - 8x + 4 = 0$

$\approx \{0.5, 3.5\}$

b. from the graph: $2x^2 - 8x + 4 = 4$

$\approx \{0, 4\}$

Answers to exercise 15b could vary, one possible solution is given for reference.

15. a. $2x^2 - 8x + 4 = -4$

b. $2x^2 - 8x + 4 < -4$

c. 3 solutions are not possible

17.

x	y	differences	
1	6	**1st**	**2nd**
		9	
2	15		4
		13	
3	28		4
		17	
4	45		4
		21	
5	66		

quadratic: $2a = 4$, $a = 2$; when $x = 0$
$y = 1$, $c = 1$; $9 = 3(2) + b$, $b = 3$
$y = 2x^2 + 3x + 1$

19.

x	y	differences	
1	−1	**1st**	**2nd**
		3	
2	2		0
		3	
3	5		0
		3	
4	8		0
		3	
5	11		

linear: $m = 3$; when $x = 0$
$y = -4$, $b = -4$; $y = 3x - 4$

21.

x	y	differences	
1	15	**1st**	**2nd**
		−3	
2	12		−2
		−5	
3	7		−2
		−7	
4	0		−2
		−9	
5	−9		

quadratic: $2a = -2$, $a = -1$; when $x = 0$
$y = 16$, $c = 16$; $3(-1) + b = -3$, $b = 0$
$y = -x^2 + 16$

23.

x	y	differences	
1	7	**1st**	**2nd**
		2	
2	9		−6
		−4	
3	5		−12
		−16	
4	−11		−18
		−34	
5	−45		

neither quadratic or linear

25. a. $y \approx 99.93 - 0.000\,989x$

b. $y \approx 0.000\,000\,002\,76x^2 - 0.001\,07x + 100.2$

c.

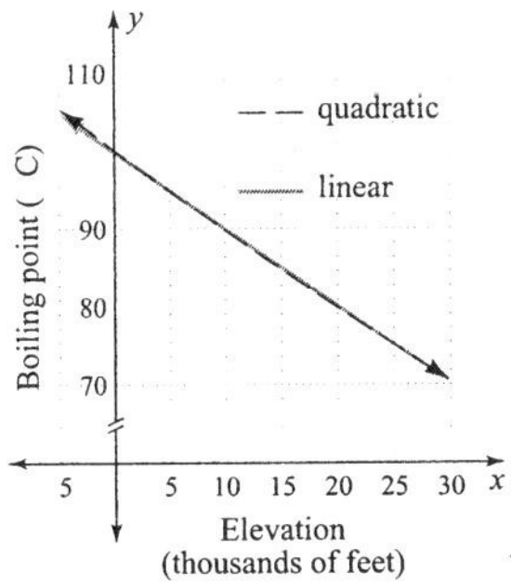

for the interval both are good fits

d. linear: $y \approx 101.2°$ C; quadratic: $y \approx 101.6°$ C

e. The linear equations does not fit this new fact. The parabola fits the new fact but then turns upward after about 190,000 ft, which is also not realistic.

27. There are two points given: starting point, (0, 0); vertex, (30, 100); end point is found from symmetry, (60, 0); using the x-intercepts we get: $y = a(x - 0)(x - 60)$; substitute the vertex to find a: $100 = a(30)(30 - 60)$; $100 = -900a$; $a = -\frac{1}{9}$ or $a \approx -0.111$

$y = -0.111(x)(x - 60)$; $y = -0.111(x^2 - 60x)$; $\mathbf{y = -0.111x^2 + 6.667}$

29. a. $x^3 - y^3$ is a ***binomial***

b. $\sqrt{x} - \sqrt{y}$ is not a polynomial

c. $x^2 + x^{-1}$ is not a polynomial

d. $x\sqrt{3} + \sqrt{3}$ is a ***binomial***

e. $3x^3 - 3x^2 + 3x$ is a ***trinomial***

f. $|x + 1|$ is not a polynomial

31. a. $(x - 3)(x + 3) = x^2 - 9$

b. $(2x - 5)(2x - 5) =$
$(2x)^2 + 2(2x)(-5) + (-5)^2 =$
$4x^2 - 20x + 25$

c. $(x - 1)(x^2 + x + 1) = x^3 + 1$

d. $(n + 4)(n + 4) =$
$n^2 + 2(n)(4) + 4^2 =$
$n^2 + 8n + 16$

e. $(2x - 3)(3x + 4) =$
$6x^2 + 8x - 9x - 12 =$
$6x^2 - x - 12$

f. $(x^2 + x - 1)(x^2 - x - 1) =$
$x^4 - x^3 - x^2 + x^3 - x^2 - x - x^2 + x + 1 =$
$x^4 - 3x^2 + 1$

33. a. $x^2 + 3x - 4 =$
$x^2 - x + 4x - 4 =$
$x(x - 1) + 4(x - 1) =$
$(x - 1)(x + 4)$

b. $2x^2 - 3x =$
$x(2x - 3)$

33. c. $2x^2 + x - 3 =$
$2x^2 - 2x + 3x - 3 =$
$2x(x - 1) + 3(x - 1) =$
$(x - 1)(2x + 3)$

d. $9x^2 + 12x + 4 =$
$9x^2 + 6x + 6x + 4 =$
$3x(3x + 2) + 2(3x + 2) =$
$(3x + 2)(3x + 2) = (3x + 2)^2$

e. $x^2 - x =$
$x(x - 1)$

f. $3x^2 + 6x + 3 =$
$3(x^2 + 2x + 1) = 3(x + 1)^2$

35. a. $(2x + 3)(2x - 3) =$
$4x^2 - 9$
difference of squares

b. $(2x - 3)(2x - 3) =$
$4x^2 - 12x + 9$
perfect square trinomial

c. $(2x + 3)(2x + 3) =$
$4x^2 + 12x + 9$
perfect square trinomial

d. $(3x - 2)(2x + 3) =$
$6x^2 + 9x - 4x - 6 =$
$6x^2 + 5x - 6$

37. $x^2 + 10x + \underline{(\frac{10}{2})^2} = (x + \underline{\frac{10}{2}})^2$
$x^2 + 10x + \underline{\mathbf{25}} = (x + \underline{\mathbf{5}})^2$

39. $x^2 + \underline{2\sqrt{169}}\,x + 169 = (x + \underline{\sqrt{169}}\,)^2$
$x^2 + \underline{\mathbf{26}}x + 169 = (x + \underline{\mathbf{13}})^2$

41. $x^2 - 49 = (x + 7)(x - 7)$

43. $4x^2 - 4x + 1 = (2x - 1)^2$

45. $x^2 + 16$ cannot be factored

47. $4x^2 + 12x + 9 = (2x + 3)^2$

49. $(a - b)(a^2 - 2ab + b^2) =$
$a^3 - 2a^2b + ab^2 - a^2b + 2ab^2 - b^3 =$
$a^3 - 3a^2b + 3ab^2 - b^3$

51. $(a + b)(a^2 - ab + b^2) =$
$a^3 + b^3$

53. $a^3 - 8 =$
$(a - 2)(a^2 + 2a + 4)$

55. $x^3 + 27 =$
$(x + 3)(x^2 - 3x + 9)$

57. $x^3 - 1000 =$
$(x - 10)(x^2 + 10x + 100)$

59. The area of the shaded section is the difference between the area of the large triangle and the area of the small triangle, $A - a$. First find the two areas, then subtract.

$A = \frac{\sqrt{3}}{4}X^2,\ a = \frac{\sqrt{3}}{4}x^2;$

Shaded area $= \frac{\sqrt{3}}{4}X^2 - \frac{\sqrt{3}}{4}x^2 = \frac{\sqrt{3}}{4}(X + x)(X - x)$

61. $x^2 + 3x - 4 = 0$
$(x + 4)(x - 1) = 0$
$x + 4 = 0$ or $x - 1 = 0$
$x = -4$ or $x = 1$
$\{-4, 1\}$

63. $x^2 + 16 = 8x$
$x^2 - 8x + 16 = 0$
$(x - 4)^2 = 0$
$x - 4 = 0$
$x = 4$ (double root); $\{4\}$

65. $9x^2 + 12x + 4 = 0$
$(3x + 2)^2 = 0$
$3x + 2 = 0$
$x = -\frac{2}{3}$
$\{-\frac{2}{3}\}$

67. $2x^2 + x = 3$
$2x^2 + x - 3 = 0$
$2x^2 - 2x + 3x - 3 = 0$
$2x(x - 1) + 3(x - 1) = 0$
$(x - 1)(2x + 3) = 0$
$x - 1 = 0$ or $2x + 3 = 0$
$x = 1$ or $x = -1.5$
$\{-1.5, 1\}$

69. $6x^2 + 5x + 2 = 8$
$6x^2 + 5x - 6 = 0$
$6x^2 + 9x - 4x - 6 = 0$
$3x(2x + 3) - 2(2x + 3) = 0$
$(2x + 3)(3x - 2) = 0$
$2x + 3 = 0$ or $3x - 2 = 0$
$x = -1.5$ or $x = \frac{2}{3}$
$\{-1.5, \frac{2}{3}\}$

71. $24(x + 6) = -x^2$
$x^2 + 24x + 144 = 0$
$(x + 12)^2 = 0$
$x + 12 = 0$
$x = -12$ (double root)
$\{-12\}$

73. $4x(x - 1) = 9 - 4x$
$4x^2 - 4x + 4x - 9 = 0$
$4x^2 - 9 = 0$
$(2x + 3)(2x - 3) = 0$
$2x + 3 = 0$ or $2x - 3 = 0$
$x = -1.5$ or $x = 1.5$
$\{\pm 1.5\}$

75. $y = a(x - 4)(x + 2)$
for any real number a

77.
$3 = a(1-4)(1+2)$
$3 = a(-3)(3)$
$3 = -9a$
$a = -\frac{1}{3}$
$y = -\frac{1}{3}(x-4)(x+2)$
$y = -\frac{1}{3}(x^2+2x-4x-8)$
$y = -\frac{1}{3}(x^2-2x-8)$
$y = -\frac{1}{3}x^2 + \frac{2}{3}x + 2\frac{2}{3}$

79.
$0 = a(x+2)(x-4)$
a is any real number

81.
$x = -\frac{2}{5}$ or $x = \frac{1}{5}$
$5x = -2$ or $5x = 1$
$5x + 2 = 0$ or $5x - 1 = 0$
$0 = a(5x+2)(5x-1)$
a is any real number

83. The graph of $y = x^2 - 9$ crosses the x-axis twice; the graph of $y = x^2 + 9$ does not cross the x-axis.

85. The two factors of $(x+4)^2$ are the same; this is called a ***double root.***

87. The graph of $y = (x+2)^2$ has the same shape as $y = x^2$ but is shifted left 2 units. The vertex and the x-intercept are at (0, 0) for $y = x^2$ and at (–2, 0) for $y = (x+2)^2$; the y-intercept f $y = x^2$ is also at (0, 0), while the y-intercept of $y = (x+2)^2$ is at (0, 4).

Chapter 4 Test

1. a. One solution is shown in the table; $y = -10$ when $x = -4$; we can tell that the other solution will be between $x = -2$ and $x = 0$. Add $x = -1$ to table, $y = (-1)^2 + 5(-1) - 6$, $y = -10$; the other solution is $x = -1$. **{–4, –1}**

b. One solution is shown in the table; $y = 8$ when $x = 2$; we can tell that the other solution will be between $x = -6$ and $x = -8$. Add $x = -7$ to table, $y = (-7)^2 + 5(-7) - 6$, $y = 8$; the other solution is $x = -7$. **{–7, 2}**

c. One solution is shown in the table; $y = 0$ when $x = -6$; we can tell that the other solution is between $x = 0$ and $x = 2$. Add $x = 1$ to the table, $y = (1)^2 + 5(1) - 6$, $y = 0$; the other solution is $x = 1$. **{–6, 1}**

d. It appears from the symmetry of the table that lowest point will be between $x = -2$ and $x = -4$; add $x = -3$ to the table, $y = (-3)^2 + 5(-3) - 6$, $y = -12$. Add $x = -2.5$ to the table; $y = (-2.5)^2 + 5(-2.5) - 6$, $y = -12.25$. This point is half-way between two equal y values, it is the lowest point on the graph. Therefore, $y = -20$ is not possible. { }

e. From part d above, the axis of symmetry is the x-value of the vertex. **$x = -2.5$**

f. The vertex is given in d above **(–2.5, –12.25)**

g. The domain is $\Re$, the range is $y \geq -12.25$.

2. a.

Value	differences	
−6	1st	2nd
	8	
2		2
	10	
12		2
	12	
24		2
	14	
38		

Next number is $38 + (14 + 2) = 54$
quadratic; $2a = 2, a = 1$;
$3(1) + b = 8, b = 5$;
$-6 = 1(1)^2 + 5(1) + c, c = -12$
$y = x^2 + 5x - 12$

b.

Value	differences	
23	1st	2nd
	−8	
15		0
	−8	
7		0
	−8	
−1		0
	−8	
−9		

Next number is $-9 - 8 = -17$
linear; $m = -8$; $23 = -8(1) + b$,
$b = 31$; $y = -8x + 31$

3. $(x - 6)(x - 2) =$
$x^2 - 2x - 6x + 12 =$
$x^2 - 8x + 12$; other

4. $(x - 12)(x + 1) =$
$x^2 + x - 12x - 12 =$
$x^2 - 11x - 12$; other

5. $(x + 12)(x + 1) =$
$x^2 + x + 12x + 12 =$
$x^2 + 13x + 12$; other

6. $(2x - 3)(x + 4) =$
$2x^2 + 8x - 3x - 12 =$
$2x^2 + 5x - 12$; other

7. $(x - 2)^2 =$
$x^2 + 2(x)(-2) + (-2)^2 =$
$x^2 - 4x + 4$; perfect square trinomial

8. $(2x + 1)(2x - 1) =$
$4x^2 - 1$; difference of squares

9. $(x - 2)(x^2 + 2x + 4) =$
$x^3 - 8$; other

10. $2x^2 - 7x + 6 =$
$2x^2 - 4x - 3x + 6 =$
$2x(x - 2) - 3(x - 2) =$
$(x - 2)(2x - 3)$

11. $x^2 - 6x + 9 =$
$(x - 3)^2$

12. $x^2 - 49 =$
$(x - 7)(x + 7)$

13. $12x^2 + 30x - 18 =$
$6(2x^2 + 5x - 3) =$
$6(2x^2 + 6x - x - 3) =$
$6[2x(x + 3) - 1(x + 3)] =$
$6(x + 3)(2x - 1)$

14. $x^3 - 27 =$
$(x - 3)(x^2 + 3x + 9)$

15. $x^3 + 125 =$
$(x + 5)(x^2 - 5x + 25)$

16. $2x^2 - 7x + 6 = 0$
$2x^2 - 4x - 3x + 6 = 0$
$2x(x - 2) - 3(x - 2) = 0$
$(x - 2)(2x - 3) = 0$
$x - 2 = 0$ or $2x - 3 = 0$
$x = 2$ or $x = 1.5$
$\{1.5, 2\}$

17. $x^2 = 4(3x + 7)$
$x^2 - 12x - 28 = 0$
$(x + 2)(x - 14) = 0$
$x + 2 = 0$ or $x - 14 = 0$
$x = -2$ or $x = 14$
$\{-2, 14\}$

18. $4x^2 = 2x + 30$
$4x^2 - 2x - 30 = 0$
$2x^2 - x - 15 = 0$
$2x^2 - 6x + 5x - 15 = 0$
$2x(x - 3) + 5(x - 3)$
$(x - 3)(2x + 5) = 0$
$x - 3 = 0$ or $2x + 5 = 0$
$x = 3$ or $x = -2.5$
$\{-2.5, 3\}$

19. $y = a(x + 3)(x - 4)$
$2 = a(0 + 3)(0 - 4)$
$2 = a(3)(-4)$
$2 = -12a$
$a = -\frac{1}{6}$
$y = -\frac{1}{6}(x + 3)(x - 4)$
$y = -\frac{1}{6}(x^2 - 4x + 3x - 12)$
$y = -\frac{1}{6}x^2 + \frac{1}{6}x + 2$

20. a. From calculator linear regression: $y = -0.298x + 59.6$; $-0.298(44) + 59.6 = 46.5$; suggests the voter turnout will continue to decline to 46.5% in 2008.

b. From calculator quadratic regression: $y = 0.0111x^2 - 0.697x + 61.7$:
$0.0111(44)^2 - 0.697(44) + 61.7 = 52.5$; suggests that voter turnout will have bottomed out in 1996 and will gradually rise to 52.5% in 2008.

21. Answers may vary but should resemble:

For the path of water in a drinking fountain or the path of a thrown ball, the x-intercepts are the starting and ending points and the vertex is the highest point.

For the cable in a suspension bridge, the x-intercepts are where the cable attaches to the end supports, or where the end supports are attached to the ground and the vertex is the lowest point.

The y-intercepts would represent the initial height of the person throwing the ball or the nozzle on the water fountain. On the bridge the y-intercept might be the height where the cable attaches to the end support.

Cumulative Review: Chapters 1 to 4

1. $8 - 3(x - 2) - (x - 3) =$

$8 - 3x + 6 - x + 3 =$

$17 - 4x$

3. $2y = 3x + 4 \Rightarrow y = 1.5x + 2$; $m = 1.5$, if function passes through the origin, $b = 0$

$y = 1.5x$

5. a. $c = nt^2$

$n = \dfrac{c}{t^2}$

b. $s = 2a + d(n - 1)$

$s - 2a = d(n - 1)$

$\dfrac{s - 2a}{d} = n - 1$

$n = \dfrac{s - 2a}{d} + 1 \text{ or } n = \dfrac{s - 2a + d}{d}$

7. a. $f(-1) = 3(-1)^2 - 5(-1) + 2$

$f(-1) = 3 + 5 + 2 = 10$

b. $f(0) = 3(0)^2 - 5(0) + 2$

$f(0) = 2$

c. $f(2) = 3(2)^2 - 5(2) + 2$

$f(2) = 3(4) - 10 + 2$

$f(2) = 12 - 8$

$f(2) = 4$

d. $f(x + 1) = 3(x + 1)^2 - 5(x + 1) + 2$

$f(x + 1) = 3(x^2 + 2x + 1) - 5x - 5 + 2$

$f(x + 1) = 3x^2 + 6x + 3 - 5x - 5 + 2$

$f(x + 1) = 3x^2 + x$

9. $y = a(x - 3)(x + 4)$

$-4 = a(2 - 3)(2 + 4)$ $-4 = a(-1)(6)$ $-4 = -6a$ $a = \frac{2}{3}$

$y = \frac{2}{3}(x - 3)(x + 4)$ $y = \frac{2}{3}(x^2 + 4x - 3x - 12)$ $y = \frac{2}{3}(x^2 + x - 12)$

$y = \frac{2}{3}x^2 + \frac{2}{3}x - 8$

11.

x	4	2	7	**1**	10	**8**
***f*(x)**	**19**	7	52	4	103	67

$f(x) = x^2 + 3$; most helpful were $f(10)$ and $f(2)$

13. $|2x - 1| > 6$

$2x - 1 > 6$ or $-(2x - 1) > 6$

$2x > 7$ or $-2x > 5$

$x > 3.5$ or $x < -2.5$

15. $A + B + C = 180$ (1)

$A + 3B = C$ (2)

$B = 4A$ (3)

substitute (3) into (2) $A + 3(4A) = C$ $13A = C$ (4)

substitute (3) and (4) into (1) $A + 4A + 13A = 180$

$18A = 180$, $\mathbf{A = 10°}$ $4(10) = B$, $\mathbf{B = 40°}$ $13(10) = C$, $\mathbf{C = 130°}$

17. a. $x^2 - 2x - 3 = 0$ $\{-1, 3\}$ **b.** $x^2 - 2x - 3 = -3$ $\{0, 2\}$

19. a. $(5x - 3)(3x - 5) =$

$15x^2 - 25x - 9x + 15 =$

$15x^2 - 34x + 15$

b. $(3 - x)(3 + x) =$

$9 - x^2$

Exercises 5.1

1. $x^2 = 225$

$x = \pm\sqrt{225}$

$x = \pm 15$

3. $x^2 = 121$

$x = \pm\sqrt{121}$

$x = \pm 11$

5. $x^2 = 10{,}000$

$x = \pm\sqrt{10{,}000}$

$x = \pm 100$

7. $x^2 = -36$

no real number

9. $15^2 = 225$

11. $\sqrt{49} = 7$

13. $11^2 = 121$

15. $\sqrt{0.01} = 0.1$

17. $\sqrt{1.96} = 1.4$

19. $\sqrt{75} = \sqrt{25 \cdot 3} = \sqrt{25} \cdot \sqrt{3} = 5\sqrt{3}$

21. $\sqrt{40} = \sqrt{4 \cdot 10} = \sqrt{4} \cdot \sqrt{10} = 2\sqrt{10}$

23. $\sqrt{200} = \sqrt{100 \cdot 2} = \sqrt{100} \cdot \sqrt{2} = 10\sqrt{2}$

25. $\sqrt{150} = \sqrt{25 \cdot 6} = \sqrt{25} \cdot \sqrt{6} = 5\sqrt{6}$

27. $\sqrt{80} = \sqrt{16 \cdot 5} = \sqrt{16} \cdot \sqrt{5} = 4\sqrt{5}$

29. $\sqrt{32} = \sqrt{16 \cdot 2} = \sqrt{16} \cdot \sqrt{2} = 4\sqrt{2}$

31. $1 + \sqrt{2} \approx 2.414$

33. $\dfrac{1-\sqrt{3}}{2} \approx -0.366$

35. $\dfrac{2+\sqrt{8}}{2} = \dfrac{2+\sqrt{4 \cdot 2}}{2} = \dfrac{2+2\sqrt{2}}{2} = \dfrac{2(1+\sqrt{2})}{2} = 1+\sqrt{2} \approx 2.414$

37. $\dfrac{3-\sqrt{27}}{6} = \dfrac{3-\sqrt{9 \cdot 3}}{3 \cdot 2} = \dfrac{3-3\sqrt{3}}{3 \cdot 2} = \dfrac{3(1-\sqrt{3})}{3(2)} = \dfrac{1-\sqrt{2}}{2} \approx -0.366$

39. $\dfrac{4+\sqrt{8}}{4} = \dfrac{4+2\sqrt{2}}{4} = \dfrac{2(2+\sqrt{2})}{2(2)} = \dfrac{2+\sqrt{2}}{2}$

41. $\dfrac{2+\sqrt{2}}{2}$, simplified

43. $\dfrac{4\sqrt{8}}{4}=\sqrt{8}=\sqrt{4}\sqrt{2}=2\sqrt{2}$

45. a. $\sqrt{0.81}=\sqrt{\dfrac{81}{100}}=\dfrac{\sqrt{81}}{\sqrt{100}}=\dfrac{9}{10}=0.9$

b. $\sqrt{0.0025}=\sqrt{\dfrac{25}{10000}}=\dfrac{\sqrt{25}}{\sqrt{10000}}=\dfrac{5}{100}=0.5$

c. $\sqrt{\dfrac{1}{36}}=\dfrac{\sqrt{1}}{\sqrt{36}}=\dfrac{1}{6}$

d. $\sqrt{\dfrac{121}{49}}=\dfrac{\sqrt{121}}{\sqrt{49}}=\dfrac{11}{7}$

e. $\sqrt{2.25}=\sqrt{\dfrac{9}{4}}=\dfrac{\sqrt{9}}{\sqrt{4}}=\dfrac{3}{2}=1.5$

f. $\sqrt{640000}=\sqrt{64}\cdot\sqrt{10000}=8(100)=800$

g. $\dfrac{\sqrt{48}}{\sqrt{3}}=\dfrac{\sqrt{16}\cdot\sqrt{3}}{\sqrt{3}}=\sqrt{16}=4$

h. $\dfrac{\sqrt{162}}{\sqrt{2}}=\dfrac{\sqrt{81}\cdot\sqrt{2}}{\sqrt{2}}=\sqrt{81}=9$

47. a. The graphs are identical for $y \geq 0$; $y = \sqrt{x}$ is a function, $x = y^2$ is not.

b. $4 = 2^2, 2 = \sqrt{4}$; $9 = 3^2, 3 = \sqrt{9}$; $16 = 4^2, 4 = \sqrt{16}$

c. $y = \sqrt{x}$ has the restriction $y \geq 0$; $x = y^2$ does not.

d. x cannot be a negative number is y is a real number.

49.

$$\frac{\sqrt{125}+\sqrt{45}-\left(\sqrt{75}+\sqrt{27}\right)}{\sqrt{5}-\sqrt{3}}=\frac{\sqrt{25\cdot5}+\sqrt{9\cdot5}-\left(\sqrt{25\cdot3}+\sqrt{9\cdot3}\right)}{\sqrt{5}-\sqrt{3}}=$$

$$\frac{5\sqrt{5}+3\sqrt{5}-\left(5\sqrt{3}+3\sqrt{3}\right)}{\sqrt{5}-\sqrt{3}}=\frac{8\sqrt{5}-8\sqrt{3}}{\sqrt{5}-\sqrt{3}}=\frac{8\left(\sqrt{5}-\sqrt{3}\right)}{\sqrt{5}-\sqrt{3}}=8$$

51. $\dfrac{6-2\sqrt{3}}{2}\neq 3-2\sqrt{3}$; Must factor 2 from the numerator before changing to lowest terms.

53. a. $\left(\sqrt{8}\right)^2+\left(\sqrt{6}\right)^2=x^2$

$8+6=x^2$

$x^2=14$

$x=\sqrt{14}$

b. $x^2+5^2=9^2$

$x^2=81-25$

$x^2=56$

$x=\sqrt{56}=2\sqrt{14}$

55. a. $x^2+3^2=6^2$

$x^2=36-9$

$x^2=27$

$x=\sqrt{27}=3\sqrt{3}$

b. $x^2+4^2=8^2$

$x^2=64-16$

$x^2=48$

$x=\sqrt{48}=4\sqrt{3}$

57. a. $x^2 = 8^2 + 8^2$

$x^2 = 2(8^2)$

$x = 8\sqrt{2}$

b $x^2 = 12^2 + 12^2$

$x^2 = 2(12^2)$

$x = 12\sqrt{2}$

c. $8^2 = 2(x^2)$

$32 = x^2$

$x = \sqrt{32} = 4\sqrt{2}$

59. If the legs are equal in length, multiply by $\sqrt{2}$ to get the length of the hypotenuse.

61. $12^2 + 16^2 = 20^2$?

$144 + 256 = 400$?

$400 = 400$

right triangle

63. $4.5^2 + 6^2 = 7.5^2$?

$20.25 + 36 = 56.25$?

$56.25 = 56.25$

right triangle

65. $8^2 + 9^2 = 12^2$?

$64 + 81 = 144$?

$145 \neq 144$

not a right triangle

67. $16^2 + 30^2 = 34^2$?

$256 + 900 = 1156$?

$1156 = 1156$

right triangle

69. $\left(\sqrt{5}\right)^2 + \left(\sqrt{11}\right)^2 = 4^2$?

$5 + 11 = 16$?

$16 = 16$

right triangle

71. a. $0.97 \text{ ft} \cdot \dfrac{12\,in.}{1\,ft.} = 11.6 \text{ in.}$

b. $0.05 \text{ ft} \cdot \dfrac{12\,in.}{1\,ft.} = 0.6 \text{ in.}$

73. $2(12^2) = d^2$

$12\sqrt{2} = d$

$d \approx 16.97$ ft

$d = 16$ ft 11.6 in.

75. $13^2 + 19^2 = d^2$

$\sqrt{530} = d$

$d \approx 23.02$ ft

$d = 23$ ft 0.3 in.

77. $25^2 + 16^2 = x^2$

$\sqrt{881} = x$

$x = 29.7$ ft

79. distance from base of wall = 16 ÷ 4 = 4; ladder length = $\sqrt{16^2 + 4^2}$
ladder length = 16.5 ft.

81. A triangle is a right triangle if and only if the sum of the squares of the two shorter sides is equal to the square of the longest side.

Exercises 5.2

1. a. $x = -5, x = -1$

b. $x = -5, x = -1$

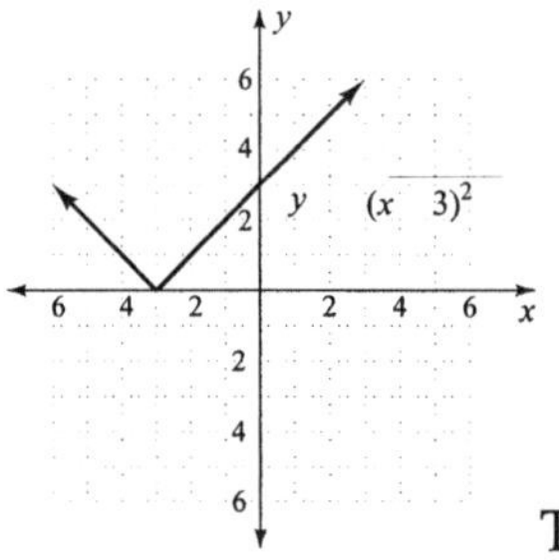

c. The graph is the same as that of $y = |x + 3|$. The $\sqrt{w^2} = w$.

3.

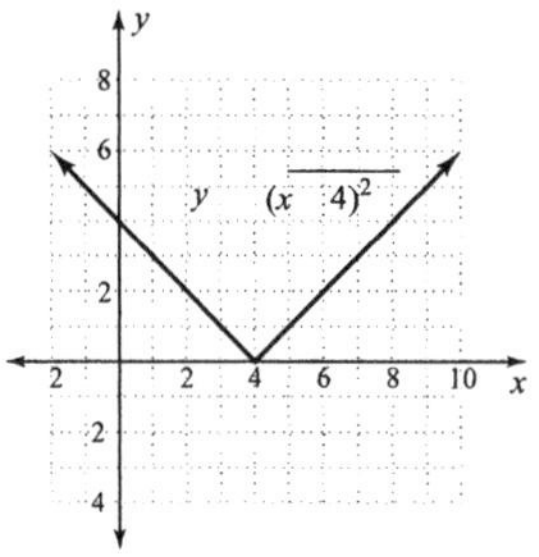

5.

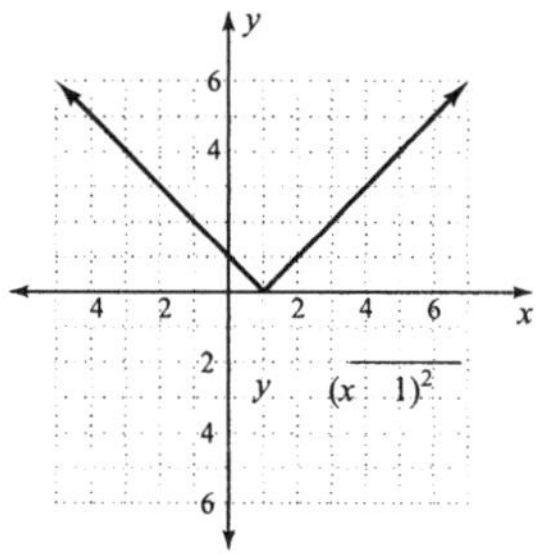

7.
$x^2 + 4x + 4 = 9$
$(x + 2)^2 = 9$
$\sqrt{(x+2)^2} = \sqrt{9}$
$|x + 2| = 3$
$x + 2 = \pm 3$
$x + 2 = 3$ or $x + 2 = -3$
$x = 1$ or $x = -5$
$\{-5, 1\}$

9.
$x^2 - 10x + 25 = 36$
$(x - 5)^2 = 36$
$\sqrt{(x-5)^2} = \sqrt{36}$
$|x - 5| = 6$
$x - 5 = \pm 6$
$x - 5 = 6$ or $x - 5 = -6$
$x = 11$ or $x = -1$
$\{-1, 11\}$

11. $x^2 + 8x + 16 = 9$
$(x + 4)^2 = 9$
$\sqrt{(x+4)^2} = \sqrt{9}$
$|x + 4| = 3$
$x + 4 = \pm 3$
$x + 4 = 3$ or $x + 4 = -3$
$x = -1$ or $x = -7$
$\{-7, -1\}$

13. $x^2 - 4x + 4 = 16$
$(x - 2)^2 = 16$
$\sqrt{(x-2)^2} = \sqrt{16}$
$|x - 2| = 4$
$x - 2 = \pm 4$
$x - 2 = 4$ or $x - 2 = -4$
$x = 6$ or $x = -2$
$\{-2, 6\}$

15. $4x^2 - 1 = 0$
$4x^2 = 1$
$\sqrt{4x^2} = \sqrt{1}$
$|2x| = 1$
$2x = \pm 1$
$2x = 1$ or $2x = -1$
$x = \frac{1}{2}$ or $x = -\frac{1}{2}$
$\{\pm\frac{1}{2}\}$

17. $4x^2 + 4x + 1 = 0$
$(2x + 1)^2 = 0$
$2x + 1 = 0$
$x = -\frac{1}{2}$
$\{-\frac{1}{2}\}$

19. $4x^2 - 16 = 0$
$x^2 - 4 = 0$
$x^2 = 4$
$\sqrt{x^2} = \sqrt{4}$
$|x| = 2$
$x = \pm 2$
$\{\pm 2\}$

21. $x^2 = 225$
$|x| = 15$
$x = \pm 15$
$\{\pm 15\}$

23. $x^2 - 121 = 0$
$x^2 = 121$
$|x| = 11$
$x = \pm 11$
$\{\pm 11\}$

25. $3x^2 - 48 = 0$
$3x^2 = 48$
$x^2 = 16$
$|x| = 4$
$x = \pm 4$
$\{\pm 4\}$

27. $5x^2 - 45 = 0$
$x^2 - 9 = 0$
$x^2 = 9$
$x = \pm 3$
$\{\pm 3\}$

29. $16x^2 - 48 = 0$
$x^2 - 3 = 0$
$x^2 = 3$
$x = \pm 1.7$
$\{\pm 1.7\}$

31. $9x^2 - 45 = 0$
$x^2 - 5 = 0$
$x^2 = 5$
$x = \pm 2.2;\ \{\pm 2.2\}$

33. $2(20^2) = c^2$
$\sqrt{2(20^2)} = c$
$c = 20\sqrt{2}$ m

35. $2(\sqrt{2})^2 = c^2$
$2(2) = c^2$
$c^2 = 4$
$c = 2$ ft

37. $2(\sqrt{18})^2 = c^2$
$2(18) = c^2$
$c^2 = 36$
$c = 6$ in.

Exercises 39 to 43 use the formula for h from Example 7 in the text: $h = \dfrac{x\sqrt{3}}{2}$

39. $x = 8;\ h = \dfrac{8\sqrt{3}}{2} = 4\sqrt{3}$

41. $x = 5;\ h = \dfrac{5\sqrt{3}}{2}$

43. $b = x;\ h = \dfrac{x\sqrt{3}}{2};\ A = \dfrac{1}{2}(x)\left(\dfrac{x\sqrt{3}}{2}\right);\ A = \dfrac{x^2\sqrt{3}}{4}$

45. $x^2 + x^2 = h^2,\ 2x^2 = h^2,\ h = x\sqrt{2}$

47. a. $A = \dfrac{\pi d^2}{4};\ 4A = \pi d^2;\ d^2 = \dfrac{4A}{\pi};\ d = \sqrt{\dfrac{4A}{\pi}};\ d = 2\sqrt{\dfrac{A}{\pi}}$

b. $\dfrac{\pi}{4} = 0.7854$; numbers form a square in the upper left of a calculator key pad.

49. $d = 2\sqrt{\dfrac{5026}{\pi}};\ d \approx 80$ in.; $80\text{ in.}\left(\dfrac{1\,ft}{12\,in.}\right) \approx 6.7$ ft

51. $100 = 4\pi r^2;\ r^2 = \dfrac{100}{4\pi};\ r = \sqrt{\dfrac{25}{\pi}};\ r \approx 2.8$ in.

53. a. $h = 0,\ v_0 = 0:\ 0 = -\frac{1}{2}gt^2 + h_0;\ \frac{1}{2}gt^2 = h_0;\ t^2 = \dfrac{2h_0}{g};\ t = \sqrt{\dfrac{2h_0}{g}}$

b. $t = \sqrt{\dfrac{2(555)}{32.2}};\ t \approx 5.9$ sec

c. $t = \sqrt{\dfrac{2(49.05)}{9.81}};\ t \approx 3.2$ sec

d. $t = \sqrt{\dfrac{2(646.4)}{9.81}};\ t \approx 11.5$ sec

55. a. $d = \sqrt{\dfrac{3h}{2}};\ d^2 = \dfrac{3h}{2};\ 2d^2 = 3h;\ h = \dfrac{2d^2}{3}$

b. $d = \sqrt{\dfrac{3(500)}{2}};\ d \approx 27.4$ miles

c. $d = \sqrt{\dfrac{3(500)}{8}};\ d \approx 13.7$ miles

d. $h = \dfrac{2(9)^2}{3};\ h \approx 54$ ft

e. $h = \dfrac{8(9)^2}{3};\ h \approx 216$ ft

f. The curvature of the moon is greater so we see ***half*** as far on the moon for a given height and must be ***four*** times higher on the moon to see the same distance.

57. The) is placed at the end of the expression.

Mid-Chapter 5 Test

1. a. $\sqrt{75} = \sqrt{25 \cdot 3} = \sqrt{25} \cdot \sqrt{3} = 5\sqrt{3}$

b. $\sqrt{18} = \sqrt{9 \cdot 2} = \sqrt{9} \cdot \sqrt{2} = 3\sqrt{2}$

c. $\sqrt{32} = \sqrt{16 \cdot 2} = \sqrt{16} \cdot \sqrt{2} = 4\sqrt{2}$

d. $\sqrt{20} = \sqrt{4 \cdot 5} = \sqrt{4} \cdot \sqrt{5} = 2\sqrt{5}$

2. a. $2\sqrt{2} = \sqrt{4} \cdot \sqrt{2} = \sqrt{4 \cdot 2} = \sqrt{8}$

b. $5\sqrt{2} = \sqrt{25} \cdot \sqrt{2} = \sqrt{25 \cdot 2} = \sqrt{50}$

c. $4\sqrt{3} = \sqrt{16} \cdot \sqrt{3} = \sqrt{16 \cdot 3} = \sqrt{48}$

d. $6\sqrt{2} = \sqrt{36} \cdot \sqrt{2} = \sqrt{36 \cdot 2} = \sqrt{72}$

e. $3\sqrt{3} = \sqrt{9} \cdot \sqrt{3} = \sqrt{9 \cdot 3} = \sqrt{27}$

f. $2\sqrt{3} = \sqrt{4} \cdot \sqrt{3} = \sqrt{4 \cdot 3} = \sqrt{12}$

3. a. $\sqrt{2500} = \sqrt{25} \cdot \sqrt{100} = 5 \cdot 10 = 50$

b. $\sqrt{0.0025} = \sqrt{\dfrac{25}{10000}} = \dfrac{\sqrt{25}}{\sqrt{10000}} = \dfrac{5}{100} = 0.05$

c. $\sqrt{2.25} = \dfrac{\sqrt{225}}{\sqrt{100}} = \dfrac{15}{10} = 1.5$

d. $\sqrt{22500} = \sqrt{225} \cdot \sqrt{100} = 15 \cdot 10 = 150$

4. a. $\sqrt{\frac{49}{64}} = \frac{\sqrt{49}}{\sqrt{64}} = \frac{7}{8}$

b. $\frac{\sqrt{24}}{\sqrt{6}} = \frac{\sqrt{4} \cdot \sqrt{6}}{\sqrt{6}} = 2$

c. $\frac{\sqrt{48}}{\sqrt{27}} = \frac{4\sqrt{3}}{3\sqrt{3}} = \frac{4}{3}$

d. $\frac{1}{\sqrt{81}} = \frac{1}{9}$

5. $x^2 = 16$ has two solutions $\{\pm 4\}$; $x = \sqrt{16}$ has one solution $\{4\}$.

6. a. ≈ 1.816

b. ≈ 0.048

7. Closing parentheses is needed after the $\sqrt{(6)}$ in the numerator, not after the 3 in the denominator.

8. a. $7.5^2 + 10^2 = 12.5^2$?
$56.25 + 100 = 156.25$?
$156.25 = 156.25$
right triangle

b. $8^2 + 15^2 = 17^2$?
$64 + 225 = 289$?
$289 = 289$
right triangle

c. $6^2 + 8^2 = 12^2$?
$36 + 64 = 144$?
$100 \neq 144$
not a right triangle

9. a. $x^2 + 5^2 = 13^2$
$x^2 = 169 - 25$
$x^2 = 144$
$x = 12$

b. $x^2 + 4^2 = 20^2$
$x^2 = 400 - 16$
$x^2 = 384$
$x = \sqrt{384} = \sqrt{64 \cdot 6} = 8\sqrt{6}$

c. $2.5^2 + 6^2 = x^2$
$6.25 + 36 = x^2$
$42.25 = x^2$
$x = 6.5$

10. a. $|x - 2| = 3$
$x - 2 = 3$ or $x - 2 = -3$
$x = 5$ or $x = -1$
$\{-1, 5\}$

b. $|x + 5| = 2$
$x + 5 = 2$ or $x + 5 = -2$
$x = -3$ or $x = -7$
$\{-7, -3\}$

11. a. $x^2 = 121$

$x = \pm 11$

$\{\pm 11\}$

b. $x^2 = 0.04$

$x = \pm 0.2$

$\{\pm 0.2\}$

c. $(x + 1)^2 = 4$

$|x + 1| = 2$

$x + 1 = 2$ or $x + 1 = -2$

$x = 1$ or $x = -3$

$\{-3, 1\}$

d. $(x + 4)^2 = 25$

$|x + 4| = 5$

$x + 4 = 5$ or $x + 4 = -5$

$x = 1$ or $x = -9$

$\{-9, 1\}$

e. $x^2 + 4x + 4 = 9$

$(x + 2)^2 = 9$

$|x + 2| = 3$

$x + 2 = 3$ or $x + 2 = -3$

$x = 1$ or $x = -5$

$\{-5, 1\}$

f. $x^2 + 6x + 9 = 1$

$(x + 3)^2 = 1$

$|x + 3| = 1$

$x + 3 = 1$ or $x + 3 = -1$

$x = -2$ or $x = -4$

$\{-4, -2\}$

12. $2\left(\sqrt{8}\right)^2 = h^2;\ h^2 = 16;\ h = 4$

13. a. The height is an axis of symmetry. ***true***

b. The height cuts the base into two equal parts. ***true***

c. The height is longer than the side of the original triangle. ***false***

$h^2 = 6^2 + 3^2,\ h^2 = 36 + 9,\ h^2 = 45, h = \sqrt{45} = 3\sqrt{3}$

14. $h = -\frac{1}{2}gt^2$

$\dfrac{-2h}{g} = t^2$

$t = \sqrt{\dfrac{-2h}{g}}$

15. $d = \sqrt{\dfrac{4A}{\pi}}$

$d^2 = \dfrac{4A}{\pi}$

$A = \dfrac{\pi d^2}{4}$

Exercises 5.3

1. $(x + 3)^2 = x^2 + 2(3x) + 3^2 = x^2 + 6x + 9$

3. $(x - 6)^2 = x^2 + 2(-6x) + (-6)^2 = x^2 - 12x + 36$

5. $(x - \frac{1}{2})^2 = x^2 + 2(-\frac{1}{2}x) + (-\frac{1}{2})^2 = x^2 - x + \frac{1}{4}$

7. $(2x-3)^2 = (2x)^2 + 2(2x)(-3) + (-3)^2 = 4x^2 - 12x + 9$

9. $x^2 - 4x + 4 = x^2 + 2(-2x) + (-2)^2 = (x-2)^2$

11. $x^2 + 8x + 16 = x^2 + 2(4x) + 4^2 = (x+4)^2$

13. $x^2 + 18x + 81 = x^2 + 2(9x) + 9^2 = (x+9)^2$

15. $x^2 - 16x + 64 = x^2 + 2(-8x) + (-8)^2 = (x-8)^2$

17.

	a	+7
a	a^2	7a
+7	7a	**49**

$(a+7)^2 = a^2 + 14a + 49$

19.

	x	$-\frac{7}{2}$
x	x^2	$-\frac{7}{2}x$
$-\frac{7}{2}$	$-\frac{7}{2}x$	$\frac{49}{4}$

$(x - \frac{7}{2})^2 = x^2 - 7x + \frac{49}{4}$

21. $x^2 + 22x = 23$

$22 \div 2 = 11$

$11^2 = 121$

$x^2 + 22x + 121 = 23 + 121$

$(x+11)^2 = 144$

23. $x^2 + 12x = 13$

$12 \div 2 = 6$

$6^2 = 36$

$x^2 + 12x + 36 = 13 + 36$

$(x+6)^2 = 49$

25. $x^2 - 11x = -18$

$-11 \div 2 = -\frac{11}{2}$

$\left(-\frac{11}{2}\right)^2 = \frac{121}{4}$

$x^2 - 11x + \frac{121}{4} = -18 + \frac{121}{4}$

$(x - \frac{11}{2})^2 = \frac{49}{4}$

27. $x^2 + 15x = -14$

$15 \div 2 = \frac{15}{2}$

$\left(\frac{15}{2}\right)^2 = \frac{225}{4}$

$x^2 + 15x + \frac{225}{4} = -14 + \frac{225}{4}$

$(x + \frac{15}{2})^2 = \frac{169}{4}$

29. $x^2 + 6x - 8 = 0$

$a = 1, b = 6, c = -8$

$$x = \frac{-6 \pm \sqrt{6^2 - 4(1)(-8)}}{2(1)}$$

$$x = \frac{-6 \pm \sqrt{68}}{2};\ x = \frac{-6 \pm 2\sqrt{17}}{2}$$

irrational

31. $x^2 - 6x + 8 = 0$

$a = 1, b = -6, c = 8$

$$x = \frac{-(-6) \pm \sqrt{(-6)^2 - 4(1)(8)}}{2(1)}$$

$$x = \frac{6 \pm \sqrt{4}}{2};\ x = \frac{6 \pm 2}{2} = 3 \pm 1$$

$x = 3 + 1$ or $x = 3 - 1$; $x = 4$ or $x = 2$

$\{2, 4\}$; rational

33. $4x^2 - 4x + 1 = 0$

$a = 4, b = -4, c = 1$

$$x = \frac{-(-4) \pm \sqrt{(-4)^2 - 4(4)(1)}}{2(4)}$$

$$x = \frac{4 \pm \sqrt{0}}{8}$$

$x = \frac{1}{2}$; rational

35. $2x^2 + 3x + 1 = 0$

$a = 2, b = 3, c = 1$

$$x = \frac{-3 \pm \sqrt{3^2 - 4(2)(1)}}{2(2)};\ x = \frac{-3 \pm \sqrt{1}}{4}$$

$x = \frac{-3+1}{4} = -\frac{1}{2}$ or $x = \frac{-3-1}{4} = -1$

$\{-1, -\frac{1}{2}\}$; rational

37. $2x^2 + 3x + 2 = 0$

$a = 2, b = 3, c = 2$

$$x = \frac{-3 \pm \sqrt{3^2 - 4(2)(2)}}{2(2)}$$

$$x = \frac{-3 \pm \sqrt{-7}}{4}$$

not in set of real numbers

39. $2x^2 - 5x + 3 = 0$

$a = 2, b = -5, c = 3$

$$x = \frac{-(-5) \pm \sqrt{(-5)^2 - 4(2)(3)}}{2(2)}$$

$$x = \frac{-5 \pm \sqrt{1}}{4}$$

$x = \frac{-5+1}{4} = -1$ or $x = \frac{-5-1}{4} = -\frac{3}{2}$

$\{-\frac{3}{2}, -1\}$; rational

41. $2x^2 + 20x + 50 = 0$ $a = 2, b = 20, c = 50$ $x = \frac{-20 \pm \sqrt{20^2 - 4(2)(50)}}{2(2)}$

$x = \frac{-20 \pm \sqrt{0}}{4}$ $x = -5$; rational

43. $x = \frac{-(-4) - \sqrt{(-4)^2 - 4(1)(-5)}}{2(1)} = \frac{4 - \sqrt{36}}{2} = \frac{4-6}{2} = -1$; $a = 1, b = -4, c = -5$

$x^2 - 4x - 5 = 0$

45. $x = \dfrac{-5+\sqrt{5^2-4(2)(-12)}}{2(2)} = \dfrac{-5+\sqrt{121}}{4} = \dfrac{-5+11}{4} = \dfrac{3}{2}$; a = 2, b = 5, c = −12

$2x^2 + 5x - 12 = 0$

47. $x = \dfrac{-1+\sqrt{1^2-4(3)(-4)}}{2(3)} = \dfrac{-1+\sqrt{49}}{6} = \dfrac{-1+7}{6} = 1$; a = 3, b = 1, c = −4

$3x^2 + x - 4 = 0$

49. $x = \frac{-b}{2a}$, the leading term of the quadratic formula when it is separated into two terms.

51. a = 2, b = 3; $x = \frac{-3}{2(2)} = -\frac{3}{4} = -0.75$; $f(-0.75) = 2(-0.75)^2 + 3(-0.75) + 5 = 3.875$

vertex is (−0.75, 3.875)

53. a = 5, b = 3; $x = \frac{-3}{2(5)} = -0.3$; $f(-0.3) = 5(-0.3)^2 + 3(-0.3) - 3 = -3.45$

vertex is (−0.3, −3.45)

55. $-0.12x^2 + 2.4x - 8 = 0$

a = −0.12, b = 2.4, c = −8

$x = \dfrac{-2.4 \pm \sqrt{2.4^2 - 4(-0.12)(-8)}}{2(-0.12)}$

$x = \dfrac{-2.4 \pm \sqrt{1.92}}{-0.24}$

$x = \dfrac{-2.4 + \sqrt{1.92}}{-0.24}$ or $x = \dfrac{-2.4 - \sqrt{1.92}}{-0.24}$

$x \approx 4.226$ or $x \approx 15.774$

57. $-0.12x^2 + 2.4x - 10 = 0$

a = −0.12, b = 2.4, c = −10

$x = \dfrac{-2.4 \pm \sqrt{2.4^2 - 4(-0.12)(-10)}}{2(-0.12)}$

$x = \dfrac{-2.4 \pm \sqrt{0.96}}{-0.24}$

$x = \dfrac{-2.4 + \sqrt{0.96}}{-0.24}$ or $x = \dfrac{-2.4 - \sqrt{0.96}}{-0.24}$

$x \approx 5.918$ or $x \approx 14.082$

59. a. The y-intercept is the initial height of 32.8 ft.

b. The part of the parabola to the left of the y-axis has no meaning in the problem setting.

c. The diver is at 32.8 ft. at 0 sec. and ≈ 0.4 sec.

In exercise 61 assume the same initial height as in Example 11 in the text.

61. $h_o = 32.8$ ft. $v_o = 4$ ft/sec; $0 = -16.1t^2 + 4t + 32.8$; a = −16.1, b = 4, c = 32.8

$t = \dfrac{-4 \pm \sqrt{4^2 - 4(-16.1)(32.8)}}{2(-16.1)}$; $t = \dfrac{-4+\sqrt{2128.32}}{-32.2}$ or $t = \dfrac{-4-\sqrt{2128.32}}{-32.2}$

$t \approx -1.3$ (discard negative) or **t ≈ 1.55 sec.**

63. $v_o = 0$, $h_o = 35$ m, $0 = -\frac{1}{2}(9.81)t^2 + 0t + 35$; $0 = -4.905t^2 + 35$; $a = -4.905$, $b = 0$, $c = 35$

$$t = \frac{0 \pm \sqrt{0 - 4(-4.905)(35)}}{2(-4.905)};\ t = \frac{\pm\sqrt{686.7}}{-9.81};\ t \approx -2.7 \text{ (discard negative) or } \mathbf{t \approx 2.7 \text{ sec.}}$$

65. $v_o = 1.5$ m/sec.; $0 = -4.905t^2 + 1.5t + 35$; $a = -4.905$, $b = 1.5$, $c = 35$

$$t = \frac{-1.5 \pm \sqrt{1.5^2 - 4(-4.905)(35)}}{2(-4.905)};\ t = \frac{-1.5 + \sqrt{688.95}}{-9.81} \text{ or } t = \frac{-1.5 - \sqrt{688.95}}{-9.81}$$

$t \approx -2.5$ (discard negative) or $\mathbf{t \approx 2.8}$ **sec.**

67. a. $\sqrt{16x^2} = 4x$, $\sqrt{(w^2)} = |w|$, not w

b. $\sqrt{(16 + x^2)} = (4 + x)$; there is no square root of sums property

c. $\sqrt{(4x^2 - 9y^2)} = (2x - 3y)$; there is not square root of differences property

d. $\sqrt{(4x^2 - 4x + 1)} = |2x - 1|$; true $(4x^2 - 4x + 1) = (2x - 1)^2$

69.

$ax^2 + bx + c = 0$	***standard form***
$4a^2x^2 + 4abx + 4ac = 0$	***multiply each side by 4a***
$4a^2x^2 + 4abx = -4ac$	***subtract 4ac from both sides***
$4a^2x^2 + 4abx + b^2 = b^2 - 4ac$	***add b^2 to both sides***
$(2ax + b)^2 = b^2 - 4ac$	***factor left side***
$\sqrt{(2ax + b)^2} = \pm\sqrt{b^2 - 4ac}$	***take square root of both sides***
$2ax + b = \sqrt{b^2 - 4ac}$ or $2ax + b = -\sqrt{b^2 - 4ac}$	***simplify left side, giving both solutions***
$2ax = -b + \sqrt{b^2 - 4ac}$ or $2ax = -b - \sqrt{b^2 - 4ac}$	***subtract b from both sides***
$x = \frac{-b + \sqrt{b^2 - 4ac}}{2a}$ or $x = \frac{-b - \sqrt{b^2 - 4ac}}{2a}$	***divide both sides by 2a***

Exercises 5.4

	Inequality	Interval	Words	Line Graph
1.	**$x < -3$ or $x > 2$**	**$(-\infty, -3)$ or $(2, +\infty)$**	Set of numbers less than –3 or greater than 2	
3.	$x \geq 4$	$[4, +\infty)$	**Set of numbers greater than or equal to 4**	given in text
5.	$x > -3$	$(-3, +\infty)$	**Set of number greater than –3**	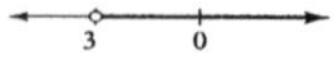
7.	$x \leq 4$	$(-\infty, 4]$	**Set of numbers less than or equal to 4**	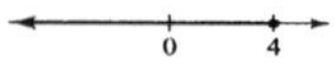

9. $x^2 - 5x + 6 > 0$
$(x - 2)(x - 3) > 0$
intercepts are $x = 2$ and $x = 3$
parabola opens upward,
$f(x) > 0$ when $x < 2$ or $x > 3$

11. $x^2 - 5x - 14 \leq 0$
$(x - 7)(x + 2) \leq 0$
intercepts are $x = 7$ and $x = -2$
graph opens upward,
$f(x) \leq 0$ for $-2 \leq x \leq 7$

13. $x^2 - 5x - 6 \geq 0$
$(x - 6)(x + 1) \geq 0$
intercepts are $x = 6$ and $x = -1$
graph opens upward,
$f(x) > 0$ for $x \leq -1$ or $x \geq 6$

15. $x^2 + 6x + 9 > 0$
$(x + 3)^2 > 0$
intercept is $x = -3$
graph opens upward,
$f(x) > 0$ for all $\Re$, $x \neq -3$

17. $4x^2 - 4x + 1 < 0$
$(2x - 1)^2 < 0$
intercept is $x = 0.5$
graph opens upward,
$f(x) < 0$, never, function ≥ 0

19. $x^2 + 2x - 15 < 0$
$(x + 5)(x - 3) < 0$
intercepts are $x = -5$ and $x = 3$
graph opens upward,
$f(x) < 0$ for $-5 < x < 3$

21. $x^2 + 10x - 24 \geq 0$
$(x + 12)(x - 2) \geq 0$
intercepts are $x = -12$ and $x = 2$
graph opens upward,
$f(x) \geq 0$ for $x \leq -12$ or $x \geq 2$

23. $x^2 - 3x - 18 \leq 0$
$(x - 6)(x + 3) \leq 0$
intercepts are $x = 6$ and $x = -3$
graph opens upward,
$f(x) \leq 0$ for $-3 \leq x \leq 6$

25. $x^2 + 17x - 18 > 0$
$(x + 18)(x - 1) > 0$
intercepts are $x = -18$ and $x = 1$
graph opens upward,
$f(x) > 0$ for $x < -18$ or $x > 1$

27. $x^2 + 8x + 16 \geq 0$
$(x + 4)^2 \geq 0$
intercept is $x = -4$
graph opens upward,
$f(x) \geq 0$ for all $\Re$

29. $x^2 - 14x + 49 < 0$
$(x - 7)^2 < 0$
intercept is $x = 7$
graph opens upward,
$f(x) < 0$ never, always ≥ 0

31. $x^2 - x > x + 3$
$x^2 - 2x - 3 > 0$
$(x - 3)(x + 1) > 0$
intercepts are $x = 3$ and $x = -1$
$x^2 - x > x + 3$, for $x < -1$ or $x > 3$
$(-\infty, -1)$ or $(3, +\infty)$

33. $x^2 + 2x - 9 \leq 1 - x$
$x^2 + 3x - 10 \leq 0$
$(x + 5)(x - 2) \leq 0$
intercepts are $x = -5$ and $x = 2$
$x^2 + 2x - 9 \leq 1 - x$, for $-5 \leq x \leq 2$
$[-5, 2]$

35. $x^2 + x < x - 1$
$x^2 < -1$
not possible in real numbers
graph both sides, no intersection
$x^2 + x$ is always above $x - 1$
{ }

37. $x^2 - x \geq x - 1$
$x - 2x + 1 \geq 0$
$(x - 1)^2 \geq 0$
intercept is $x = 1$
$x^2 - x \geq x - 1$, for all $\Re$
$(-\infty, +\infty)$

39. $2x^2 - 2 < x^2 - x$
$x^2 + x - 2 < 0$
$(x + 2)(x - 1) < 0$
intercepts are $x = -2$ and $x = 1$
$2x^2 - 2 < x^2 - x$, for $-2 < x < 1$
$(-2, 1)$

41. $3x^2 - 3 \geq 4x - x^2$
$4x^2 - 4x - 3 \geq 0$
$(2x - 3)(2x + 1) \geq 0$
intercepts are $x = 1.5$ and $x = -0.5$
$3x^2 - 3 \geq 4x - x^2$, for $x \leq -0.5$ or $x \geq 1.5$
$(-\infty, -0.5]$ or $[1.5, +\infty)$

Answers for Exercise 43 are taken from the graph, answers are recorded for reference.

43. a. $\approx 12 < x < 100$ **b.** $\approx 0 < x < 12$ or $100 < x < 112$

c. ≈ 28 ft.

d. No; the graph shows the ball's path, both axes are in feet.

e. 50 ft. **f.** $\approx 25 < x < 85$

Exercise 45 refers to text examples, a calendar is handy for exercise 45.

45. a. seasonal products such as ski equipment or heating oil

b. $\approx$ November 1 to March 31

c. December 29 **d.** October 28

e. x-intercepts are at $\approx$ 120 and 245, y-coordinate of vertex is 80,000; $b = 125$, $h = 80{,}000$
$A = \frac{2}{3}(125)(80{,}000)$; $A \approx 6{,}666{,}666$; revenue $\approx$ \$6,600,000

f. $b = 365$, $h = 17{,}500$; $A = \frac{2}{3}(365)(17{,}500)$, $A \approx 4{,}258{,}333$; cost $\approx 4{,}250{,}000$

g. Yes, there will be an overall profit for the year.

Chapter 5 Review

1. a. $\sqrt{75} = \sqrt{25 \cdot 3} = 5\sqrt{3}$ **b.** $\sqrt{8} = \sqrt{4 \cdot 2} = 2\sqrt{2}$

c. $\sqrt{32} = \sqrt{16 \cdot 2} = 4\sqrt{2}$

3. a. $\dfrac{3 - 3\sqrt{6}}{3} = \dfrac{3(1 - \sqrt{6})}{3} = 1 - \sqrt{6}$ **b.** $\dfrac{3 + 3\sqrt{6}}{3} = \dfrac{3(1 + \sqrt{6})}{3} = 1 + \sqrt{6}$

5. a. $\sqrt{14400} = \sqrt{144 \cdot 100} = 12 \cdot 10 = 120$ **b.** $\sqrt{1.44} = \sqrt{\dfrac{144}{100}} = \dfrac{12}{10} = 1.2$

c. $\sqrt{0.0121} = \sqrt{\dfrac{121}{10{,}000}} = \dfrac{11}{100} = 0.11$

d. $\sqrt{0.000\,000\,01} = \sqrt{\dfrac{1}{100{,}000{,}000}} = \dfrac{1}{10{,}000} = 0.0001$

7. a. $7.5^2 + 10^2 = 12.5^2$ **b.** $18^2 + 24^2 = 30^2$
$56.25 + 100 = 156.25$ $324 + 576 = 900$

c. $(\sqrt{5})^2 + (\sqrt{8})^2 = (\sqrt{13})^2$ **d.** $4^2 + (\sqrt{20})^2 = 6^2$
$5 + 8 = 13$ $16 + 20 = 36$

9. $2x^2 = z^2;\ z = x\sqrt{2};\ \dfrac{z}{x} = \sqrt{2}$

11. False, there is no graph in the 3rd or 4th quadrant.

13. $C = 2\pi\sqrt{\dfrac{a^2+b^2}{2}};\ \dfrac{C}{2\pi} = \sqrt{\dfrac{a^2+b^2}{2}};\ \dfrac{C^2}{4\pi^2} = \dfrac{a^2+b^2}{2};\ \dfrac{C^2}{2\pi^2} = a^2+b^2;$

$\dfrac{C^2}{2\pi^2} - b^2 = a^2;\ a = \sqrt{\dfrac{C^2}{2\pi^2} - b^2};\ or\ \ a = \sqrt{\dfrac{C^2 - 2\pi^2 b^2}{2\pi^2}}$

15. a. $\left(x + \dfrac{b}{2a}\right)^2 = x^2 + 2x\dfrac{b}{2a} + \left(\dfrac{b}{2a}\right)^2 = x^2 + \dfrac{bx}{a} + \dfrac{b^2}{4a^2}$

b. $\dfrac{b^2}{4a^2} - \dfrac{c}{a} = \dfrac{b^2}{4a^2} - \dfrac{c(4a)}{a(4a)} = \dfrac{b^2 - 4ac}{4a^2}$ **c.** $\sqrt{\left(x + \dfrac{b}{2a}\right)^2} = \left|x + \dfrac{b}{2a}\right|$

17. $3x^2 = 75 \quad x^2 = 25 \quad x = \pm 5$

19. $-4.905x^2 + 6x + 10 = 11 \quad -4.905x^2 + 6x - 1 = 0$

$x = \dfrac{-6 \pm \sqrt{6^2 - 4(-4.905)(-1)}}{2(-4.905)} \qquad x = \dfrac{-6 \pm \sqrt{16.38}}{-9.81}$

$x \approx 0.199$ or $x \approx 1.024$

21. $(x-5)^2 = 1$

$|x-5| = 1$

$x - 5 = \pm 1$

$x - 5 = 1$ or $x - 5 = -1$

$x = 6$ or $x = 4$

23. $9x^2 - 18x - 25 = 47$

$9x^2 - 18x - 72 = 0$

$9(x^2 - 2x - 8) = 0$

$(x-4)(x+2) = 0$

$x - 4 = 0$ or $x + 2 = 0$

$x = 4$ or $x = -2$

25. $4x^2 + 4x + 1 = 25$

$(2x+1)^2 = 25$

$|2x+1| = 5$

$2x + 1 = \pm 5$

$2x + 1 = 5$ or $2x + 1 = -5$

$x = 2$ or $x = -3$

27. $3x^2 - 6x + 3 = 0$

$3(x^2 - 2x + 1) = 0$

$(x-1)^2 = 0$

$x - 1 = 0$

$x = 1$

29. $5x^2 + 7x + 4 = 0$; $x = \dfrac{-7 \pm \sqrt{7^2 - 4(5)(4)}}{2(5)}$; $x = \dfrac{-7 \pm \sqrt{-31}}{10}$

no real-number solutions; no x-intercepts; negative number under the radical sign

31. a. $h_o = 0$, $v_o = 72$ ft/sec: $h = -16t^2 + 72t$

b. $-16t^2 + 72t = 0$; $-16t(t - 4.5) = 0$; $-16t = 0$ or $t - 4.5 = 0$; $t = 0$ or $t = 4.5$ sec.
(0, 0), (4.5, 0); first is when ball is thrown, second is when it comes back down.

c. $80 = -16t^2 + 72t$

$16t^2 - 72t + 80 = 0$
$16(t^2 - 4.5t + 5) = 0$
$(t - 2)(t - 2.5) = 0$
$t - 2 = 0$ or $t - 2.5 = 0$
$t = 2$ or $t = 2.5$
ball goes through 80 ft on the way up and down.

d. x-coordinate of vertex is $\dfrac{-b}{2a} = \dfrac{-72}{-32} = 2.25$

$-16(2.25)^2 + 72(2.25) = 81$
highest point is 81 ft.

e. $-16t^2 + 72t > 80$, (see *c* for intercepts) 2 sec $< t <$ 2.5 sec

f. $-16t^2 + 72t < 32$
$0 < 16t^2 - 72t + 32$
$0 < 16(t^2 - 4.5t + 2)$
$0 < (t - 4)(t - 0.5)$
intercepts are $t = 4$ and $t = 0.5$
parabola turns down
$0 < t < 0.5$ and $4 < t < 4.5$ (from when thrown to 32 feet and again from 32 feet to when the ball hits the ground at 4.5 sec. - see *b*) ball has traveled 64 feet.

g. The ball travels **2** feet in 0.5 seconds at the top of the parabola and **32** feet in 0.5 seconds near the ground. Speed slows at the top.

33. a. x-coordinate is $\dfrac{-3}{2(1)} = -\dfrac{3}{2}$; y-coordinate is $y = (-\frac{3}{2})^2 + 3(-\frac{3}{2}) + 4 = \frac{7}{4}$;
vertex is $(-\frac{3}{2}, \frac{7}{4})$

b. x-coordinate is $\dfrac{-(-2)}{2(0.5)} = 2$; y-coordinate is $y = 0.5(2^2) - 2(2) + 1 = -1$
vertex is $(2, -1)$

c. x-coordinate is $\dfrac{-(-4)}{2(3)} = \dfrac{2}{3}$; y-coordinate is $y = 3(\frac{2}{3})^2 - 4(\frac{2}{3}) = -\frac{4}{3}$
vertex is $(\frac{2}{3}, -\frac{4}{3})$

35. $x^2 + 6x - 8 \le 0$

use quadratic formula to find intercepts

$$x = \frac{-6 \pm \sqrt{36 - 4(1)(8)}}{2(1)}$$

$x = 1.123$ or $x = -7.123$

graph turns upward

function is ≤ 0 for $-7.123 \le x \le 1.123$

37. $1 - 4x^2 \le 0$

$(1 + 2x)(1 - 2x) \le 0$

$x = \pm\frac{1}{2}$

graph turns downward

function is ≤ 0 for $x \le -\frac{1}{2}$ or $x \ge \frac{1}{2}$

39. $3x^2 + 5x - 3 > 19$

$3x^3 + 5x - 22 > 0$

$(3x + 11)(x - 2) > 0$

$x = -3.667$ or $x = 2$

graph turns upward function is > 19 for $x < -3.667$ or $x > 2$

41. $x^2 + 5x + 2 > x - 3$

$x^2 + 4x + 5 > 0$

use quadratic formula to find intercepts

$$x = \frac{-4 \pm \sqrt{16 - 4(1)(5)}}{2(1)}; \quad x = \frac{-4 \pm \sqrt{-4}}{2}$$

no intercepts, graph both sides; $x^2 + 5x + 2 > x - 3$ for all x, $\Re$

43. $x^2 + 3 < 2 - x^2$

$2x^2 + 1 < 0$

$2x^2 < -1$

no real number solution

45. $x^2 + 5x + 2 \le 2 - x$

$x^2 + 6x \le 0$

$x(x + 6) \le 0$

$x = 0$ or $x = -6$

$x^2 + 5x + 2 \le 2 - x$ for $-6 \le x \le 0$

47. $0.0375(x - 20)^2 + 10 \le 20$

$0.0375(x^2 - 40x + 400) + 10 \le 20$

$0.0375x^2 - 1.5x + 15 + 10 \le 20$

$0.0375x^2 - 1.5x + 5 \le 0$

$$x = \frac{-(-1.5) \pm \sqrt{(-1.5)^2 - 4(0.0375)(5)}}{2(0.0375)}; \; x = \frac{1.5 \pm \sqrt{1.5}}{0.075}; \; x \approx 3.7 \text{ or } x \approx 36.3$$

cable can be inspected on the interval $3.7 \le x \le 36.3$

Chapter 5 Test

1. $x = \dfrac{-b \pm \sqrt{b^2 - 4ac}}{2a}$

2. a. $\sqrt{45} = \sqrt{9 \cdot 5} = 3\sqrt{5}$

b. $\sqrt{98} = \sqrt{49 \cdot 2} = 7\sqrt{2}$

3. $\dfrac{6+\sqrt{18}}{12} = \dfrac{6+3\sqrt{2}}{12} = \dfrac{3(2+\sqrt{2})}{3(4)} = \dfrac{2+\sqrt{2}}{4}$

4.
$x^2 = 196$
$\sqrt{x^2} = \sqrt{196}$
$|x| = 14$
$x = \pm 14$
$\{\pm 14\}$

5.
$(x-4)^2 = 25$
$\sqrt{(x-4)^2} = \sqrt{25}$
$|x-4| = 5$
$x-4 = 5$ or $x-4 = -5$
$x = 9$ or $x = -1$
$\{-1, 9\}$

6.
$x^2 + 10x + 5^2 = 5^2 + 24$
$(x+5)^2 = 49$
$\sqrt{(x+5)^2} = \sqrt{49}$
$|x+5| = 7$
$x+5 = 7$ or $x+5 = -7$
$x = 2$ or $x = -12$
$\{-12, 2\}$

7.
$0.04x^2 - 169 = 0$
$0.04x^2 = 169$
$\sqrt{0.04x^2} = \sqrt{169}$
$|0.2x| = 13$
$0.2x = \pm 13$
$x = \pm 65$
$\{\pm 65\}$

8.
$x^2 + 2x + 1 = 16$
$(x+1)^2 = 16$
$\sqrt{(x+1)^2} = \sqrt{16}$
$|x+1| = 4$
$x+1 = \pm 4$
$x+1 = 4$ or $x+1 = -4$
$x = 3$ or $x = -5$
$\{-5, 3\}$

9.
$|x-2| = 1$
$x - 2 = \pm 1$
$x-2 = 1$ or $x-2 = -1$
$x = 3$ or $x = 1$
$\{1, 3\}$

10. $a = 2, b = -7, c = 6$

$$x = \frac{-(-7) \pm \sqrt{(-7)^2 - 4(2)(6)}}{2(2)}$$

$$x = \frac{7 \pm \sqrt{49 - 48}}{4}$$

$$x = \frac{7 \pm \sqrt{1}}{4}$$

$$x = \frac{7+1}{4} \text{ or } x = \frac{7-1}{4}$$

$x = 2$ or $x = 1.5$

11. a. $9^2 + x^2 = (2x)^2$

$81 + x^2 = 4x^2;\ 81 = 3x^2$

$x^2 = 27;\ x = \sqrt{27} = 3\sqrt{3}$

b. $2x^2 = 7^2$

$x^2 = \frac{49}{2}$

$x = \frac{7}{\sqrt{2}} = \frac{7\sqrt{2}}{\sqrt{2}\sqrt{2}} = \frac{7\sqrt{2}}{2}$

12. $x^2 - x - 2 = 0$

$(x - 2)(x + 1) = 0$

$x = 2$ or $x = -1$

intercepts $\{-1, 2\}$

axis of symmetry is $x = \frac{-(-1)}{2(1)} = \frac{1}{2}$

$y = (0.5)^2 - 0.5 - 2$

$y = -2.25$

vertex is $(0.5, -2.25)$

13. $x^2 + 5x - 6 \le 0$

$(x + 6)(x - 1) \le 0$

$x = -6$ or $x = 1$

graph opens upward

$x^2 + 5x - 6 \le 0$ for $-6 \le x \le 1$

14. $x^2 + 5x - 6 > 0$

$(x + 6)(x - 1) > 0$

$x = -6$ or $x = 1$

graph opens upward

$x^2 + 5x - 6 > 0$ for $x < -6$ or $x > 1$

15. $x^2 + x < x + 1$

$x^2 - 1 < 0$

$(x + 1)(x - 1) < 0$

$x = \pm 1$

$x^2 + x < x + 1$ for $-1 < x < 1$

16. $4 - x^2 < x^2 + 6$

$-2 < 2x^2$

$-1 < x^2$

true for all x,

$4 - x^2 < x^2 + 6$ for

all real numbers

17. $5x^2 + 15x - 16 \ge 4 - 4x - x^2$

$6x^2 + 19x - 20 \ge 0$

$(6x - 5)(x + 4) \ge 0$

$x = 0.833$ or $x = -4$

$5x^2 + 15x - 16 \ge 4 - 4x - x^2$ for

$x \le -4$ or $x \ge 0.833$

18. a.

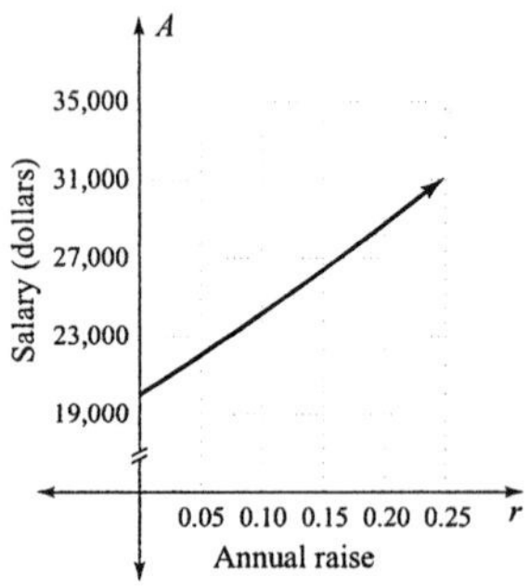

r	$A = 20{,}000(1 + r)^2$
0	$20{,}000(1 + 0)^2 = 20{,}000$
0.02	$20{,}000(1 + 0.02)^2 = 20{,}808$
0.04	$20{,}000(1 + 0.04)^2 = 21{,}632$
0.06	$20{,}000(1 + 0.06)^2 = 22{,}472$
0.07	$20{,}000(1 + 0.07)^2 = 22{,}898$
0.08	$20{,}000(1 + 0.08)^2 = 23{,}328$
0.10	$20{,}000(1 + 0.10)^2 = 24{,}200$

b. 7%

c. $\approx 22.5\%$

19.

$$v = \sqrt{2gs}$$

$$v^2 = 2gs$$

$$s = \frac{v^2}{gs}$$

20. Let x = steps for one side of the block

$$2x^2 = 212^2$$

$$x^2 = \frac{44{,}944}{2} = 22{,}472$$

$x = \sqrt{22{,}472}$; $x \approx 150$ steps for one side

≈ 300 steps to go around 2 sides

21. $x = \sqrt{4}$; $x = 2$ (principal square root)

$\sqrt{x^2} = 2$; $|x| = 2$; $x = \pm 2$; $\sqrt{w^2} = |w|$; if $|w| = z$ then $w = \pm z$

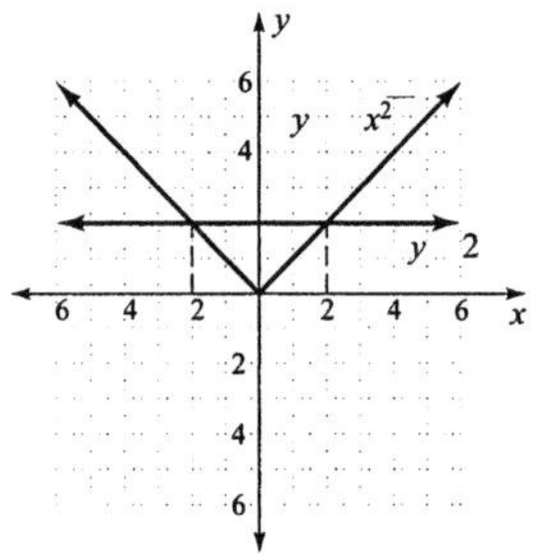

Exercises 6.1

1. The coefficient on x^2 in $y = -0.00375x^2 + 1.5x + 0$ is negative, indicating that the graph will turn down.

3. The absolute value of a is 0.00375 which is less than the absolute value of -1, so the graph will be flatter.

5. The y-intercept is zero, $c = 0$.

7. If a quadratic graph passes through the origin the y-intercept is zero and $c = 0$.

9. The equation of a parabolic cable supporting a bridge would have a positive coefficient on x^2; $y = \frac{4}{125}x^2$.

11.

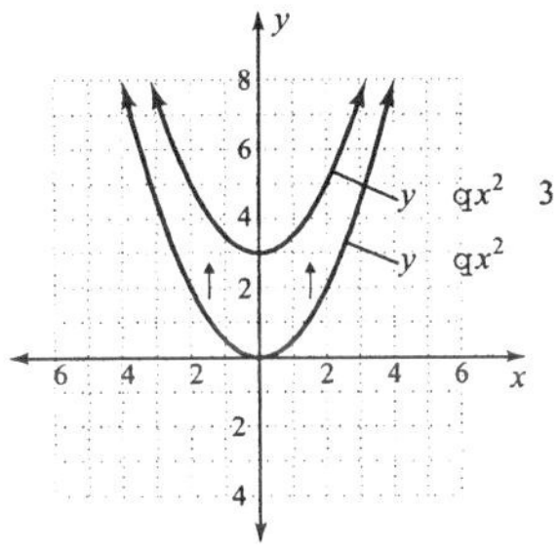

13.

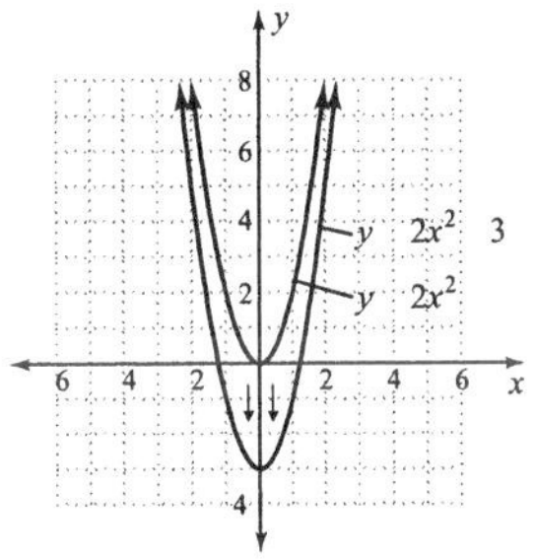

15. $f(x) = 2x^2$, $g(x) = \pi x^2$
$\pi > 2$, $g(x)$ is steeper

17. $j(x) = 4x^2$, $g(x) = \pi x^2$
$4 > \pi$, $j(x)$ is steeper

19. $f(x) = \frac{2}{3}x^2$, $h(x) = \frac{3}{4}x^2$
$\frac{3}{4} > \frac{2}{3}$, $h(x)$ is steeper

21. $p(x) = 2.41x^2$, $q(x) = 2.288x^2$
$2.41 > 2.288$, $p(x)$ is steeper

23. A coefficient on x^2 larger than 1 gives a graph ***steeper*** than the graph of $y = x^2$.

25. & 27.

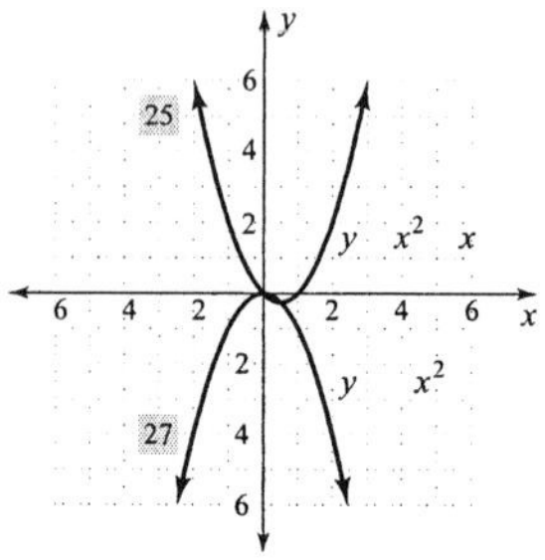

29. & 31.

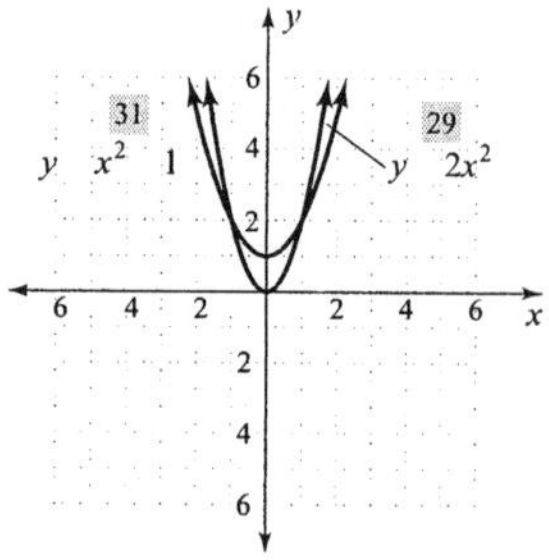

33. & 35.

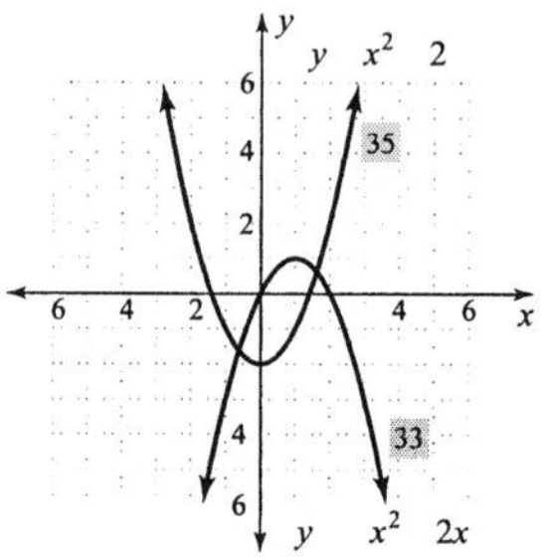

37. & 39.

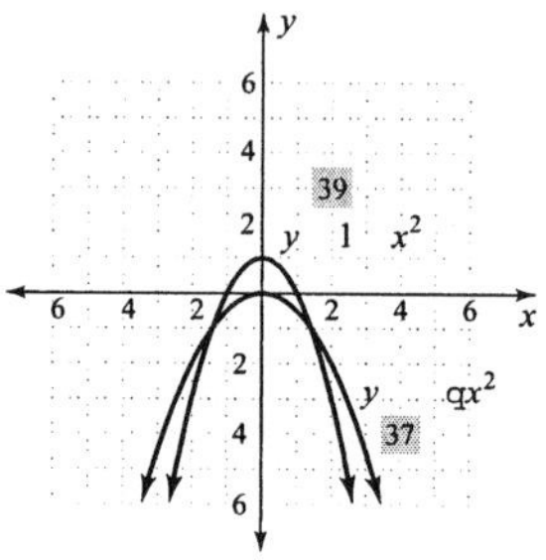

41. & 43.

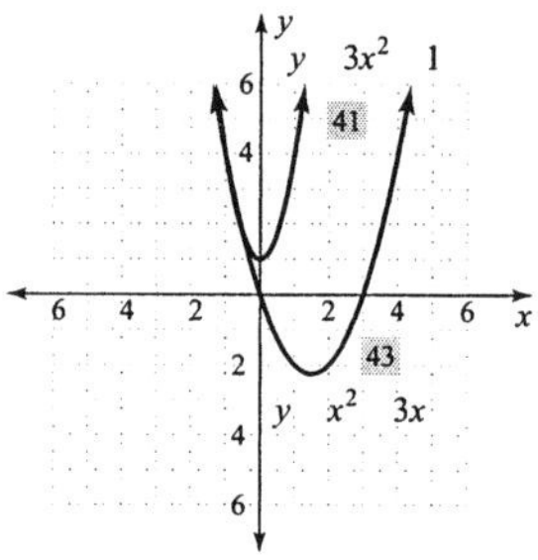

45.

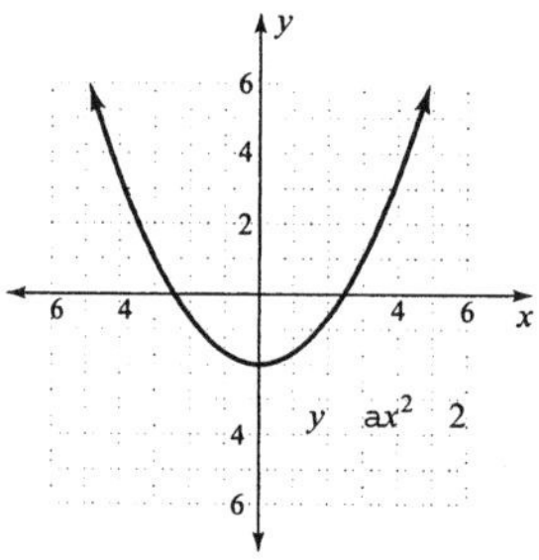

47. The coefficient on x^2 in $g(x)$ is 0.5, $h(x)$, $m(x)$, and $q(x)$ also have $a = 0.5$.

49. $f(x)$, $g(x)$, $h(x)$, $j(x)$, $m(x)$ and $q(x)$ all have parabolic graphs that turn up.

51. $j(x) = 2(x - 1)^2 = 2x^2 - 4x + 2$; the y-intercept is $(0, 2)$.

53. In the vertical motion equation, if motions starts at h = 0 then $h_o = 0$ and equation is:

$h = \frac{-g}{2}t^2 + v_o t$, a = $\frac{-g}{2}$, b = v_o; x-coordinate of the vertex is $\frac{-v_o}{2\frac{-g}{2}} = \frac{v_o}{g}$; y - coordinate

of the vertex is $\frac{-g}{2}\left(\frac{v_o}{g}\right)^2 + v_o\left(\frac{v_o}{g}\right) = \frac{-v_o^2}{2g} + \frac{v_o^2}{g} = \frac{-v_o^2}{2g} + \frac{2v_o^2}{2g} = \frac{v_o^2}{2g}$

vertex in terms of v_o and g is $\left(\frac{v_o}{g}, \frac{v_o^2}{2g}\right)$

55. The two possibilities are a graph with the same shape opening up or opening down. The coefficient on x^2 is the same, the y-intercept is –3; $y = x^2 - 3$ or $y = -x^2 - 3$.

57. a. If the origin is 20 feet directly below the cannon all the points on the water spray graph shift up 20. The cannon is at (0, 20), the vertex is (200, 170) and the fire is (400, 20).

b. The new equation has a y-intercept of 20 instead of 0. The equation will be:

$y = -0.00375x^2 + 1.5x + 20$

Exercises 6.2

1. a. $\sqrt{-20} = i\sqrt{20} = i\sqrt{4 \cdot 5} = 2i\sqrt{5}$ **b.** $\sqrt{-40} = i\sqrt{40} = i\sqrt{4 \cdot 10} = 2i\sqrt{10}$

c. $\sqrt{-72} = i\sqrt{72} = i\sqrt{36 \cdot 2} = 6i\sqrt{2}$

3. a. $\sqrt{-3} = i\sqrt{3}$ **b.** $\sqrt{-54} = i\sqrt{54} = i\sqrt{9 \cdot 6} = 3i\sqrt{6}$

c. $\sqrt{-27} = i\sqrt{27} = i\sqrt{9 \cdot 3} = 3i\sqrt{3}$

5. a. $16 = 16 + 0i$ **b.** $\sqrt{-16} = i\sqrt{16} = 4i = 0 + 4i$

c. $\sqrt{36} + \sqrt{-6} = 6 + i\sqrt{6}$

7. a. $\frac{6+\sqrt{-12}}{2} = \frac{6+2i\sqrt{3}}{2} = \frac{2(3+i\sqrt{3})}{2} = 3 + i\sqrt{3}$

b. $\frac{2-\sqrt{-24}}{4} = \frac{2-2i\sqrt{6}}{4} = \frac{2}{4} - \frac{2}{4}i\sqrt{6} = \frac{1}{2} - \frac{1}{2}i\sqrt{6}$

9. $2 - 3i$ **11.** $i + 1 = 1 + i$, conjugate is: $1 - i$

13. $2i - 3 = -3 + 2i$, conjugate is: $-3 - 2i$ **15.** $4 + 2i$

17. $2i + 3i - 4i - 5i + 6i + 7i - 8i - 9i = -8i$

19. $6 + 4i - 7i + 5 - 12i =$
$6 + 5 + 4i - 7i - 12i =$
$11 - 15i$

21. $-a + bi - (a + bi) =$
$-a + bi - a - bi =$
$-a - a + bi - bi = -2a$

23. $3 - 4(5 - i) =$
$3 - 20 + 4i =$
$-17 + 4i$

25. $(6 + 4i) - (3 - 2i) =$
$6 + 4i - 3 + 2i =$
$6 - 3 + 4i + 2i =$
$3 + 6i$

27. $(7 - 2i) + (6 - 4i) =$
$7 - 2i + 6 - 4i =$
$7 + 6 - 2i - 4i =$
$13 - 6i$

29. $4i - 3(5 - i) =$
$4i - 15 + 3i =$
$-15 + 4i + 3i =$
$-15 + 7i$

31. a. $(4 - 3i)(4 + 3i) =$
$16 - 9i^2 =$
$16 - 9(-1) =$
$16 + 9 = 25$

b. $(3 + 4i)(3 - 4i) =$
$9 - 16i^2 =$
$9 - 16(-1) =$
$9 + 16 = 25$

33. a. $(\sqrt{2} + i)(\sqrt{2} - i) =$
$2 - i^2 =$
$2 - (-1) =$
$2 + 1 = 3$

b. $(1 - i\sqrt{2})(1 + i\sqrt{2}) =$
$1 - 2i^2 =$
$1 - 2(-1) =$
$1 + 2 = 3$

35. $a = 4, b = 7, c = -2$
$7^2 - 4(4)(-2) =$
$49 + 32 = 81$
$4x^2 + 7x - 2 = 0$ has
2 real solutions

37. $a = 5, b = 4, c = 1$
$4^2 - 4(5)(1) =$
$16 - 20 = -4$
$5x^2 + 4x + 1 = 0$ has
2 complex solutions

39. $a = 4, b = 4, c = 1$
$4^2 - 4(4)(1) =$
$16 - 16 = 0$
$4x^2 + 4x + 1 = 0$ has
1 real solution

41. $a = 2, b = 6, c = 5$
$6^2 - 4(2)(5) =$
$36 - 40 = -4$
$2x^2 + 6x + 5 = 0$ has
2 complex solutions

43. $a = \frac{1}{2}$, $b = 0$, $c = -3$

$0 - 4(\frac{1}{2})(-3) = 6$

$\frac{1}{2}x^2 - 3 = 0$ has

2 real solutions

45. $x^2 - 8 = 0$

$x^2 = 8$

$x = \pm\sqrt{8}$

$x = \pm 2\sqrt{2} \approx \{\pm 2.8\}$

47. $x^2 + 2x + 4 = 0$

$$x = \frac{-2 \pm \sqrt{2^2 - 4(1)(4)}}{2(1)}$$

$$x = \frac{-2 \pm \sqrt{-12}}{2}$$

$$x = \frac{-2 \pm 2i\sqrt{3}}{2} = -1 \pm i\sqrt{3}$$

$x \approx \{-1 \pm 1.7i\}$

49. $x^2 - 4x + 8 = 0$

$$x = \frac{-(-4) \pm \sqrt{(-4)^2 - 4(1)(8)}}{2(1)}$$

$$x = \frac{4 \pm \sqrt{-16}}{2} = \frac{4 \pm 4i}{2}$$

$x = 2 \pm 2i$

51. $x^2 + 3x - 4 = 0$

$(x + 4)(x - 1) = 0$

$x + 4 = 0$ or $x - 1 = 0$

$x = -4$ or $x = 1$

$\{-4, 1\}$

53. $x^2 - 4x + 4 = 0$

$(x - 2)^2 = 0$

$x - 2 = 0$; $x = 2$

$\{2\}$

55. $(x + 1)^2 + 1 = 0$

$(x + 1)^2 = -1$

$x + 1 = \pm i$

$x = -1 \pm i$

$\{-1 \pm i\}$

57. $2x^2 + 10x + 13 = 0$

$$x = \frac{-10 \pm \sqrt{10^2 - 4(2)(13)}}{2(2)}$$

$$x = \frac{-10 \pm \sqrt{-4}}{4} = \frac{-10 \pm 2i}{4}$$

$$x = -\frac{5}{2} \pm \frac{i}{2} \approx \{-2.5 \pm 0.5i\}$$

59.

Factor	$\mathbf{x^2}$	**+2x**	**+4**
x	x^3	$\mathbf{+2x^2}$	**+4x**
−2	$\mathbf{-2x^2}$	**−4x**	−8

$x^3 - 8 = (x - 2)(x^2 + 2x + 4)$

61. $x^3 + 8 = 0$

$(x + 2)(x^2 - 2x + 4) = 0$

$x + 2 = 0;\ x^2 - 2x + 4 = 0$

$x = -2$ or

$x = \dfrac{-(-2) \pm \sqrt{(-2)^2 - 4(1)(4)}}{2(1)}$

$x = \dfrac{2 \pm 2i\sqrt{3}}{2};\ x = 1 \pm i\sqrt{3}$

$\{-2,\ 1 \pm i\sqrt{3}\} \approx \{-2,\ 1 \pm 1.7i\}$

63. $x^4 - 81 = 0$

$(x^2 - 9)(x^2 + 9) = 0$

$x^2 - 9 = 0$ or $x^2 + 9 = 0$

$(x + 3)(x - 3) = 0$ $x = \pm 3$ or

$x^2 + 9 = 0,\ x^2 = -9,\ x = \pm 3i$

$\{\pm 3, \pm 3i\}$

65. $x^3 + 3x^2 + 2x = 0$

$x(x + 2)(x + 1) = 0$

$x = 0$ or $x + 2 = 0$ or $x + 1 = 0$

$x = -2$ or $x = -1$

$\{-2, -1, 0\}$

67. $x^3 - 27 = 0$

$(x - 3)(x^2 + 3x + 9) = 0$

$x - 3 = 0$ or $x^2 + 3x + 9 = 0$

$x = 3$ or $x = \dfrac{-3 \pm \sqrt{3^2 - 4(1)(9)}}{2(1)}$

$x = \dfrac{-3 \pm 3i\sqrt{3}}{2} =$

$\{3,\ -\tfrac{3}{2} \pm i\tfrac{3\sqrt{3}}{2}\} \approx \{3, -1.5 \pm 2.6i\}$

69. $x^3 - 4x^2 = 0$

$x^2(x - 4) = 0$

$x^2 = 0$ or $x - 4 = 0$

$x = 0$ (dbl root) or $x = 4$

$\{0$ (dbl root), $4\}$

71. $x^2 - 8x^5 = 0$

$x^2(1 - 8x^3) = 0$

$x^2 = 0$ or $1 - 8x^3 = 0$

$x = 0$ (dbl root) or $(1 - 2x)(1 + 2x + 4x^2) = 0$

$1 - 2x = 0$ or $1 + 2x + 4x^2 = 0$

$x = \tfrac{1}{2}$, or $x = \dfrac{-2 \pm \sqrt{2^2 - 4(4)(1)}}{2(4)};\ x = \dfrac{-2 \pm 2i\sqrt{3}}{8} = \dfrac{-1 \pm i\sqrt{3}}{4}$

$\{0$ (dbl root), $\tfrac{1}{2},\ -\tfrac{1}{4} \pm i\tfrac{\sqrt{3}}{4}\} \approx \{0$ (dbl root), $0.5,\ -0.25 \pm 0.43i\}$

Mid-Chapter 6 Test

1. The –2 changes the graph to open down and makes it steeper.

2. The –4 shifts the graph down 4 units, the vertex and y-intercept are (0, –4) instead of (0, 0), the x-intercepts are (–2, 0), (2, 0) instead of (0, 0).

3. & 4.

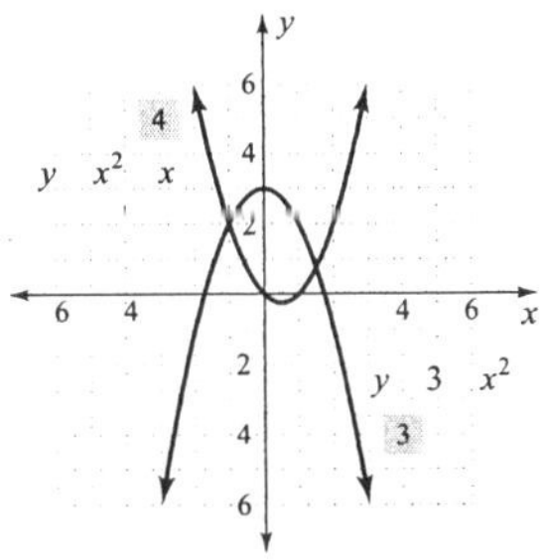

5 & 6.

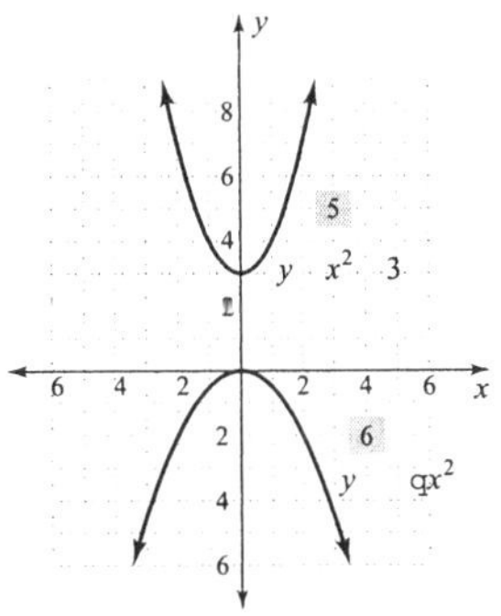

7. a. $\sqrt{-16} = 0 + 4i$ **b.** $4 = 4 + 0i$

c. $3 + \sqrt{-4} = 3 + 2i$ **d.** $\sqrt{-56} = 0 + 2i\sqrt{14}$

8. a. $\dfrac{4-\sqrt{-48}}{2} = \dfrac{4-4i\sqrt{3}}{2} = 2 - 2i\sqrt{3}$ **b.** $\dfrac{9+\sqrt{-18}}{6} = \dfrac{9+3i\sqrt{2}}{6} = \dfrac{3+i\sqrt{2}}{2}$

9. a. $(3 + 2i) - (6 + 2i) =$
$3 + 2i - 6 - 2i = -3$

b. $(4 - 2i) - 2(3 - 4i) =$
$4 - 2i - 6 + 8i = -2 + 6i$

10. a. $(3 - 2i)(3 + 2i) =$
$9 - 4(-1) = 13$

b. $(2 - i)(2 + i) =$
$4 - (-1) = 5$

c. $(2 - 3i)(2 + 3i) =$
$4 - 9(-1) = 13$

11. a. Both are 13; reversing a and b in $(a + bi)(a - bi)$ gives the same real-number product.

b. $(1 + 2i)(1 - 2i) = 1 + 4 = 5$; $(2 + i)(2 - i) = 4 + 1 = 5$

12. $a = -1, b = 2, c = 3$; $2^2 - 4(-1)(3) = 16$, two real number solutions;
$-x^2 + 2x + 3 = 0$; $(x + 1)(-x + 3) = 0$; $x = -1$ or $x = 3$; $\{-1, 3\}$

13. $(x-2)(x^2+2x+4)=0$; $x-2=0$ or $x^2+2x+4=0$; $x=2$ or $x=\dfrac{-2\pm\sqrt{2^2-4(1)(4)}}{2(1)}$

$x=\dfrac{-2\pm 2i\sqrt{3}}{2}=-1\pm i\sqrt{3}$; $\{2, -1\pm i\sqrt{3}\}\approx\{2, -1\pm 1.7i\}$

14. $x=\dfrac{-1\pm\sqrt{1^2-4(1)(1)}}{2(1)}$

$x=\dfrac{-1\pm i\sqrt{3}}{2}$

15. $x^2(1-9x^2)=0$; $x^2=0$ or $1-9x^2=0$

$x=0$ (dbl root) or $(1+3x)(1-3x)=0$

$x=\pm\frac{1}{3}$; $\{0 \text{ (dbl root)}, \pm\frac{1}{3}\}$

Exercises 6.3

1. a. graph is shifted 2 units left
$y=(x+2)^2$

b. graph is shifted 3 units right
$y=(x-3)^2$

3. a. graph is shifted up 2 units
$y=x^2+2$

b. graph is shifted down 1 unit
$y=x^2-1$

5. a. graph is shifted down 1 unit
$y=x^2-1$

b. graph is shifted down 3 units
$y=-x^2-3$

7. a. graph is shifted left 1 unit and down 2 units
$y=(x+1)^2-2$

b. graph is shifted right 2 units and up 1 unit
$y=(x-2)^2+1$

9. Shift $y=x^2$ left 2 units.

11. Shift $y=x^2$ right 4 units.

13. Shift $y=x^2$ right 3 units and up 4 units.

15. Shift $y=x^2$ left 4 units and up 3 units.

17. $y=x^2$ shifted right 1 unit, new vertex is at (1, 0)

19. $y=x^2$ shifted left 3 and up 4, new vertex is at (–3, 4)

21. $y=x^2$ shifted right 2 and down 3 new vertex is at (2, –3)

23. $y = a(x-3)^2 + 1$

$3 = a(-1-3)^2 + 1$

$3 = 16a + 1$

$2 = 16a,\ a = \frac{1}{8}$

$y = \frac{1}{8}(x-3)^2 + 1$

25. $y = a(x+1)^2 - 1$

$7 = a(1+1)^2 - 1$

$7 = 4a - 1$

$8 = 4a,\ a = 2$

$y = 2(x+1)^2 - 1$

27. $y = a(x-3)^2 - 2$

$-10 = a(5-3)^2 - 2$

$-10 = 4a - 2$

$-8 = 4a,\ \ a = -2$

$y = -2(x-3)^2 - 2$

29. One possible point is (10, 0); quadratic regression gives: $y = -0.24x^2 + 2.4x$

31. a. A (0, –25), C (20, –15)

b. $y = a(x-20)^2 - 15;$

$0 = a(0-20)^2 - 15$

$15 = 400a;\ \ a = 0.0375$

$y = 0.0375(x-20)^2 - 15$ or $y = 0.0375x^2 - 1.5x$

c. y-intercepts indicate a vertical shift of 25 feet.

33. Vertex is (30, 100); $y = a(x-30)^2 + 100$

$0 = a(0-30)^2 + 100$

$-100 = 900a;\ \ a = -\frac{1}{9}$

$y = -\frac{1}{9}(x-30)^2 + 100$ or $y = -\frac{1}{9}x^2 + \frac{20}{3}x$

35. $y = x^2 + 10x + 25$

perfect square trinomial

$y = (x+5)^2$

37. $y = x^2 - 6x + 9$

perfect square trinomial

$y = (x-3)^2$

39. $y = x^2 + 10x + 30$

$y - 30 = x^2 + 10x$

$y - 30 + 25 = x^2 + 10x + 25$

$y - 5 = (x+5)^2$

$y = (x+5)^2 + 5$

41. $y = x^2 - 6x + 8$

$y - 8 = x^2 - 6x$

$y - 8 + 9 = x^2 - 6x + 9$

$y + 1 = (x-3)^2$

$y = (x-3)^2 - 1$

43. a. $h = \dfrac{-10}{2(-0.025)} = 200$

$k = -0.025(200)^2 + 10(200) + 25$

$k = 1025$

$y = -0.025(x - 200)^2 + 1025$

b. $h = \dfrac{-(-5)}{2(-0.125)} = -20$

$k = -0.125(-20)^2 - 5(-20) + 10$

$k = 60$

$y = -0.125(x + 20)^2 + 60$

c. $h = \dfrac{-(-6)}{2(0.15)} = 20$

$k = 0.15(20)^2 - 6(20) + 15$

$k = -45$

$y = 0.15(x - 20)^2 - 45$

d. $h = \dfrac{-4.5}{2(0.75)} = -3$

$k = 0.75(-3)^2 + 4.5(-3) - 5$

$k = -11.75$

$y = 0.75(x + 3)^2 - 11.75$

e. $h = \dfrac{-(-2.7)}{2(0.9)} = 1.5$; $k = 0.9(1.5)^2 - 2.7(1.5) - 15$

$k = -17.025$; $y = 0.9(x - 1.5)^2 - 17.025$

Exercises 6.4

1. $0 = x(x - 2)$; $x = 0$ or $x = 2$

x-intercepts: (0, 0) and (2, 0)

$(0 + 2) \div 2 = 1$; $0 + 1 = 1$

axis of symmetry: $x = 1$

$y = 1^2 - 2(1)$; $y = -1$

vertex: (1, –1)

3. $0 = (-x - 7)(x - 3)$; $x = -7$ or $x = 3$

x-intercepts: (–7, 0) and (3, 0)

$(|-7| + 3) \div 2 = 5$; $-7 + 5 = -2$

axis of symmetry: $x = -2$

$y = -(-2)^2 - 4(-2) + 21$; $y = 25$

vertex: (–2, 25)

5. $x = \dfrac{-8}{2(1)} = -4$

$y = (-4)^2 + 8(-4) + 15$; $y = -1$

vertex: (–4, –1)

7. $x = \dfrac{-(-4)}{2(1)} = 2$

$y = 2^2 - 4(2) + 5$; $y = 1$

vertex: (2, 1)

9. $x = \dfrac{-(-1)}{2(2)} = \dfrac{1}{4}$

$y = 2(\frac{1}{4})^2 - (\frac{1}{4}) - 3$; $y = -\frac{25}{8}$

vertex: $(\frac{1}{4}, -\frac{25}{8})$ or (0.25, –3.125)

11. $x = \dfrac{-6}{2(3)} = -1$

$y = 3(-1)^2 + 6(-1) - 2$; $y = -5$

vertex: (–1, –5)

13. $x = \dfrac{-3}{2(-1)} = \dfrac{3}{2} = 1.5$

$y = -(1.5)^2 + 3(1.5) - 2$; $y = 0.25$

vertex: (1.5, 0.25)

15. $x = \dfrac{-\frac{4}{3}}{2(-\frac{2}{225})} = 75$

$y = -\frac{2}{225}(75)^2 + \frac{4}{3}(75) + 10$;

$y = 60$

vertex: (75, 60)

a. clearance is y value of vertex: 60 ft.

b. clearance is 60 – 10 = 50 ft.

17. $x = \dfrac{-(-0.4)}{2(0.001)} = 200$

$y = 0.001(200)^2 - 0.4(200) + 50$

$y = 10$

vertex: (200, 10)

a. from y value of vertex to roadbed is 10 ft.

b. the y-intercept is (0, 50), mid-way is y = 25, the distance to the vertex is 15 ft.

Exercises 19 - 25 use the equation from Example 2 in the text: $y = -\dfrac{32.2}{v^2}x^2 + x$

19. Factor to find the x-intercepts and find the distance between them; use the quadratic formula to find the x-intercepts; double the horizontal distance to the vertex.

Using the x-intercept method: 0 = x(–0.005x + 1), x = 0 or x = 200; distance between them is 200 feet.

21. $v = 90;\ y = -\dfrac{32.2}{90^2}x^2 + x$;

at the vertex, $x = -\dfrac{1}{2(\frac{32.2}{90^2})} = 125.8$

$y = -\frac{32.2}{90^2}(125.8)^2 + 125.8$

$y \approx 62.9$ ft;

ball travels 2(125.8) = 251.6 ft.

23. $v = 200;\ y = -\dfrac{32.2}{200^2}x^2 + x$;

at the vertex; $x = -\dfrac{1}{2(\frac{32.2}{200^2})} = 621.1$

$y = -\frac{32.2}{200^2}(621.1)^2 + 621.1$

$y \approx 310.6$ ft., does not reach 500 ft.

25.

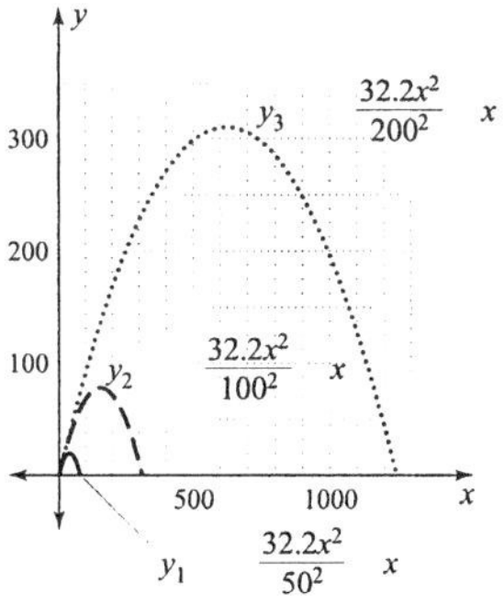

a. When the initial velocity is doubled, the height of the vertex above the ground is ***quadrupled.***

b. When the initial velocity is doubled, the distance the ball travels is ***quadrupled.***

Exercises 27 to 31 use the formula: $h = -0.5gt^2 + v_o t + h_o$

27. $v_o = 40$ ft/sec; $h_o = 5.5$ ft; $g = 32.2$ ft/sec^2; $h = -0.5(32.2)t^2 + 40t + 5.5$

max. height is h-coordinate of vertex; t-coordinate $= \dfrac{-40}{2(-16.1)} \approx 1.242$;

$h = -16.1(1.242)^2 + 40(1.242) + 5.5$; **$h \approx 30.3$ ft**

29. $v_o = 115$ ft/sec; $h_o = 0$; $h = -16.1t^2 + 115t$;

$t = \dfrac{-115}{2(-16.1)} \approx 3.57$; $h = -16.1(3.57)^2 + 115(3.57)$; $h \approx 205.4$

fireworks burst at (3.6 sec., 205.4 ft), the vertex of the equation

31. The t-coordinate of the vertex, $t = \dfrac{-v_o}{2(-0.5g)}, \left(\dfrac{-b}{2a}\right)$, does not contain h_o (c).

33. Point L is where x = 2000: $y = \frac{7}{400{,}000}(2000)^2 - \frac{1}{25}(2000) + 700 \approx$ **690 ft.**

Storm drain is at the vertex: $x = \dfrac{-(-\frac{1}{25})}{2(\frac{7}{400{,}000})} \approx 1143$;

$y = \frac{7}{400{,}000}(1143)^2 - \frac{1}{25}(1143) + 700 \approx 677$

Storm drain is at **(1143, 677)**

35. Point M is where y = 0 or at the y-intercept. From the equation elevation at M is **1188 ft**

$x = \dfrac{-(-\frac{1}{20})}{2(\frac{11}{480{,}000})} \approx 1091$; $y = \frac{11}{480{,}000}(1091)^2 - \frac{1}{20}(1091) + 1188 \approx 1161$

Storm drain is at **(1091, 1161)**

37. $\frac{1}{60{,}000}x^2 - \frac{3}{100}x + 1000 < 986.6$, multiply both sides by 60,000 to clear fractions

$x^2 - 1800x + 60{,}000{,}000 < 59{,}196{,}000$

$x^2 - 1800x + 804{,}000 < 0$; use any method to solve for x-intercepts

$x < 977$ and $x > 822$; length of puddle is distance between intercepts $977 - 822 = 155$

puddle is approximately 155 ft long.

39. $l + 2w = 60$; $l = 60 - 2w$

$A = lw$; substituting: $A = (60 - 2w)(w) = 60w - 2w^2$

Maximum area is at vertex: $w = \dfrac{-60}{2(-2)} = 15$; $60(15) - 2(15)^2$ **$A = 450$ ft^2**

For maximum area **$w = 15$ ft**, $l = 60 - 2(15)$; **$l = 30$ ft**

41. $2l + 3w = 120;\ l = 60 - 1.5w$

$A = lw;\ A = (60 - 1.5w)(w) = 60w - 1.5w^2$

Maximum area is at the vertex: $w = \dfrac{-60}{2(-1.5)} = 20;\ 60(20) - 1.5(20)^2$ $\mathbf{A = 600\ ft^2}$

For maximum area $\mathbf{w = 20}$ **ft.**; $l = 60 - 1.5(20)$, $\mathbf{l = 30}$ **ft.**

43. The other x-intercept will be the distance $|h - x|$ from the vertex on the opposite side.

Chapter 6 Review Exercises

1. The graph of $y = 3x^2$ is steeper than the graph of $y = x^2$.

3. The graph of $y = x^2 - 9$ is shifted down 9 units from $y = x^2$.

5. 7. 9.

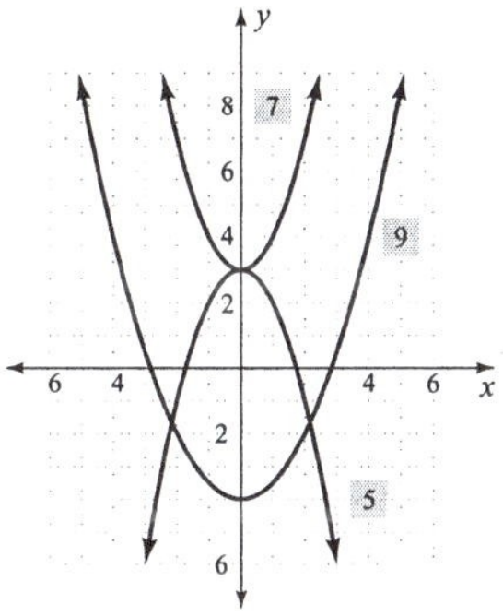

11. a. Olympic diver; input variable t indicates vertical motion equation; $a = -4.905$ is a steep downward curve, $c = 10$ meters is diver's initial height, $b = 5$ is a small initial velocity.

b. fireworks; input variable t indicates vertical motion equation; $a = -4.905$ is a steep downward curve, $c = 0$ indicates the ground as a starting point, $b = 115$ is a large initial velocity.

c. bridge cable; $a = 0.03125$ is a shallow upward curve, $c = 60$ feet indicates starting point above the ground.

d. golf ball; $a = -0.0005812$ is a very shallow downward curve.

13. a. $\sqrt{-16} = i\sqrt{16} = 4i$ **b.** $\sqrt{-50} = i\sqrt{2 \cdot 25} = 5i\sqrt{2}$

15. a. $\dfrac{8+6i}{2} = \dfrac{2(4+3i)}{2} = 4 + 3i$ **b.** $\dfrac{6+3i}{3} = \dfrac{3(2+i)}{3} = 2 + i$

17. a. $(3 + 2i) - (6 + 2i) =$
$3 + 2i - 6 - 2i = -3$

b. $(4 - 2i) - 2(3 - 4i) =$
$4 - 2i - 6 + 8i = -2 + 6i$

19. a. $(4 + 3i)(4 + 3i) =$
$16 + 24i + 9i^2 =$
$16 + 24i + 9(-1) =$
$16 - 9 + 24i = 7 + 24i$

b. $(4 - 3i)(4 + 3i) =$
$16 - 9i^2 =$
$16 - 9(-1) =$
$16 + 9 = 25$

c. $(3 - 4i)(3 + 4i) =$
$9 - 16i^2 =$
$9 - 16(-1) =$
$9 + 16 = 25$

21. If $b^2 - 4ac$ is negative, ***c*** ***(then the quadratic equation has two complex number solutions).***

23. If $b^2 - 4ac$ is positive, ***b*** ***(then the quadratic equation has two real number roots).***

25. $b^2 - 4ac > 0$, ***b*** ***(the graph of the quadratic equation intersects the x-axis twice).***

27. $(1)^2 - 4(-2)(3) = 1 + 24 = 25$; 2 real number solutions; 2 x-intercepts

29. $(3)^2 - 4(2)(4) = 9 - 32 = -23$; 2 complex number solutions; no x-intercepts

31. $(6)^2 - 4(1)(9) = 36 - 36 = 0$; 1 real number solution (a double root); 1 x-intercept

33. $-x^2 + 2x - 1 = 0$

$$x = \frac{-2 \pm 0}{2(-1)} = \frac{-2}{-2} = 1$$

$\{1\}$

34. $2x^2 + 3x + 4 = 0$

$$x = \frac{-3 \pm \sqrt{-23}}{2(2)} = \frac{-3 \pm i\sqrt{23}}{4}$$

$$\left\{\frac{-3}{4} + i\frac{\sqrt{23}}{4}, \frac{-3}{4} - i\frac{\sqrt{23}}{4}\right\}$$

35. $3x^2 + 2x + 1 = 0$

$$\frac{-2 \pm \sqrt{-8}}{2(3)} = \frac{-2 \pm 2i\sqrt{2}}{2(3)} = \frac{-1 \pm i\sqrt{2}}{3}; \left\{\frac{-1}{3} + i\frac{\sqrt{2}}{3}, \frac{-1}{3} - i\frac{\sqrt{2}}{3}\right\}$$

In exercise 37 some answers may vary

37. a. not possible to have no solutions
b. y= 5 is one possible answer
c. y = 4 is one answer
d. y = 2 is one possible answer
e. 4 solutions are not possible

39. a. from the graph {–3, –1, 2}
b. from the graph ≈ {–3.5, 0, 1.5}

41.
$x^3 - 27 = 0$
$(x - 3)(x^2 + 3x + 9) = 0$
$x - 3 = 0;\ x = 3$
$x^2 + 3x + 9 = 0$
$x = \dfrac{-3 \pm \sqrt{3^2 - 4(1)(9)}}{2(1)}$
$x = \dfrac{-3 \pm \sqrt{-27}}{2}$
$x = -\frac{3}{2} \pm \frac{3i\sqrt{3}}{2}$
$\{3,\ -\frac{3}{2} \pm \frac{3i\sqrt{3}}{2}\}$

43.
$x^3 + x^2 + x = 0$
$x(x^2 + x + 1) = 0$
$x = 0$
$x^2 + x + 1 = 0$
$x = \dfrac{-1 \pm \sqrt{1^2 - 4(1)(1)}}{2(1)} = \dfrac{-1 \pm \sqrt{-3}}{2}$
$x = -\frac{1}{2} \pm i\frac{\sqrt{3}}{2}$
$\{0,\ -\frac{1}{2} \pm i\frac{\sqrt{3}}{2}\}$

45.
$2x^3 - 2x^2 = 0$
$2x^2(x - 1) = 0$
$2x^2 = 0;\ x = 0$
$x - 1 = 0;\ x = 1$
{0 (double root), 1}

47.
$x^4 - 16 = 0$
$(x^2 + 4)(x^2 - 4) = 0$
$x^2 + 4 = 0;\ x^2 = -4,\ x = \pm 2i$
$x^2 - 4 = 0;\ x^2 = 4,\ x = \pm 2$
$\{\pm 2, \pm 2i\}$

49.

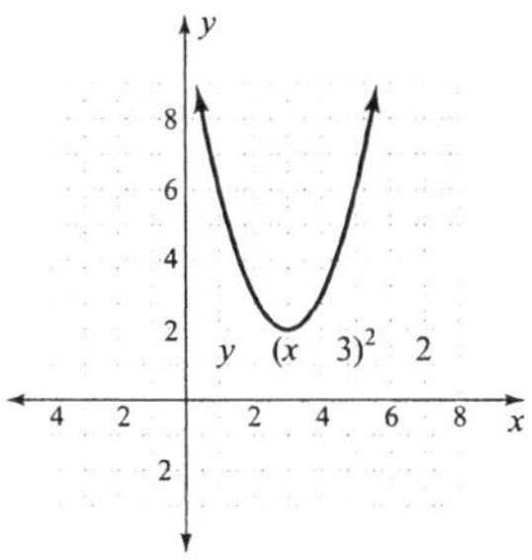

51.

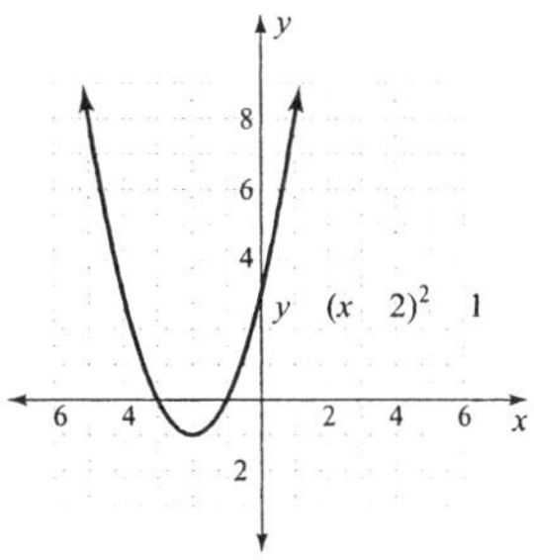

53.
$y = a(x - 8)^2 + 2$
$0 = a(0 - 8)^2 + 2$
$-2 = 64a,\ a = -\frac{1}{32}$
$y = -\frac{1}{32}(x - 8)^2 + 2$

55.
$y = a(x - 0)^2 + 0$
$2 = a(8)^2$
$a = \frac{1}{32}$
$y = \frac{1}{32}x^2$

57. $y = a(x + 2)^2 + 3$

$-3 = a(2 + 2)^2 + 3$

$-3 = 16a + 3$

$a = -\frac{3}{8}$

$y = -\frac{3}{8}(x + 2)^2 + 3$

59. vertex is (100, 250)

$y = a(x - 100)^2 + 250$

$0 = a(0 - 100)^2 + 250$

$-250 = 10{,}000a;\ \ a = -0.025$

$y = -0.025(x - 100)^2 + 250$

61. $y = x^2 + 4x + 9$

$y - 9 + 4 = x^2 + 4x + 4$

$y - 5 = (x + 2)^2$

$y = (x + 2)^2 + 5$

63. $x^2 - x - 2 = 0$

$(x - 2)(x + 1) = 0$

$x - 2 = 0;\ \ x = 2$

$x + 1 = 0;\ \ x = -1$

$\{-1, 2\}$

vertex is half-way between intercepts at $x = 0.5$

$y = (0.5)^2 - 0.5 - 2;\ y = -2.25;$ vertex: $(0.5, -2.25)$

65. $2x^2 - 4x + 5 = 0$

$x = \dfrac{-(-4) \pm \sqrt{(-4)^2 - 4(2)(5)}}{2(2)};\ x = \dfrac{4 \pm \sqrt{-24}}{4};$ no x-intercepts

vertex: $x = 1;\ y = 2(1)^2 - 4(1) + 5;\ y = 3$

vertex: (1, 3)

67. a. $v = 132$ ft/sec. $y = -\frac{32.2}{132^2}x^2 + x;\ \ x = \dfrac{-1}{2\left(\frac{32.2}{132^2}\right)} \approx 271,\ y = \frac{32.2}{132^2}(271)^2 + 271 \approx 135$

vertex: (271, 135)

b. $v = 220$ ft/sec. $y = -\frac{32.2}{220^2}x^2 + x;\ \ x = \dfrac{-1}{2\left(\frac{32.2}{220^2}\right)} \approx 752,\ y = \frac{32.2}{220^2}(752)^2 + 752 \approx 376$

vertex (752, 376)

c. $v = 400$ ft/sec. $y = -\frac{32.2}{400^2}x^2 + x;\ \ x = \dfrac{-1}{2\left(\frac{32.2}{400^2}\right)} \approx 2484,\ y = \frac{32.2}{400^2}(2484)^2 + 2484 \approx 1242$

vertex (2484, 1242)

d. $v = 500$ ft/sec. $y = -\frac{32.2}{500^2}x^2 + x;\ \ x = \dfrac{-1}{2\left(\frac{32.2}{500^2}\right)} \approx 3882,\ y = \frac{32.2}{500^2}(3882)^2 + 3882 \approx 1941$

69. a. $x = \dfrac{-\frac{20}{3}}{2(-\frac{1}{9})} = 30$, $y = -\frac{1}{9}(30)^2 + \frac{20}{3}(30) = $ **100 ft.**

b. $-\frac{1}{9}x^2 + \frac{20}{3}x < 50$

$-\frac{1}{9}x^2 + \frac{20}{3}x - 50 < 0$

$$x = \frac{-\frac{20}{3} \pm \sqrt{(\frac{20}{3})^2 - 4(-\frac{1}{9})(-50)}}{2(-\frac{1}{9})}$$

$x = \dfrac{-\frac{20}{3} \pm \sqrt{\frac{200}{9}}}{-\frac{2}{9}}$, $x \approx 8.8$ or $x \approx 51.2$ ft.

graph turns down $x < 8.8$ ft or $x > 51.2$ ft

71. a. $x = \dfrac{-(-\frac{1}{50})}{2(\frac{1}{48,000})} \approx 480$; $y = \frac{1}{48,000}(480)^2 - \frac{1}{50}(480) + 700 \approx 695.2$

drain location (480, 695.2)

b. elevation of puddle is $695.2 + 0.3 = 695.5$;

$695.5 = \frac{1}{48,000}x^2 - \frac{1}{50}x + 700$;

$0 = \frac{1}{48,000}x^2 - \frac{1}{50}x + 4.5$ clear fractions

$0 = x^2 - 960x + 216,000$

$0 = (x - 360)(x - 600)$; $x = 360$ or $x = 600$

$600 - 360 = 240$; puddle is 240 ft. long

73. $2l + 2w = 80$; $l = 40 - w$

$A = lw$; $A = (40 - w)w = 40w - w^2$

vertex: $w = \dfrac{-40}{2(-1)} = 20$; $A = 40(20) - (20)^2 = 400$; (20, 400)

The rectangular shape with the maximum area is a square: $l = w$; $4w = 80$; $w = 20$

$A = 20^2 = 400$

Chapter 6 Test

1. $y = x^2 + 2$ is shifted up 2 units from $y = x^2$; it has no x-intercepts; the vertex moves from (0, 0) to (0, 2)

2. The negative sign in $y = -x^2$ turn the graph of $y = x^2$ upside down.

3. The first equation, $y = x - \frac{32.2}{2500}x^2$, could describe the path of a ball because the graph opens down.

4. a. $\sqrt{-36} = i\sqrt{36} = 6i$

b. $\sqrt{-75} = i\sqrt{3 \cdot 25} = 5i\sqrt{3} \approx 8.7i$

5. a. $\dfrac{5+10i}{5} = \dfrac{5(1+2i)}{5} = 1+2i$

b. $\dfrac{8+4i}{2} = \dfrac{2(4+2i)}{2} = 4+2i$

6. a. $(1 + 5i)(1 - 5i) =$

$1 - 25i^2 = 1 - 25(-1) =$

$1 + 25 = 26$

b. $(5 - i)(5 + i) =$

$25 - i^2 = 25 - (-1) =$

$25 + 1 = 26$

c. $(1 - 5i)(1 - 5i) =$

$1 - 10i + 25i^2 =$

$1 - 10i + 25(-1) =$

$1 - 25 - 10i = -24 - 10i$

7. $5(2 - 3i) - 3(6 - 7i) =$

$10 - 15i - 18 + 21i =$

$-8 + 6i$

8. a. $2^2 - 4(-1)(3) = 4 + 12 = 16$

b. The discriminant is a positive real number, there will be two real number solutions and two x-intercepts.

c. $-x^2 + 2x + 3 = 0$

$(-x - 1)(x - 3) = 0$

$-x - 1 = 0;\ x = -1$ or $x - 3 = 0;\ x = 3$

$\{-1, 3\}$

9. $(x - 1)(x^2 + x + 1) = 0$

$x - 1 = 0$ or $x^2 + x + 1 = 0$

$x = 1$ or $x = \dfrac{-1 \pm \sqrt{1^2 - 4(1)(1)}}{2(1)};\ x = \dfrac{-1 \pm \sqrt{-3}}{2};\ x = \dfrac{-1 \pm i\sqrt{3}}{2}$

$\{1,\ -\frac{1}{2} \pm i\frac{\sqrt{3}}{2}\}$

10. The graph of $y = (x - 3)^2 - 1$ is $y = x^2$ shifted right 3 units and down 1 unit.

11. $y = a(x+1)^2 - 8;$
$19 = a(2+1)^2 - 8$
$27 = 9a;\ a = 3$
$y = 3(x+1)^2 - 8$ or
$y = 3(x^2 + 2x + 1) - 8$
$y = 3x^2 + 6x + 3 - 8$
$y = 3x^2 + 6x - 5$

12. $y = x^2 + 4x - 1$
$y + 1 = x^2 + 4x$
$y + 1 + 4 = x^2 + 4x + 4$
$y + 5 = (x+2)^2$
$y = (x+2)^2 - 5$

13. a. $x = \dfrac{-6}{2(-4.905)} \approx 0.612;\ y = -4.905(0.612)^2 + 6(0.612) + 3 \approx 4.83$

(0.61 sec, 4.83 m)

b. $0 = -4.905t^2 + 6t + 3,$

$$t = \frac{-6 \pm \sqrt{6^2 - 4(-4.905)(3)}}{2(-4.905)} = \frac{-6 \pm \sqrt{94.86}}{-9.81} \approx \frac{-6 \pm 9.74}{-9.81}$$

$t = -0.38$ (discard negative) or $t = 1.6$

1.6 sec. to hit water

14. a. $y = lw$

b. $2w + l = 15;\ l = 15 - 2w;\ y = (15 - 2w)w;\ y = 15w - 2w^2$

c. $w = \dfrac{-15}{2(-2)} = \dfrac{15}{4},\ y = 15(\tfrac{15}{4}) - 2(\tfrac{15}{4})^2;\ y = \tfrac{225}{4} - \tfrac{225}{8} = \tfrac{450}{8} - \tfrac{225}{8} = \tfrac{225}{8}$

vertex: $(\tfrac{15}{4}, \tfrac{225}{8})$; area, y, is largest at the peak of the graph.

Cumulative Review 1 to 6

1. $8(3\tfrac{1}{4}) = 8(3) + 8(\tfrac{1}{4}) = 24 + 2 = 26$

3. $5 - 3(x-2) = 3(x-1)$
$5 - 3x + 6 = 3x - 3$
$11 - 3x = 3x - 3$
$11 + 3 = 3x + 3x$
$14 = 6x$
$x = \tfrac{14}{6} = \tfrac{7}{3}$

5. a. $\sqrt{(-2)^2} = \sqrt{4} = 2$

b. $\sqrt{-2^2} = \sqrt{-4} = i\sqrt{4} = 2i$

c. $\sqrt{-2} = i\sqrt{2}$

d. $(\sqrt{-2})^2 = -2$

7. $(3 - i)(3 + i) = 9 - i^2 = 9 - (-1) = 9 + 1 = 10$

9. a. $(x + 1)^2 = x^2 + 2x + 1$

b. $(a + b)^2 = a^2 + 2ab + b^2$

11. a.

y	1st diff	2nd diff
3		
	−3	
0		4
	1	
1		4
	5	
6		4
	9	
15		

Next number 9 + 4 + 15 = 28

quadratic

$2a = 4$; $a = 2$; $3(2) + b = -3$

$b = -9$; $c - 3 - 4 = 3$; $c = 10$

$y = 2x^2 - 9x + 10$

b.

y	1st diff	2nd diff
8		
	3	
11		0
	3	
14		0
	3	
17		0
	3	
20		

Next number 20 + 3 = 23

linear

$m = 3$; $b = 8 - 3 = 5$

$y = 3x + 5$

c.

y	1st diff	2nd diff
3		
	1	
4		2
	3	
7		1
	4	
11		3
	7	
18		

each number is the sum of the two previous numbers, next number is 11 + 18 = 29

sequence is neither quadratic or linear

13. $x^2 - 2x - 3 \geq -3$

$x^2 - 2x \geq 0$

$x(x - 2) \geq 0$; $x \leq 0$ or $x \geq 2$

15. $x^2 - 4x + 4 \geq 8 - x$

$x^2 - 3x - 4 \geq 0$

$(x - 4)(x + 1) \geq 0$

$x \leq -1$ or $x \geq 4$

17. $|x + 6| < 5$

$x + 6 < 5$ or $-(x + 6) < 5$

$x < -1$ or $-x - 6 < 5$; $-6 - 5 < x$; $-11 < x$

$-11 < x < -1$

19. $2a - 3b = -8$; $-2(2a - 3b) = -2(-8)$

$-4a + 6b = 16$

$4a + 5b = 17$

$0 + 11b = 33$; $\mathbf{b = 3}$

$2a - 3(3) = -8$; $2a = 1$; $\mathbf{a = \frac{1}{2}}$

21.

$	**rate**	**QV**
450	0.22	450(0.22) = 99
1200	x	1200x
1650	0.10	1650(0.10) = 165

$1200x + 99 = 165$; $1200x = 66$; $x = 0.055$

interest rate = 5.5%

23. $x^2 + 3x + \underline{(\frac{9}{4})} = (x + \frac{3}{2})^2$

Exercises 7.1

1. $\frac{186\,cm^2}{6\,cm} = \frac{6\,cm \cdot 31\,cm}{6\,cm} = 31\,cm$

3. $\frac{36\,in.}{1728\,in.^3} = \frac{36\,in.}{36\,in. \cdot 48\,in^2} = \frac{1}{48\,in^2}$

5. $\frac{1500\,ft \cdot lbs}{25\,ft} = \frac{25\,ft \cdot 60\,lbs}{25\,ft} = 60\,lbs$

7. $\frac{1200\,kw\,hr}{24\,hr} = \frac{24\,hr \cdot 50\,kw}{24\,hr} = 50\,kw$

9. $300\,mL \cdot \frac{1\,L}{1000\,mL} = \frac{300\,L}{1000} = 0.3\,L$

11. $25\,ft \cdot \frac{12\,in.}{1\,ft} \cdot \frac{1\,m}{39.37\,in} = \frac{300\,m}{39.37} \approx 7.62\,m$

13. $150\,ft^3 \cdot \left(\frac{1\,yd}{3\,ft}\right)^3 = \frac{150\,ft^3 \cdot 1\,yd^3}{9\,ft^3} \approx 5.6\,yd^3$

15. $100\,in^3 \cdot \left(\frac{1\,ft}{12\,in}\right)^3 = \frac{100\,in^3 \cdot 1\,ft^3}{1728\,in^3} \approx 0.058\,ft^3$

17. $200\,mL \cdot \frac{1\,kg}{1000\,mL} \cdot \frac{1000\,g}{1\,kg} = 200\,g$

19. a. $\frac{\frac{1}{2}\,ft}{2\,in} \cdot \frac{12\,in}{1\,ft} = \frac{6}{2} = \frac{3}{1}$

b. $\frac{3000\,g}{6\,kg} \cdot \frac{1\,kg}{1000\,g} = \frac{3000}{6000} = \frac{1}{2}$

c. $\frac{32\,oz}{6\,lb} \cdot \frac{1\,lb}{16\,oz} = \frac{32}{96} = \frac{1}{3}$

21. a. $\frac{300\,mL}{30\,L} \cdot \frac{1\,L}{1000\,mL} = \frac{300}{30{,}000} = \frac{1}{100}$

b. $\frac{2\,yr}{180\,mo} \cdot \frac{12\,mo}{1\,yr} = \frac{24}{180} = \frac{2}{15}$

c. $\frac{40\,\text{min}}{\frac{1}{4}\,hr} \cdot \frac{1\,hr}{60\,\text{min}} = \frac{40}{15} = \frac{8}{3}$

23. $\frac{55\,ft}{1\,\text{sec}} \cdot \frac{1\,mi}{5280\,ft} \cdot \frac{60\,\text{sec}}{1\,\text{min}} \cdot \frac{60\,\text{min}}{1\,hr} = \frac{55 \cdot 60 \cdot 60\,mi}{5280\,hr} = 37.5\,mi/hr$

25. $\frac{1\,gal}{5\,mi} \cdot \frac{55\,mi}{1\,hr} \cdot \frac{\$1.35}{1\,gal} \cdot \frac{10\,hr}{1\,day} = \frac{55 \cdot \$1.35 \cdot 10}{5\,day} = \$148.50/day$

27. $1\,yr \cdot \dfrac{12\,mo}{1\,yr} \cdot \dfrac{adult}{150\,mo} \cdot \dfrac{500\,mg}{adult} = \dfrac{12 \cdot 500\,mg}{150} = 40\,mg$

29. a. $14x \cdot \dfrac{3}{x} = \dfrac{5}{14} \cdot 14x$

$14 \cdot 3 = 5x$

$x = 8.4$

b. $25 \cdot 15 \cdot \dfrac{4}{25} = \dfrac{x}{15} \cdot 25 \cdot 15$

$15 \cdot 4 = 25x$

$x = 2.4$

c. $24x \cdot \dfrac{7}{24} = \dfrac{16}{x} \cdot 24x$

$7x = 16 \cdot 24$

$x \approx 54.9$

31. $\dfrac{1}{12} = \dfrac{15}{x}$

$12x \cdot \dfrac{1}{12} = \dfrac{15}{x} \cdot 12x$

$x = 15 \cdot 12$ in

$x = 180\,in \cdot \dfrac{1\,ft}{12\,in} = 15\,ft$

33. $\dfrac{a}{12} = \dfrac{5}{9} \qquad \dfrac{b}{9} = \dfrac{8}{12}$

$9a = 60 \qquad 12b = 72$

$a = \frac{60}{9} = 6\frac{2}{3} \qquad b = 6$

35. $\dfrac{d}{6} = \dfrac{4\sqrt{3}}{4}$

$d = 6\sqrt{3}$

$f = \sqrt{\left(6\sqrt{3}\right)^2 + 6^2}$

$f = 12$

37. $\dfrac{e}{10} = \dfrac{4}{4\sqrt{3}}$

$e = \dfrac{10\sqrt{3}}{3}$

$f = \sqrt{10^2 + \left(\dfrac{10\sqrt{3}}{3}\right)^2}$

$f = \sqrt{\dfrac{400}{3}} = \dfrac{20}{\sqrt{3}} = \dfrac{20\sqrt{3}}{3}$

39. $\dfrac{d}{\sqrt{5}} = \dfrac{4\sqrt{3}}{4}$; $d = \sqrt{15}$

$f = \sqrt{\left(\sqrt{15}\right)^2 + \left(\sqrt{5}\right)^2}$

$f = \sqrt{20} = 2\sqrt{5}$

41. $n^2 = 2^2 + 2^2$; $n = \sqrt{8} = 2\sqrt{2}$

$\dfrac{x}{2} = \dfrac{y}{2} = \dfrac{8}{n}$

$x = y = \dfrac{16}{2\sqrt{2}} = \dfrac{8\sqrt{2}}{\sqrt{2}\cdot\sqrt{2}} = \dfrac{8\sqrt{2}}{2} = 4\sqrt{2}$

43. Total length of $b_1 = 18 + b_2$

$\dfrac{25}{18 + b_2} = \dfrac{5}{b_2}$

$25b_2 = 5(18 + b_2)$

$25b_2 = 90 + 5b_2$

$20b_2 = 90$

$b_2 = 4.5$ ft

45. Total length of $b_1 = 12 + 5 = 17$ ft

$\dfrac{20}{17} = \dfrac{h_2}{5}$

$20(5) = 17h_2$

$100 = 17h_2$

5.9 ft $\approx h_2$

47. $\dfrac{6\,ft}{2\,yd}\cdot\dfrac{1\,yd}{3\,ft} = \dfrac{6}{6}$ needed unit conversion

49. $15x\dfrac{4}{15} = \dfrac{6}{x}15x$; $4x = 90$ did not multiply each side by 15x

Exercises 7.2

1. a. x = weight in lb.; y = distance stretched in inches; $m = \dfrac{6-4}{15-10} = \dfrac{2}{5}$;

$y = \frac{2}{5}x + b$; $4 = \frac{2}{5}(10) + b$; $4 = 4 + b$; $b = 0$; $y = \frac{2}{5}x$

b. relationship is a direct variation **c.** intercept is zero

3. a. x = number of credit hours; y = cost in \$; $m = \dfrac{345-185}{4-2} = 80$;

$y = 80x + b$; $185 = 80(2) + b$; $185 = 160 + b$; $b = 25$; $y = 80x + 25$

b. relationship is not proportional **c.** b = \$25 student fees

5. a. x = purchase price in \$; y = tax in \$; $m = \dfrac{2.25-1.20}{30-16} = 0.075$;

$y = 0.075x + b$; $1.20 = 0.075(16) + b$; $1.20 = 1.20 + b$; $b = 0$; $y = 0.075x$

b. relationship is a direct variation **c.** intercept is zero

7. k = $\frac{165\,mi}{3\,hr} = 55\,mph$; 55 mph (8 hr) = 440 mi

9. k = $\frac{\$115.92}{8\,CD}$ = \$14.49/CD; \$14.49/CD (10 CD) = \$144.90

11. k = $\frac{4.5\,oz}{(3.5\,in)^2} \approx 0.367$ oz/in^2; 0.367 oz/in^2 (1.5 in)$^2 \approx 0.827$ oz

13. Linear equations that can be written as y = mx **are** linear variations.

15. If a set of data is proportional, then its equation is of the form **y = mx.**

17. a. A = $\frac{1}{2}$bh; k = $\frac{1}{2}$ **b.** A = *lw*; k = 1 **c.** V = $\frac{1}{3}\pi r^2 h$; k = $\frac{1}{3}\pi$

19. C = kd; k = π

21. A = kr^2; k = π

23. A = k*lw*; k = 1

25. V = khr^2; k = π

27. C = kd; $\frac{k8}{k3} = \frac{8}{3}$; Circumference of 8-foot diameter circle is $\frac{8}{3}$ the circumference of the 3-foot diameter circle

29. E = kh^2; $\frac{k10^2}{k2^2} = \frac{100}{4} = 25$; The 10-foot wave has 25 times the energy of the 2-foot wave.

31. L = ks^2; $\frac{k40^2}{k25^2} = \frac{1600}{625} = 2.56$; The length of the 40-mph skid mark will be 2.56 times the length of the 25 mph skid mark.

33. V = $\frac{1}{3}$bh; b ≈ 53,095 m^2; h ≈ 146.6 m; V = $\frac{1}{3}$(53,095)(146.6) m^3; V ≈ 2,594,576 m^3

Exercises 7.3

1. x = no. of problems, y = no. of points per problem; k = 100 points; $y = \frac{100}{x}$

3. x = trash in yd^3/yr; y = time in yrs.; k = 1,000,000 yd^3; $y = \dfrac{1{,}000{,}000}{x}$

5. x = copper usage in metric tons/yr; y = time in yrs; k = 650,000,000 metric tons; $y = \dfrac{650{,}000{,}000}{x}$

7. $160 \cdot 35 = 5600$; $200 \cdot 28 = 5600$; inverse variation; $k = 5600\dfrac{tons}{day} \cdot years$

9. $160 \cdot \$35 = \$5600$; $200 \cdot \$43.75 = \$8750$; not inverse variation

$\$35 \div 160 = \$0.21875$; $\$43.75 \div 200 = \$0.21875$; direct variation; $k = \$0.21875$ per item

11. $15 \cdot 6 = 90$; $25 \cdot 3.6 = 90$; inverse variation; $k = 90$ mi

13. $2 \cdot 6 = 12$; $6 \cdot 2 = 12$; inverse variation; $k = 12$ worker · days

15. $2 \cdot \$2200 = \$4400$; $3 \cdot \$3300 = \$9900$; not inverse variation

$\$2200 \div 2 = \$1100$; $\$3300 \div 3 = \$1100$; direct variation; $k = \$1100$ per semester

17. $35 \cdot 6 = 4x$

x = 52.5 lb

k = 35 lb · 6 ft = 210 lb·ft

19. $80 \cdot 0.75 = x \cdot 1$

x = 60 people

k = 60 people · 1 cup/person = 60 cups

21. $71 \cdot 1{,}820{,}000 = 1{,}300{,}000x$

x = 99.4 yr

k = 71 yr · 1,820,000 metric tons/yr

k ≈ 129,000,000 metric tons

23. $4 \cdot 900 = 3x$

x = 1200 ft/gal

k = 4 in · 900 ft/gal

k = 3600 in·ft/gal · 1ft/12 in

k = 300 ft^2/gal

25.

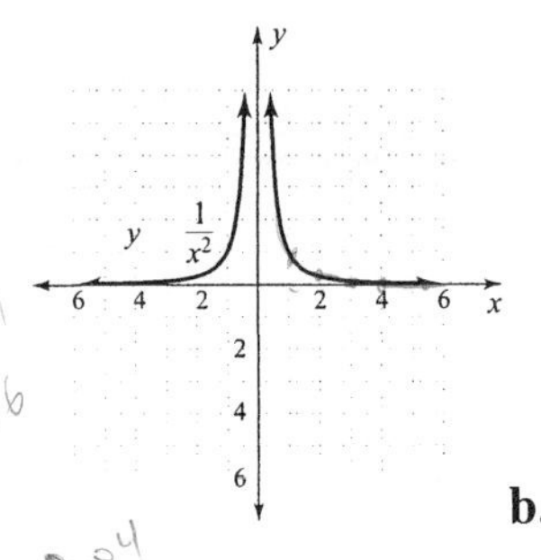

a. $f(0)$ is undefined; division by zero

b. x = 0

c. The graph approaches the x-axis

27.

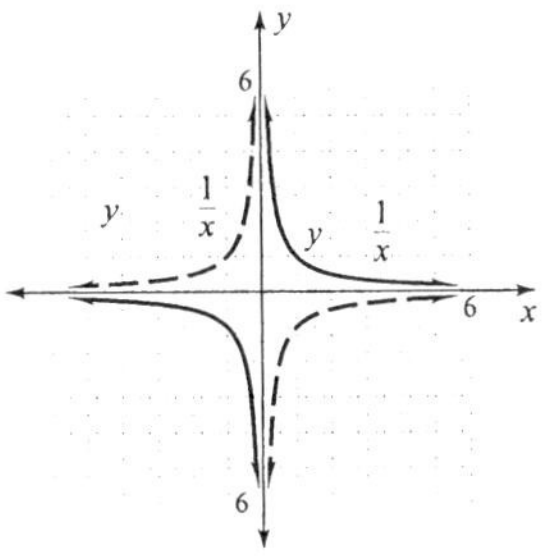

The graphs are reflected over the x-axis

29. $y_1 = \frac{k}{x_1}$; $y_2 = \frac{k}{x_2}$; solve 2nd equation for k and substitute into the 1st;

$k = x_2 y_2$; $y_1 = \frac{x_2 y_2}{x_1}$; Multiply both sides by x_1; $x_1 y_1 = x_2 y_2$

31. Multiply both sides by $x_1 x_2$; $x_1 x_2 \frac{y_1}{x_1} = \frac{y_2}{x_2} x_1 x_2$; $x_2 y_1 = x_1 y_2$; the subscripts are not alike on each side of the equation

33. No, if there were steps the equations would be the same variation. The first is inverse variation; the second is direct variation.

35. **b.** $H = \frac{M}{r^3}$; distance cubed term is in the denominator

37. $\frac{\frac{k}{2^2}}{\frac{k}{6^2}} = \frac{36}{4} = 9$ times greater

39. a. (10 lbs, 3 inches), (x lbs, 1 inch) **b.** $10 \cdot 3 = 1x$; $x = 30$ lbs

c. $k = 10$ lb. $\cdot$ 3 in. $= 30$ lb·in

Exercises 7.4

1. $x + 3 = 0$; $x = -3$

3. $4 - a = 0$; $a = 4$

5. $x^2 - 3x + 28 = 0$ $(b^2 - 4ac) = (-3)^2 - 4(1)(28) = -103$

no real roots; defined for all $\Re$

7. **a.** $-x - y$ **b.** $x - y$ **c.** $x - y$

9. **a.** $x - 3, x \neq 3;\ \dfrac{x-3}{x-3} = 1$ **b.** $-2 - x, x \neq -2;\ \dfrac{x+2}{-2-x} = -1$

c. $b - a, b \neq a;\ \dfrac{b-a}{b-a} = 1$ **d.** $x - 3, x \neq 3;\ \dfrac{3-x}{x-3} = -1$

11. **a.** $\dfrac{16a^2}{10a} = \dfrac{2\cdot 8\cdot a\cdot a}{2\cdot 5\cdot a} = \dfrac{8a}{5}, a \neq 0$ **b.** $\dfrac{2x^2y}{10xy^2} = \dfrac{2\cdot x\cdot x\cdot y}{2\cdot 5\cdot x\cdot y\cdot y} = \dfrac{x}{5y}, x \neq 0, y \neq 0$

c. $\dfrac{-49x^3y^5}{63x^2y^2} = \dfrac{-7\cdot 7\cdot x^2\cdot x\cdot y^2\cdot y^3}{7\cdot 9\cdot x^2\cdot y^2} = -\dfrac{7xy^3}{9}, x \neq 0, y \neq 0$

d. $\dfrac{64a^3b^2c}{-48a^5b^3} = \dfrac{2\cdot 4\cdot 8\cdot a^3\cdot b^2\cdot c}{-3\cdot 2\cdot 8\cdot a^3\cdot a^2\cdot b^2\cdot b} = -\dfrac{4c}{3a^2b};\ a \neq 0, b \neq 0$

13. **a.** $\dfrac{xy}{2x+y}$; simplified; $y \neq -2x$ **b.** $\dfrac{xy}{2x+xy} = \dfrac{xy}{x(2+y)} = \dfrac{y}{x+y};\ x \neq 0, y \neq -2$

15. **a.** $\dfrac{3-x}{(x+3)(x-3)} = \dfrac{-(x-3)}{(x+3)(x-3)} = -\dfrac{1}{x+3};\ x \neq \pm 3$

b. $\dfrac{x^2-9}{x^2+5x+6} = \dfrac{(x+3)(x-3)}{(x+3)(x+2)} = \dfrac{x-3}{x+2};\ x \neq -3, -2$

17. **a.** $\dfrac{2ac+4bc}{4ad+8bd} = \dfrac{2c(a+2b)}{4d(a+2b)} = \dfrac{c}{2d};\ a \neq -2b, d \neq 0;$

b. $\dfrac{6x^2+3x}{12x^2-6x} = \dfrac{3x(2x+1)}{6x(2x-1)} = \dfrac{2x+1}{2(2x-1)};\ x \neq 0, \frac{1}{2}$

19. **a.** $\dfrac{x^2+x-6}{2-x} = \dfrac{(x+3)(x-2)}{-(x-2)} = -(x+3);\ x \neq 2$

b. $\dfrac{x-3}{6-2x} = \dfrac{x-3}{-2(x-3)} = -\dfrac{1}{2};\ x \neq 3$

21. **a.** $\dfrac{1}{x}\cdot\dfrac{x^2}{1} = \dfrac{x\cdot x}{x} = x$ **b.** $\dfrac{1}{a} \div \dfrac{a^2b^2}{1} = \dfrac{1}{a}\cdot\dfrac{1}{a^2b^2} = \dfrac{1}{a^3b^2}$

c. $\dfrac{a}{b} \div \dfrac{a^2b}{b^2} = \dfrac{a}{b}\cdot\dfrac{b^2}{a^2b} = \dfrac{ab^2}{a^2b^2} = \dfrac{1}{a}$ **d.** $\dfrac{b}{a} \div \dfrac{a^2}{b^2} = \dfrac{b}{a}\cdot\dfrac{b^2}{a^2} = \dfrac{b^3}{a^3}$

21. e. $\dfrac{x}{y} \div \dfrac{x^3}{y^2} = \dfrac{x}{y} \cdot \dfrac{y^2}{x^3} = \dfrac{y}{x^2}$

f. $\dfrac{18}{cd} \div \dfrac{4d^2}{9c} = \dfrac{18}{cd} \cdot \dfrac{9c}{4d^2} = \dfrac{9 \cdot 9}{2d^3} = \dfrac{81}{2d^3}$

23. a. $\dfrac{a^2+7a+12}{a^2-4} \div \dfrac{a^2+4a}{a-2} = \dfrac{(a+4)(a+3)}{(a+2)(a-2)} \cdot \dfrac{(a-2)}{a(a+4)} = \dfrac{a+3}{a(a+2)}$

b. $\dfrac{x^2-2x}{x} \cdot \dfrac{x^2-1}{x^2-3x+2} = \dfrac{x(x-2)}{x} \cdot \dfrac{(x+1)(x-1)}{(x-2)(x-1)} = x+1$

25. a. $\dfrac{x^2+3x}{x} \cdot \dfrac{x^2-x-6}{x^2-9} = \dfrac{x(x+3)}{x} \cdot \dfrac{(x-3)(x+2)}{(x+3)(x-3)} = x+2$

b. $\dfrac{4a^2+4a+1}{4-9a^2} \div \dfrac{4a^2+2a}{3a-2} = \dfrac{(2a+1)(2a+1)}{(2-3a)(2+3a)} \cdot \dfrac{(3a-2)}{2a(2a+1)} = \dfrac{-(2a+1)}{2a(2+3a)} = \dfrac{-2a-1}{2a(2+3a)}$

27. a. $\dfrac{x+2}{x^2-4x+4} \cdot \dfrac{x^2-2x}{x+2} = \dfrac{(x+2)}{(x-2)(x-2)} \cdot \dfrac{x(x-2)}{(x+2)} = \dfrac{x}{x-2}$

b. $\dfrac{a^2-5a}{a^2+5a} \div \dfrac{a^2-10a+25}{a} = \dfrac{a(a-5)}{a(a+5)} \cdot \dfrac{a}{(a-5)(a-5)} = \dfrac{a}{(a+5)(a-5)}$

29. a. $\dfrac{a^2-b^2}{a^2+2ab+b^2} \cdot \dfrac{a+b}{a^3-b^3} = \dfrac{(a+b)(a-b)}{(a+b)(a+b)} \cdot \dfrac{(a+b)}{(a-b)(a^2+ab+b^2)} = \dfrac{1}{a^2+ab+b^2}$

b. $\dfrac{x^2-2xy+y^2}{x^2-y^2} \cdot \dfrac{x-y}{y-x} = \dfrac{(x-y)(x-y)}{(x+y)(x-y)} \cdot \dfrac{(x-y)}{-(x-y)} = \dfrac{-(x-y)}{x+y} = \dfrac{y-x}{x+y}$

31. a. $\dfrac{a^3-b^3}{a+b} \cdot \dfrac{a^2+2ab+b^2}{a^2-b^2} = \dfrac{(a-b)(a^2+ab+b^2)}{(a+b)} \cdot \dfrac{(a+b)(a+b)}{(a+b)(a-b)} = a^2+ab+b^2$

b. $\dfrac{x^2+2xy+y^2}{x^3+y^3} \cdot \dfrac{x^2-y^2}{x+y} = \dfrac{(x+y)(x+y)}{(x+y)(x^2-xy+y^2)} \cdot \dfrac{(x+y)(x-y)}{(x+y)} = \dfrac{(x+y)(x-y)}{x^2-xy+y^2}$

33. a. $\dfrac{6+x-2x^2}{9-4x^2} \cdot \dfrac{15-7x-2x^2}{x^2-2x} = \dfrac{(2-x)(3+2x)}{(3-2x)(3+2x)} \cdot \dfrac{(5+x)(3-2x)}{x(x-2)} = -\dfrac{(5+x)}{x}$

b. $\dfrac{25-9x^2}{6x^2+7x-5} \div \dfrac{5+7x-6x^2}{2+2x-4x^2} = \dfrac{(5+3x)(5-3x)}{(3x+5)(2x-1)} \cdot \dfrac{-2(1+2x)(-1+x)}{(5-3x)(1+2x)} = \dfrac{-2(x-1)}{2x-1}$

35. a. $\dfrac{\frac{1}{a}}{\frac{1}{b}} = \dfrac{1}{a} \cdot \dfrac{b}{1} = \dfrac{b}{a}$

b. $\dfrac{\frac{1}{b}}{a} = \dfrac{1}{b} \cdot \dfrac{1}{a} = \dfrac{1}{ab}$

35. c. $\frac{\frac{1}{b}}{b} = \frac{1}{b} \cdot \frac{1}{b} = \frac{1}{b^2}$ **d.** $\frac{\frac{a}{b}}{\frac{1}{b}} = \frac{a}{b} \cdot \frac{b}{1} = a$

37. $\frac{\frac{x^2-3x+2}{x}}{\frac{x^2-1}{x^2}} = \frac{(x-2)(x-1)}{x} \cdot \frac{x^2}{(x+1)(x-1)} = \frac{x(x-2)}{x+1}$

39. $\frac{\frac{x-2}{x^2+3x-4}}{\frac{2-x}{x-1}} = \frac{(x-2)}{(x+4)(x-1)} \cdot \frac{(x-1)}{-(x-2)} = \frac{-1}{x+4}$

41. $\frac{5280\ ft}{88\ \frac{ft}{\text{sec}}} = 60\ \frac{ft}{1} \cdot \frac{\text{sec}}{ft} = 60\ \text{sec}$ **43.** $\frac{65\ \frac{mi}{hr}}{15\ \frac{mi}{gal}} = \frac{13}{3} \cdot \frac{mi}{hr} \cdot \frac{gal}{mi} = 4\tfrac{1}{3}\ gal/hr$

45. $\frac{24\ \frac{cans}{case}}{3.98\ \frac{\$}{case}} \approx 6 \cdot \frac{cans}{case} \cdot \frac{case}{\$} \approx 6\ cans/\$$

47. $\frac{\frac{8\,stitches}{1\,inch}}{\frac{1\,foot}{12\,inches}} = \frac{8\,stitches}{1\,inch} \cdot \frac{12\,inches}{1\,foot} = \frac{96\,stitches}{foot}$

49. $\frac{95\ \frac{words}{\text{min}}}{300\ \frac{words}{page}} \approx 0.3\ \frac{words}{\text{min}} \cdot \frac{page}{words} \approx 0.3\ page/\text{min}$

51. Use division to remove common unit of gallons: $\frac{1.40\frac{\$}{gal}}{7\frac{mi}{gal}} = 0.20\frac{\$}{gal} \cdot \frac{gal}{mi} = \$0.20/mi$

or $\frac{7\frac{mi}{gal}}{1.40\frac{\$}{gal}} = 5\frac{mi}{gal} \cdot \frac{gal}{\$} = 5\,mi/\$$

53. Use multiplication to remove the common unit of days: $\frac{4\,days}{1\,week} \cdot \frac{2\,hrs}{1\,day} = 8\,hrs/week$

55. Terms not factors were simplified. Equation is already simplified

57. Numerator factors to (x – 1)(x – 1); (x – 2) is not a common factor in the numerator.

59. The result has an extra factor of *a*; multiplication is not distributive over multiplication.

61. True; always works; may give fraction in numerator or denominator.

Mid-Chapter 7 Test

1. $\dfrac{32}{24}=\dfrac{8\cdot 4}{8\cdot 3}=\dfrac{4}{3}$

2. $\dfrac{3}{10}\cdot\dfrac{5}{18}=\dfrac{3}{2\cdot 5}\cdot\dfrac{5}{3\cdot 6}=\dfrac{1}{12}$

3. $\dfrac{abc}{aces}=\dfrac{b}{es}$

4. $\dfrac{3}{10}\div\dfrac{5}{18}=\dfrac{3}{2\cdot 5}\cdot\dfrac{2\cdot 9}{5}=\dfrac{27}{25}$

5. $\dfrac{12cd^2}{8c^2d}=\dfrac{3\cdot 4cd\cdot d}{2\cdot 4cd\cdot c}=\dfrac{3d}{2c}$

6. $\dfrac{(x-2)(x+3)}{(x+3)}=x-2$

7. $\dfrac{a-6}{a^2-3a-18}=\dfrac{(a-6)}{(a-6)(a+3)}=\dfrac{1}{a+3}$

8. $\dfrac{b+bc}{b}=\dfrac{b(1+c)}{b}=1+c$

9. $\dfrac{12x^2-43x-20}{x^2-16}=\dfrac{(12x+5)(x-4)}{(x+4)(x-4)}=\dfrac{12x+5}{x+4}$

10. $\dfrac{ab^2}{c^2}\div\dfrac{ac}{b}=\dfrac{ab^2}{c^2}\cdot\dfrac{b}{ac}=\dfrac{b^3}{c^3}$

11. $\dfrac{x^2-16}{x^2+6x+8}\cdot\dfrac{x+2}{x-2}=\dfrac{(x-4)(x+4)(x+2)}{(x+4)(x+2)(x-2)}=\dfrac{x-4}{x-2}$

12. $\dfrac{\frac{x^2}{x+1}}{\frac{x}{x^2-1}}=\dfrac{x^2}{(x+1)}\cdot\dfrac{(x+1)(x-1)}{x}=x(x-1)$

13. $\dfrac{x^3-64}{x^2+6x+8}\div\dfrac{4-x}{x+2}=\dfrac{(x-4)(x^2+4x+16)}{(x+4)(x+2)}\cdot\dfrac{(x+2)}{-(x-4)}=\dfrac{x^2+4x+16}{x+4}$

14. f(x) is undefined when the denominator equals zero; $x + 2 = 0$ or $2x^2 - 7x + 3 = 0$; $2x^2 - 7x + 3 = (2x - 1)(x - 3)$; $x + 2 = 0$, $\mathbf{x = -2}$; $2x - 1 = 0$, $\mathbf{x = 0.5}$; $x - 3 = 0$, $\mathbf{x = 3}$

15. $\frac{2\frac{2}{3}\,yd}{5\,ft}\cdot\frac{3\,ft}{1\,yd}=\frac{\frac{8}{3}\cdot 3}{5}=\frac{8}{5}$

16. $\frac{150\,mo}{15\,yr}\cdot\frac{1\,yr}{12\,mo}=\frac{10}{12}=\frac{5}{6}$

17. $1{,}000{,}000\,tbs\cdot\frac{1\,cup}{16\,tbs}\cdot\frac{1\,pt}{2\,cup}\cdot\frac{1\,qt}{2\,pt}\cdot\frac{1\,gal}{4\,qt}=\frac{1{,}000{,}000}{256}\,gal=3906.25\,gal$

18. $\frac{450\,\frac{mi}{hr}}{1500\,\frac{\$}{hr}}=0.3\frac{mi}{hr}\cdot\frac{hr}{\$}=0.3\,{}^{mi}\!/\!_{\$}$ or $\frac{1500\,\frac{\$}{hr}}{450\,\frac{mi}{hr}}=3.33\frac{\$}{hr}\cdot\frac{hr}{mi}=\$3.33\,/\!_{mi}$

19. $15\cdot 35\cdot\frac{a}{15}=\frac{12}{35}\cdot 15\cdot 35$

$35a=180$

$a\approx 5.14$

20. $8\cdot 14\cdot\frac{x+1}{8}=\frac{2x-1}{14}\cdot 8\cdot 14$

$14(x+1)=8(2x-1)$

$14x+14=16x-8$

$14+8=2x$

$x=11$

21. $\frac{n}{5}=\frac{9.6}{6}$ $\qquad$ $\frac{p}{6}=\frac{11}{5}$

$n=\frac{9.6\cdot 5}{6}=8$ $\qquad$ $p=\frac{11\cdot 6}{5}=13.2$

22. $m=\frac{15-(-6)}{10-(-4)}=\frac{21}{14}=\frac{3}{2}$; $15=\frac{3}{2}(10)+b$, $15=15+b$, $b=0$

$y=\frac{3}{2}x$; proportional

23. $m=\frac{1.50-0.60}{3-1}=\frac{0.90}{2}=0.45$; $0.60=0.45(1)+b$, $b=0.15$

$y=0.45x+0.15$; not proportional

24. a. 50 lb · 8 ft = 100 lb · 4 ft = 400 lb·ft $\qquad$ **b.** $\frac{400\,lb\cdot ft}{80\,lb}=5$ ft from the pivot point

25. y = number of cookies per child, x = number of children; k = 100 cookies

$y=\frac{100}{x}$

26. $\$12=k(13^2)$, $k\approx 0.071$; $p=0.071(16^2)$; $p=\$18.18$

Exercises 7.5

1. Comparing the degree of the numerator and the degree of the denominator we would predict the outcome to be of degree 2, or quadratic.

$$\begin{array}{r} x^2-2x+1 \\ (x-1)\overline{)x^3-3x^2+3x-1} \\ -(x^3-x^2) \\ \hline -2x^2+3x \\ -(-2x^2+2x) \\ \hline x-1 \\ -(x-1) \\ \hline 0 \end{array}$$

3. Comparing the degree of the numerator and the degree of the denominator we would predict the outcome to be of degree 2, or quadratic.

$$\begin{array}{r} x^2+2x+4 \\ (x-2)\overline{)x^3+0x^2+0x-8} \\ -(x^3-2x^2) \\ \hline 2x^2+0x \\ -(2x^2-4x) \\ \hline 4x-8 \\ -(4x-8) \\ \hline 0 \end{array}$$

5. Comparing the degree of the numerator and the degree of the denominator we would predict the outcome to be of degree 1, or linear.

$$\begin{array}{r} 2x+5 \\ (x-1)\overline{)2x^2+3x-5} \\ -(2x^2-2x) \\ \hline 5x-5 \\ -(5x-5) \\ \hline 0 \end{array}$$

7. Comparing the degree of the numerator and the degree of the denominator we would predict the outcome to be of degree1, or linear.

$$\begin{array}{r} 2x-5 \\ (6x+5)\overline{)12x^2-20x-25} \\ -(12x^2+10x) \\ \hline -30x-25 \\ -(-30x-25) \\ \hline 0 \end{array}$$

9. Comparing degrees we predict the outcome will be degree 3, or cubic. Dividing the answer would give an outcome of degree 2, or quadratic.

$$\begin{array}{r} 4x^3 - 4x^2 - 9x + 9 \\ (x+1)\overline{)4x^4 + 0x^3 - 13x^2 + 0x + 9} \\ \underline{-(4x^4 + 4x^3)} \\ -4x^3 - 13x^2 \\ \underline{-(-4x^3 - 4x^2)} \\ -9x^2 + 0x \\ \underline{-(-9x^2 - 9x)} \\ 9x + 9 \\ \underline{-(9x+9)} \\ 0 \end{array}$$

$$\begin{array}{r} 4x^2 - 9 \\ (x-1)\overline{)4x^3 - 4x^2 - 9x + 9} \\ \underline{-(4x^3 - 4x^2)} \\ 0 - 9x + 9 \\ \underline{-(-9x + 9)} \\ 0 \end{array}$$

$4x^2 - 9 = (2x \pm 3)$

11. Comparing degrees we predict the outcome will be degree 3, or cubic. Dividing the answer would give an outcome of degree 2, or quadratic.

$$\begin{array}{r} x^3 + 6x^2 - x - 6 \\ (x-6)\overline{)x^4 + 0x^3 - 37x^2 + 0x + 36} \\ \underline{-(x^4 - 6x^3)} \\ +6x^3 - 37x^2 \\ \underline{-(6x^3 - 36x^2)} \\ -x^2 + 0x \\ \underline{-(-x^2 + 6x)} \\ -6x + 36 \\ \underline{-(-6x + 36)} \\ 0 \end{array}$$

$$\begin{array}{r} x^2 - 1 \\ (x+6)\overline{)x^3 + 6x^2 - x - 6} \\ \underline{-(x^3 + 6x^2)} \\ 0 - x - 6 \\ \underline{-(-x - 6)} \\ 0 \end{array}$$

$x^2 - 1 = (x \pm 1)$

13. Defined for all $\Re$ except when $x - 4 = 0$ or $x = 4$

$$\frac{x^2 - 3x - 4}{x - 4} = \frac{(x-4)(x+1)}{(x-4)} = x + 1$$

15. Defined for all $\Re$ except when x – 3 = 0 or x = 3

$$\begin{array}{r} x \\ (x-3)\overline{)x^2-3x-4} \\ \underline{-(x^2-3x)} \\ 0-4 \end{array}$$

$$\frac{x^2-3x-4}{x-3} = x - \frac{4}{x-3}$$

17. Defined for all $\Re$ except when x – 2 = 0 or x = 2

$$\begin{array}{r} x-1 \\ (x-2)\overline{)x^2-3x-4} \\ \underline{-(x^2-2x)} \\ -x-4 \\ \underline{-(-x+2)} \\ -6 \end{array}$$

$$\frac{x^2-3x-4}{x-2} = x - 1 - \frac{6}{x-2}$$

19. Defined for all $\Re$ except when x – 1 = 0 or x = 1

$$\begin{array}{r} x-2 \\ (x-1)\overline{)x^2-3x-4} \\ \underline{-(x^2-x)} \\ -2x-4 \\ \underline{-(-2x+2)} \\ -6 \end{array}$$

$$\frac{x^2-3x-4}{x-1} = x - 2 - \frac{6}{x-1}$$

21. Defined for all $\Re$ except when x = 0

$$\begin{array}{r} x-3 \\ x\overline{)x^2-3x-4} \\ \underline{-(x^2-3x)} \\ 0-4 \end{array}$$

$$\frac{x^2-3x-4}{x} = x - 3 - \frac{4}{x}$$

23. Defined for all $\Re$ except when x + 1 = 0 or x = –1

$$\frac{x^2-3x-4}{x+1} = \frac{(x-4)(x+1)}{(x+1)} = x - 4$$

25. a. $x^2 - 3x - 4 = (x - 4)(x + 1)$; When the denominator is a factor of the numerator the quotient has a remainder of 0.

b.

x	$f(x) = x^2 - 3x - 4$
-1	$(-1)^2 - 3(-1) - 4 = 0$
0	$(0)^2 - 3(0) - 4 = -4$
1	$(1)^2 - 3(1) - 4 = -6$
2	$(2)^2 - 3(2) - 4 = -6$
3	$(3)^2 - 3(3) - 4 = -4$
4	$(4)^2 - 3(4) - 4 = 0$

Function value and remainder are the same.

c. 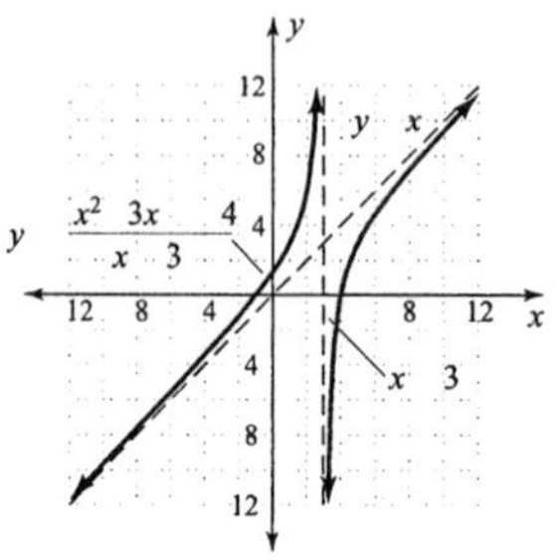

d. Nearly vertical section is where the graph approaches the restriction; $x = 3$.

e. For $x < -8$ and $x > 8$ the graph approaches the function $y = x$.

f. 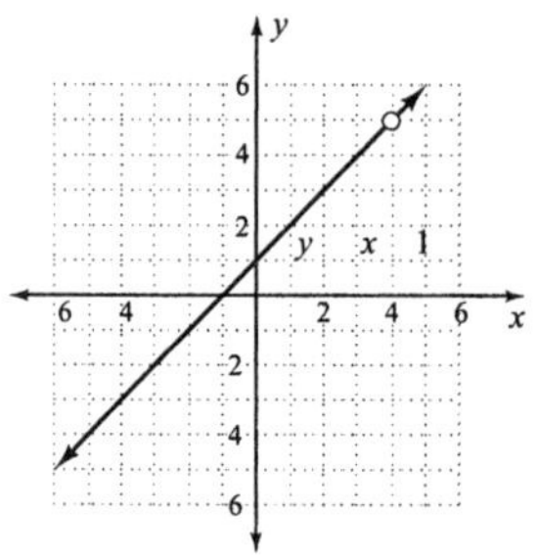 graph is linear with a hole at $x = 4$; remainder in division was 0.

g. 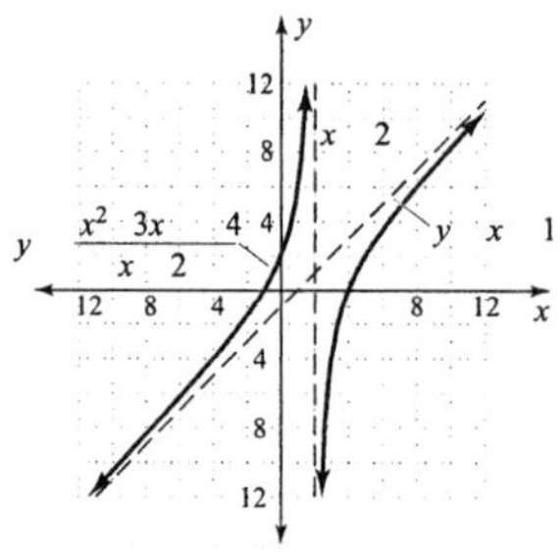 graph is nearly vertical at the restriction; $x = 2$.

27.

$$\begin{array}{r} x^2+2x+4 \\ (x-2)\overline{)x^3-0x^2-0x-1} \\ \underline{-(x^3-2x^2)} \\ 2x^2-0x \\ \underline{-(2x^2-4x)} \\ 4x-1 \\ \underline{-(4x-8)} \\ 7 \end{array}$$

$$\frac{x^3-1}{x-2}=x^2+2x+4+\frac{7}{x-2}$$

29. Denominator is a factor of the numerator.

$$\begin{array}{r} x^2+x+1 \\ (x-1)\overline{)x^3-0x^2-0x-1} \\ \underline{-(x^3-x^2)} \\ x^2-0x \\ \underline{-(x^2-x)} \\ x-1 \\ \underline{-(x-1)} \\ 0 \end{array}$$

$$\frac{x^3-1}{x-1}=x^2+x+1$$

31.

$$\begin{array}{r} x^2 \\ x\overline{)x^3-0x^2-0x-1} \\ \underline{-(x^3)} \\ 0-0x^2-0x-1 \end{array}$$

$$\frac{x^3-1}{x}=x^2-\frac{1}{x}$$

33.

$$\begin{array}{r} x^2-x+1 \\ (x+1)\overline{)x^3-0x^2-0x-1} \\ \underline{-(x^3+x^2)} \\ -x^2-0x \\ \underline{-(-x^2-x)} \\ x-1 \\ \underline{-(x+1)} \\ -2 \end{array}$$

$$\frac{x^3-1}{x+1}=x^2-x+1-\frac{2}{x+1}$$

35. Denominator is a factor of numerator

$$
\begin{array}{r}
x^2 - x + 1 \\
(x+1)\overline{)x^3 + 0x^2 + 0x + 1} \\
-\underline{(x^3 + x^2)} \\
-x^2 + 0x \\
-\underline{(-x^2 - x)} \\
x + 1 \\
-\underline{(x+1)} \\
0
\end{array}
$$

$$\frac{x^3+1}{x+1} = x^2 - x + 1$$

37.

$$
\begin{array}{r}
x^3 - x^2 + x - 1 \\
(x+1)\overline{)x^4 + 0x^3 + 0x^2 + 0x + 1} \\
-\underline{(x^4 + x^3)} \\
-x^3 + 0x^2 \\
-\underline{(-x^3 - x^2)} \\
x^2 + 0x \\
-\underline{(x^2 + x)} \\
-x + 1 \\
-\underline{(-x-1)} \\
2
\end{array}
$$

$$\frac{x^4+1}{x+1} = x^3 - x^2 + x - 1 + \frac{2}{x+1}$$

39. Denominator is a factor of numerator

$$
\begin{array}{r}
x^3 - 3x^2 + 3x - 1 \\
(x-1)\overline{)x^4 - 4x^3 + 6x^2 - 4x + 1} \\
-\underline{(x^4 - x^3)} \\
-3x^3 + 6x^2 \\
-\underline{(-3x^3 + 3x^2)} \\
3x^2 - 4x \\
-\underline{(3x^2 - 3x)} \\
-x + 1 \\
-\underline{(-x+1)} \\
0
\end{array}
$$

$$\frac{x^4 - 4x^3 + 6x^2 - 4x + 1}{x-1} = x^3 - 3x^2 + 3x - 1$$

41.

$$
\begin{array}{r}
3x^2 - 2x + 2 \\
(x+1)\overline{)3x^3 + x^2 + 0x - 1} \\
-\underline{(3x^3 + 3x^2)} \\
-2x^2 + 0x \\
-\underline{(-2x^2 - 2x)} \\
2x - 1 \\
-\underline{(2x+2)} \\
-3
\end{array}
$$

$$\frac{x^2 + 3x^3 - 1}{x+1} = 3x^2 - 2x + 2 - \frac{3}{x+1}$$

43. Hole is at $(-1, 3)$

45. Hole is at $(1, 0)$

47.

$$\begin{array}{r} x^2 - xy + y^2 \\ (x+y)\overline{)x^3 + 0x^2y + 0xy^2 + y^3} \\ -(x^3 + x^2y) \\ \hline -x^2y + 0xy^2 \\ -(-x^2y - xy^2) \\ \hline xy^2 + y^3 \\ -(xy^2 + y^3) \\ \hline 0 \end{array}$$

49. ***True***, if the divisor $(x - a)$ is a factor of the dividend, there is a hole in the graph at $x = a$.

Exercises 7.6

1. a. $\frac{3\cdot 7}{5\cdot 7} = \frac{21}{35}$ **b.** $\frac{5\cdot 6}{8\cdot 6} = \frac{30}{48}$ **c.** $\frac{8\cdot 16}{3\cdot 16} = \frac{128}{48}$

3. a. $\frac{3\cdot 5}{8\cdot 5} = \frac{15}{40}$ **b.** $\frac{5\cdot 9}{6\cdot 9} = \frac{45}{54}$ **c.** $\frac{5\cdot 8}{4\cdot 8} = \frac{40}{32}$

5. a. $\frac{4\cdot 2x}{xy\cdot 2x} = \frac{8x}{2x^2y}$ **b.** $\frac{3\cdot 2b}{a^2b\cdot 2b} = \frac{6b}{2a^2b^2}$

7. a. $\frac{x\cdot(x+1)}{(x-1)(x+1)} = \frac{x^2+x}{x^2-1}$ **b.** $\frac{a(a-b)}{(a+b)(a-b)} = \frac{a^2-ab}{a^2-b^2}$

9. a. $\frac{3}{4} - \frac{x}{4} = \frac{3-x}{4}$ **b.** $\frac{2}{5x} - \frac{7}{5x} = \frac{2-7}{5x} = \frac{-5}{5x} = -\frac{1}{x};\ x \neq 0$

c. $\frac{4}{x+2} - \frac{x^2}{x+2} = \frac{4-x^2}{x+2} = \frac{(2+x)(2-x)}{(x+2)} = 2 - x;\ x \neq -2$

d. $\frac{2x}{x^2-1} - \frac{x-1}{x^2-1} = \frac{2x-(x-1)}{x^2-1} = \frac{2x-x+1}{(x-1)(x+1)} = \frac{x+1}{(x-1)(x+1)} = \frac{1}{x-1};\ x \neq -1, 1$

11. a. common denominator is 12; $\frac{1\cdot 3}{4\cdot 3}+\frac{7}{12}=\frac{3+7}{12}=\frac{10}{12}=\frac{5\cdot 2}{6\cdot 2}=\frac{5}{6}$

b. common denominator is 2a; $\frac{2\cdot 2}{a\cdot 2}-\frac{5}{2a}=\frac{4-5}{2a}=-\frac{1}{2a}$

c. common denominator is a^2b^2; $\frac{b\cdot b^2}{a^2b^2}-\frac{c\cdot a}{ab^2\cdot a}=\frac{b^3-ac}{a^2b^2}$

d. $x^2-2x=x(x-2)$; $x^2-4=(x+2)(x-2)$; common denominator is $x(x+2)(x-2)$;

$$\frac{5(x+2)}{x(x-2)(x+2)}+\frac{3x}{x(x-2)(x+2)}=\frac{5x+10+3x}{x(x-2)(x+2)}=\frac{8x+10}{x(x-2)(x+2)}$$

e. $x^2-6x+9=(x-3)^2$; $x^2-9=(x-3)(x+3)$; common denominator is $(x+3)(x-3)^2$;

$$\frac{5x(x+3)}{(x-3)^2(x+3)}-\frac{3x(x-3)}{(x+3)(x-3)(x-3)}=\frac{5x^2+15x-3x^2+9x}{(x+3)(x-3)^2}=\frac{2x^2+24x}{(x+3)(x-3)^2}$$

13. $$\frac{5\cdot 3b}{4a\cdot 3b}+\frac{3\cdot 2a}{6b\cdot 2a}=\frac{15b+6a}{12ab}=\frac{3(2a+5b)}{3(4ab)}=\frac{2a+5b}{4ab}$$

15. $$\frac{x(x)}{(x-3)x}+\frac{1(x-3)}{x(x-3)}=\frac{x^2+x-3}{x(x-3)}$$

17. $$\frac{5x(x)}{(x-1)x}-\frac{(8+x)(x-1)}{x(x-1)}=\frac{5x^2-(x^2+7x-8)}{x(x-1)}=\frac{4x^2-7x+8}{x(x-1)}$$

19. $$\frac{x(x-3)}{(x-3)(x-3)}+\frac{3}{(x-3)^2}=\frac{x^2-3x+3}{(x-3)^2}$$

21. $$\frac{2b}{(b+1)(b-1)}-\frac{3(b+1)}{-(b-1)(b+1)}=\frac{2b+3b+3}{(b+1)(b-1)}=\frac{5b+3}{(b+1)(b-1)}$$

23. $$\frac{2b}{a(2+b)b}-\frac{3a}{b(2+b)a}=\frac{2b-3a}{ab(2+b)}$$

25. $$\frac{x(x)}{(x-3)^2(x)}+\frac{3(x-3)}{x(x-3)(x-3)}=\frac{x^2+3x-9}{x(x-3)^2}$$

27. $$\frac{(x+2)(x)}{(2x+3)(3x+2)(x)}-\frac{1(3x+2)}{x(2x+3)(3x+2)}=\frac{x^2+2x-3x-2}{x(2x+3)(3x+2)}=\frac{x^2-x-2}{x(2x+3)(3x+2)}=$$

$$\frac{(x-2)(x+1)}{x(2x+3)(3x+2)}$$

29. $$\frac{24}{24}+\frac{24x}{24}+\frac{12x^2}{24}+\frac{4x^3}{24}+\frac{x^4}{24}=\frac{x^4+4x^3+12x^2+24x+24}{24}$$

31. $$\frac{RT(v^2)}{v^2(v-b)}-\frac{a(v-b)}{v^2(v-b)}=\frac{RTv^2-av+ab}{v^2(v-b)}$$

33. $$\frac{t_2}{t_1t_2}+\frac{t_1}{t_1t_2}=\frac{t_1+t_2}{t_1t_2}$$

35. $$\frac{m_2}{m_1m_2}+\frac{m_1}{m_1m_2}=\frac{m_1+m_2}{m_1m_2}$$

37. $$\frac{1}{t}=\frac{t_2}{t_1t_2}+\frac{t_1}{t_1t_2}$$
$$\frac{1}{t}=\frac{t_1+t_2}{t_1t_2}$$
$$t=\frac{t_1t_2}{t_1+t_2}$$

39. $$\frac{1}{t}=\frac{t_2t_3}{t_1t_2t_3}+\frac{t_1t_3}{t_1t_2t_3}+\frac{t_2t_1}{t_1t_2t_3}$$
$$\frac{1}{t}=\frac{t_2t_3+t_1t_3+t_2t_1}{t_1t_2t_3}$$
$$t=\frac{t_1t_2t_3}{t_2t_3+t_1t_3+t_2t_1}$$

41. Using formula from exercise 37; $\frac{5\cdot 6}{5+6}=\frac{30}{11}\approx 2.727$ hrs; air is changed in less than 3 hrs

43. Using the formula from exercise 37; $\frac{5\cdot 9}{5+9}=\frac{45}{14}\approx 3.2$ min.

45. Using the formula from exercise 37; $\frac{4\cdot 3}{4+3}=\frac{12}{7}\approx 1.7$ hrs

47. $$\frac{7+1}{\frac{1}{7}+1}=\frac{8}{\frac{1}{7}+\frac{7}{7}}=\frac{8}{\frac{8}{7}}=\frac{8}{1}\cdot\frac{7}{8}=7$$

49. $$a=\frac{V}{\frac{4}{3}\pi b^2}=\frac{V}{\frac{4\pi b^2}{3}}=\frac{V}{1}\cdot\frac{3}{4\pi b^2}=\frac{3V}{4\pi b^2}$$

51. $$\frac{x-\frac{x}{2}}{2+\frac{x}{3}}=\frac{\frac{2x}{2}-\frac{x}{2}}{\frac{6}{3}+\frac{x}{3}}=\frac{\frac{x}{2}}{\frac{x+6}{3}}=\frac{x}{2}\cdot\frac{3}{(x+6)}=\frac{3x}{2(x+6)}$$

53. $$I=\frac{E}{R+\frac{r}{2}}=\frac{E}{\frac{2R}{2}+\frac{r}{2}}=\frac{E}{\frac{2R+r}{2}}=\frac{E}{1}\cdot\frac{2}{2R+r}=\frac{2E}{2R+r}$$

55. $$v=\frac{2x}{\frac{x}{v_1}+\frac{x}{v_2}}=\frac{2x}{\frac{v_2x}{v_1v_2}+\frac{v_1x}{v_1v_2}}=\frac{2x}{\frac{v_1x+v_2x}{v_1v_2}}=\frac{2x}{1}\cdot\frac{v_1v_2}{(v_1+v_2)x}=\frac{2v_1v_2}{v_1+v_2}$$

57. $284 \div 55 \approx 5.16$ hr; $284 \div 65 \approx 4.37$ hr; $5.16 + 4.37 \approx 9.5$ hrs total

Using the equation from exercise 55 to find average speed: $\dfrac{2(55)(65)}{55+65} \approx 59.6$ mph

59. Using the equation from exercise 55: $\dfrac{2(18)(22)}{18+22} = 19.8$ kph

63. Answer is wrong; did not convert to common denominator before adding.

65. Answer is right; common denominator is not "least common denominator" so additional simplification may be necessary.

67. a. "·" multiplication **b.** "÷" division

69. a. "+" addition **b.** "–" subtraction

71. a. "·" multiplication
b. "÷" division

73. a. "+" addition
b. "·" multiplication

75. a.
$$\frac{1}{4}+\frac{1}{10}=\frac{5}{20}+\frac{2}{20}=\frac{7}{20}$$
$$\frac{1}{4}-\frac{1}{10}=\frac{5}{20}-\frac{2}{20}=\frac{3}{20}$$
$$\frac{1}{4}\cdot\frac{1}{10}=\frac{1}{40}$$
$$\frac{1}{4}\div\frac{1}{10}=\frac{1}{4}\cdot\frac{10}{1}=\frac{10}{4}=\frac{5}{2}$$

b.
$$\frac{1}{8}+\frac{1}{10}=\frac{5}{40}+\frac{4}{40}=\frac{9}{40}$$
$$\frac{1}{8}-\frac{1}{10}=\frac{5}{40}-\frac{4}{40}=\frac{1}{40}$$
$$\frac{1}{8}\cdot\frac{1}{10}=\frac{1}{80}$$
$$\frac{1}{8}\div\frac{1}{10}=\frac{1}{8}\cdot\frac{10}{1}=\frac{10}{8}=\frac{5}{4}$$

c.
$$\frac{3}{4}+\frac{5}{6}=\frac{9}{12}+\frac{10}{12}=\frac{19}{12}$$
$$\frac{3}{4}-\frac{5}{6}=\frac{9}{12}-\frac{10}{12}=-\frac{1}{12}$$
$$\frac{3}{4}\cdot\frac{5}{6}=\frac{15}{24}=\frac{5}{8}$$
$$\frac{3}{4}\div\frac{5}{6}=\frac{3}{4}\cdot\frac{6}{5}=\frac{18}{20}=\frac{9}{10}$$

d.
$$\frac{2}{3}+\frac{5}{6}=\frac{4}{6}+\frac{5}{6}=\frac{9}{6}=\frac{3}{2}$$
$$\frac{2}{3}-\frac{5}{6}=\frac{4}{6}-\frac{5}{6}=-\frac{1}{6}$$
$$\frac{2}{3}\cdot\frac{5}{6}=\frac{10}{18}=\frac{5}{9}$$
$$\frac{2}{3}\div\frac{5}{6}=\frac{2}{3}\cdot\frac{6}{5}=\frac{12}{15}=\frac{4}{5}$$

75. e. $1\frac{1}{3}=\frac{4}{3};\ 2\frac{1}{2}=\frac{5}{2}$

$$\frac{4}{3}+\frac{5}{2}=\frac{8}{6}+\frac{15}{6}=\frac{23}{6}=3\tfrac{5}{6}$$

$$\frac{4}{3}-\frac{5}{2}=\frac{8}{6}-\frac{15}{6}=-\frac{7}{6}=-1\tfrac{1}{6}$$

$$\frac{4}{3}\cdot\frac{5}{2}=\frac{20}{6}=\frac{10}{3}=3\tfrac{1}{3}$$

$$\frac{4}{3}\div\frac{5}{2}=\frac{4}{3}\cdot\frac{2}{5}=\frac{8}{15}$$

f. $2\frac{1}{4}=\frac{9}{4};\ 1\frac{2}{3}=\frac{5}{3}$

$$\frac{9}{4}+\frac{5}{3}=\frac{27}{12}+\frac{20}{12}=\frac{47}{12}=3\tfrac{11}{12}$$

$$\frac{9}{4}-\frac{5}{3}=\frac{27}{12}-\frac{20}{12}=\frac{7}{12}$$

$$\frac{9}{4}\cdot\frac{5}{3}=\frac{45}{12}=\frac{15}{4}=3\tfrac{3}{4}$$

$$\frac{9}{4}\div\frac{5}{3}=\frac{9}{4}\cdot\frac{3}{5}=\frac{27}{20}=1\tfrac{7}{20}$$

g. $2\frac{1}{5}=\frac{11}{5};\ 1\frac{1}{4}=\frac{5}{4}$

$$\frac{11}{5}+\frac{5}{4}=\frac{44}{20}+\frac{25}{20}=\frac{69}{20}=3\tfrac{9}{20}$$

$$\frac{11}{5}-\frac{5}{4}=\frac{44}{20}-\frac{25}{20}=\frac{19}{20}$$

$$\frac{11}{5}\cdot\frac{5}{4}=\frac{55}{20}=\frac{11}{4}=2\tfrac{3}{4}$$

$$\frac{11}{5}\div\frac{5}{4}=\frac{11}{5}\cdot\frac{4}{5}=\frac{44}{25}=1\tfrac{19}{25}$$

h. $1\frac{3}{4}=\frac{7}{4};\ 2\frac{1}{7}=\frac{15}{7}$

$$\frac{7}{4}+\frac{15}{7}=\frac{49}{28}+\frac{60}{28}=\frac{109}{28}=3\tfrac{25}{28}$$

$$\frac{7}{4}-\frac{15}{7}=\frac{49}{28}-\frac{60}{28}=-\frac{11}{28}$$

$$\frac{7}{4}\cdot\frac{15}{7}=\frac{15}{4}=3\tfrac{3}{4}$$

$$\frac{7}{4}\div\frac{15}{7}=\frac{7}{4}\cdot\frac{7}{15}=\frac{49}{60}$$

Exercises 7.7

1. $\frac{3}{4}x+5=23$

$\frac{3}{4}x=23-5=18$

$x=18\cdot\frac{4}{3}=24$

3. $L=\dfrac{\pi r\theta}{180}$

$180L=\pi r\theta$

$r=\dfrac{180L}{\pi\theta},\ \theta\neq 0$

5. $x(\frac{1}{10}+\frac{1}{12})=1$

$x(\frac{6}{60}+\frac{5}{60})=1$

$x\frac{11}{60}=1$

$x=\frac{60}{11}=5\frac{5}{11}$

7. $\dfrac{2n+4}{7}=\dfrac{3n-7}{4}$

$4(2n+4)=7(3n-7)$

$8n+16=21n-49$

$16+49=21n-8n \quad 65=13n$

$n=5$

9. There are no variables in the denominators.

11. $\frac{7}{11+3} = \frac{11-7}{8}$

$\frac{7}{14} = \frac{4}{8}$

$\frac{1}{2} = \frac{1}{2}$

13. $\frac{3}{4} + \frac{1}{5} = \frac{1}{\frac{20}{19}}$

$\frac{15}{20} + \frac{4}{20} = \frac{19}{20}$

$\frac{19}{20} = \frac{19}{20}$

15. $\frac{2}{-3+1} + \frac{3}{-3} = -2$

$\frac{2}{-2} + (-1) = -2$

$-1 + (-1) = -2$

$-2 = -2$

17. $\frac{x+1}{2} = \frac{x-3}{1}$

$x + 1 = 2(x - 3)$

$x + 1 = 2x - 6$

$1 + 6 = 2x - x$

$7 = x$

19. $\frac{x-8}{5} = \frac{-6}{x+5}$

$(x + 5)(x - 8) = (-6)(5)$

$x^2 - 3x - 40 = -30$

$x^2 - 3x - 10 = 0$

$(x - 5)(x + 2) = 0$

$\{-2, 5\},\ x \neq -5$

21. $\frac{x-3}{6} = \frac{3}{2x-1}$

$(2x - 1)(x - 3) = (3)(6)$

$2x^2 - 7x + 3 = 18$

$2x^2 - 7x - 15 = 0$

$(x - 5)(2x + 3) = 0$

$\{-1.5, 5\},\ x \neq 0.5$

23. $\frac{x}{32} + \frac{x}{8} = 10$

$32 \cdot \left(\frac{x}{32} + \frac{x}{8}\right) = 10 \cdot 32$

$x + 4x = 320$

$5x = 320$

$x = 64$

25. $6x^2\left(\frac{1}{2x^2} - \frac{1}{3x}\right) =$

$\frac{6x^2}{2x^2} - \frac{6x^2}{3x} =$

$3 - 2x$

27. $2x^2\left(\frac{1}{2x^2} + \frac{3}{x}\right) =$

$\frac{2x^2}{2x^2} + \frac{6x^2}{x} =$

$1 + 6x$

29. $x(x+1)\left(\frac{1}{x+1} - \frac{2}{x}\right) =$

$\frac{x(x+1)}{x+1} - \frac{2x(x+1)}{x} =$

$x - 2(x + 1) = x - 2x - 2 = -x - 2$

31. $(x+3)(x-1)\left(\frac{1}{x+3}+\frac{2}{x-1}\right)=$

$\frac{(x+3)(x-1)}{x+3}+\frac{2(x+3)(x-1)}{x-1}=$

$(x-1)+2(x+3)=$

$x-1+2x+6=$

$3x+5$

33. $x \neq 0;\ 15x\left(\frac{3}{5}+\frac{2}{3}\right)=\left(\frac{1}{x}\right)15x$

$\frac{15x\cdot 3}{5}+\frac{15x\cdot 2}{3}=\frac{15x}{x}$

$3x\cdot 3+5x\cdot 2=15$

$9x+10x=15$

$19x=15;\ x=\frac{15}{19}$

35. $x \neq 0;\ 6x\left(\frac{1}{3}+\frac{1}{x}\right)=\left(\frac{1}{2}\right)6x$

$\frac{6x}{3}+\frac{6x}{x}=\frac{6x}{2}$

$2x+6=3x$

$6=3x-2x;\ x=6$

37. $x \neq 0;\ x\left(\frac{5}{x}+\frac{2}{x}\right)=\left(\frac{4}{x}\right)x$

$5+2=4$; contradiction

{ } or $\varnothing$

39. $x \neq 0;\ 2x\left(\frac{1}{2}-\frac{1}{x}\right)=\left(\frac{1}{2x}\right)2x$

$\frac{2x}{2}-\frac{2x}{x}=\frac{2x}{2x}$

$x-2=1;\ x=3$

41. $x \neq \frac{1}{2}, 1;\ (2x-1)(x-1)\frac{4}{2x-1}=\frac{1}{x-1}(2x-1)(x-1)$

$4(x-1)=2x-1$

$4x-4=2x-1$

$4x-2x=-1+4$

$2x=3;\ x=\frac{3}{2}$

43. $x \neq -2;\ 2(x+2)\frac{(x+1)}{2}=\frac{15}{(x+2)}(x+2)2$

$(x+2)(x+1)=30$

$x^2+3x+2=30$

$x^2+3x-28=0$

$(x+7)(x-4)=0$

$\{-7, 4\}$

45. $x \neq 1;\ 10(x-1)\frac{4}{(x-1)}=\frac{(x+2)}{10}(x-1)10$

$40=(x+2)(x-1)$

$40=x^2+x-2$

$x^2+x-42=0$

$(x+7)(x-6)=0$

$\{-7, 6\}$

47. $x \neq 0;\; x^2\left(\frac{18}{x^2}+\frac{9}{x}-2\right)=x^2\cdot 0$

$18 + 9x - 2x^2 = 0$ $\qquad$ $(6 - x)(3 + 2x) = 0$

$\{-1.5, 6\}$

49. $x \neq 2;\; (x-2)\left(\frac{1}{(x-2)}-4\right)=\frac{(3-x)}{(x-2)}(x-2)$

$1 - 4(x - 2) = 3 - x$ $\qquad$ $1 - 4x + 8 = 3 - x$

$9 - 3 = -x + 4x$ $\qquad$ $6 = 3x$

$x = 2$; violates restriction on variable; { } or ∅

51. $x \neq \pm 1;\; 2(x+1)(x-1)\left(\frac{3}{(x-1)}+\frac{5}{2(x+1)}\right)=\frac{3}{2}(x-1)(x+1)2$

$6(x + 1) + 5(x - 1) = 3(x - 1)(x + 1)$ $\qquad$ $6x + 6 + 5x - 5 = 3(x^2 - 1)$

$11x + 1 = 3x^2 - 3$ $\qquad$ $3x^2 - 11x - 4 = 0$

$(3x + 1)(x - 4) = 0$ $\qquad$ $\{-\frac{1}{3}, 4\}$

53. $x \neq 3, 6;\; (6-x)(x-3)\frac{x}{(6-x)}=\frac{2}{(x-3)}(6-x)(x-3)$

$x(x - 3) = 2(6 - x)$ $\qquad$ $x^2 - 3x = 12 - 2x$

$x^2 - x - 12 = 0$ $\qquad$ $(x + 3)(x - 4) = 0$

$\{-3, 4\}$

55. $x \neq 0, 2;\; 2x(x-2)\left(\frac{1}{(x-2)}+\frac{(x-1)}{x}\right)=\frac{2x+1}{2x}(x-2)2x$

$2x + 2(x - 2)(x - 1) = (2x + 1)(x - 2)$ $\qquad$ $2x + 2(x^2 - 3x + 2) = 2x^2 - 3x - 2$

$2x + 2x^2 - 6x + 4 = 2x^2 - 3x - 2$ $\qquad$ $2x^2 - 4x + 4 = 2x^2 - 3x - 2$

$4 + 2 = 2x^2 - 2x^2 - 3x + 4x$ $\qquad$ $6 = x$

57. $x \neq -3, 0;\; 8x(x+3)\left(\frac{x-3}{x}+\frac{x-2}{x+3}\right)=\frac{6x+1}{8x}(x+3)8x$

$8(x + 3)(x - 3) + 8x(x - 2) = (6x + 1)(x + 3)$

$8(x^2 - 9) + 8x^2 - 16x = 6x^2 + 19x + 3$ $\qquad$ $8x^2 - 72 + 8x^2 - 16x = 6x^2 + 19x + 3$

$16x^2 - 16x - 72 = 6x^2 + 19x + 3$ $\qquad$ $16x^2 - 6x^2 - 16x - 19x - 72 - 3 = 0$

$10x^2 - 35x - 75 = 0$ $\qquad$ $5(2x^2 - 7x - 15) = 0$

$2x^2 - 7x - 15 = 0$ $(2x + 3)(x - 5) = 0$

$\{-\frac{3}{2}, 5\}$

59. $x \neq \pm 1;\ 3(x+1)(x-1)\left(\frac{2}{x-1} + \frac{x-2}{x+1}\right) = \frac{4x}{3(x+1)}(x-1)(x+1)3$

$6(x + 1) + 3(x - 1)(x - 2) = 4x(x - 1)$ $6x + 6 + 3(x^2 - 3x + 2) = 4x^2 - 4x$

$6x + 6 + 3x^2 - 9x + 6 = 4x^2 - 4x$ $3x^2 - 3x + 12 = 4x^2 - 4x$

$4x^2 - 3x^2 - 4x + 3x - 12 = 0$ $x^2 - x - 12 = 0$

$(x - 4)(x + 3) = 0$ $\{-3, 4\}$

61. $x \neq 1;\ (x - 1)(1 - x)\frac{x+1}{x-1} = \frac{1}{1-x}(x - 1)(1 - x)$

$(1 - x)(x + 1) = x - 1$ $-x^2 + 1 = x - 1$

$x^2 + x - 2 = 0$ $(x + 2)(x - 1) = 0$

$\{-2\}$ (note: $x = 1$ violates restriction on variable)

63. From the equation $x \neq -3, 0$; from the graph $x = -6, -2$

65. $\frac{x^2}{1.000 - x} = 4$

$x^2 = 4(1.000 - x)$

$x^2 = 4 - 4x$

$x^2 + 4x - 4 = 0$

$a = 1, b = 4, c = -4$

$x = \frac{-4 \pm \sqrt{4^2 - 4(1)(-4)}}{2(1)}$

$x = \frac{-4 \pm \sqrt{32}}{2} = \frac{-4 \pm 4\sqrt{2}}{2}$

$x = -2 + 2\sqrt{2}$ or $x = -2 - 2\sqrt{2}$

$x \approx 0.828$ or

$x \approx -4.828$ (discard - violates restriction)

67. $\frac{(3.000 + 2x)^2}{(1.000 - x)(2.000 - x)} = 42$

$9 + 12x + 4x^2 = 42(2 - 3x + x^2)$

$4x^2 + 12x + 9 = 84 - 126x + 42x^2$

$42x^2 - 4x^2 - 126x - 12x + 84 - 9 = 0$

$38x^2 - 138x + 75 = 0$

$x = \frac{-(-138) \pm \sqrt{(-138)^2 - 4(38)(75)}}{2(38)}$

$x = \frac{138 \pm \sqrt{7644}}{76} = \frac{138 \pm 14\sqrt{39}}{76}$

$x = \frac{69 + 7\sqrt{39}}{38}$ or $x = \frac{69 - 7\sqrt{39}}{38}$

$x \approx 2.966$ (discard - violates restriction) or

$x \approx 0.665$

69. Let x = width, $1.618x$ = length; $x(1.618x) = 1.5\ \text{ft}^2$; $1.618x^2 = 1.5$; $x^2 = \frac{1.5}{1.618}$;

$x = \sqrt{\frac{1.5}{1.618}} \approx 0.963$ ft. $0.963\ \text{ft} \cdot 12\ \text{in}/1\ \text{ft} =$ **11.6 in.**; $1.618(11.6) =$ **18.7 in.**

71. $\frac{1}{6}+\frac{1}{x}=\frac{1}{2}$

$6x\left(\frac{1}{6}+\frac{1}{x}\right)=\frac{1}{2}(6x)$

x + 6 = 3x

6 = 2x; x = 3 min

73. $\frac{1}{20}+\frac{1}{x}=\frac{1}{12}$

$60x\left(\frac{1}{20}+\frac{1}{x}\right)=\frac{1}{12}(60x)$

3x + 60 = 5x

60 = 2x; x = 30 min

75. $\frac{1}{10{,}000}+\frac{1}{4000}=\frac{1}{R}$

$40{,}000R\left(\frac{1}{10{,}000}+\frac{1}{4000}\right)=\frac{1}{R}40{,}000R$

4R + 10R = 40,000

14R = 40,000; R ≈ 2857 ohms

77. $\frac{1}{5}+\frac{1}{3}+\frac{1}{3}=\frac{1}{x}$

$15x\left(\frac{1}{5}+\frac{1}{3}+\frac{1}{3}\right)=\frac{1}{x}15x$

3x + 5x + 5x = 15

13x = 15; x ≈ 1.15 min.

79. $\frac{l}{w}=\frac{5}{2};\ 2l+2w=39.9$ cm

$2l=5w;\ 5w+2w=39.9$

$7w=39.9;$ $\boldsymbol{w=5.7}$ **cm**

$2l=5(5.7)$

$\boldsymbol{l\approx 14.3}$ **cm**

81. $\frac{l}{w}=\frac{7}{3};\ 2l+2w=44.4$ cm

$l=\frac{7w}{3};\ 2(\frac{7w}{3})+2w=44.4$

$3\left(\frac{14w}{3}+2w\right)=44.4(3)$

$14w+6w=133.2$

$20w=133.2;$ $\boldsymbol{w\approx 6.7}$ **cm**

$l=\frac{7(6.7)}{3};$ $\boldsymbol{l\approx 15.5}$ **cm**

83. $abc\left(\frac{1}{a}+\frac{1}{b}\right)=\frac{1}{c}abc$

bc + ac = ab

bc = ab – ac

bc = a(b – c)

$a=\frac{bc}{b-c}$

85. $abc\left(\frac{1}{a}+\frac{1}{b}\right)=\frac{1}{c}abc$

bc + ac = ab; ac + bc = ab

c(a + b) = ab

$c=\frac{ab}{a+b}$

87. $I=\frac{E}{R+r}$; I(R + r) = E; IR + Ir = E; IR = E – Ir; $R=\frac{E-Ir}{I}=\frac{E}{I}-r$

89. Did not distribute 15x(x + 2) over $\frac{2}{x}$; should be:

15x + 30(x + 2) = 11x(x + 2)

91. Forgot the 2 in the numerator of the 2nd term; should be:

$$\frac{15x(x+2)}{x+2}+\frac{30x(x+2)}{x}=11x(x+2)$$

Chapter 7 Review Exercises

1.
$$\frac{2}{3}+\frac{4}{3}=\frac{6}{3}=2$$
$$\frac{2}{3}-\frac{4}{3}=\frac{-2}{3}$$
$$\frac{2}{3}\cdot\frac{4}{3}=\frac{8}{9}$$
$$\frac{2}{3}\div\frac{4}{3}=\frac{2}{3}\cdot\frac{3}{4}=\frac{2}{4}=\frac{1}{2}$$

3.
$$\frac{a}{b}+\frac{a}{c}=\frac{ac}{bc}+\frac{ab}{bc}=\frac{ac+ab}{bc}$$
$$\frac{a}{b}-\frac{a}{c}=\frac{ac}{bc}-\frac{ab}{bc}=\frac{ac-ab}{bc}$$
$$\frac{a}{b}\cdot\frac{a}{c}=\frac{a^2}{bc}$$
$$\frac{a}{b}\div\frac{a}{c}=\frac{a}{b}\cdot\frac{c}{a}=\frac{c}{b}$$

5. a. $\frac{27xy^2}{15x^2y}=\frac{3\cdot 9xyy}{3\cdot 5xxy}=\frac{9y}{5x}$

b. $\frac{2-x}{x-2}=\frac{-(x-2)}{x-2}=-1$

c. $\frac{a-ac}{a}=\frac{a(1-c)}{a}=1-c$

7. a. $\frac{3x^2-12}{x+2}=\frac{3(x^2-4)}{x+2}=\frac{3(x-2)(x+2)}{x+2}=3(x-2)$

b. $\frac{a-b}{(a+b)(a-b)}=\frac{1}{a+b}$

c. $8x\left(\frac{1}{4x}+\frac{1}{2}\right)=\frac{8x}{4x}+\frac{8x}{2}=2+4x$

9. a. $\frac{x^2-5x-6}{x^2-4x-5}=\frac{(x+1)(x-6)}{(x+1)(x-5)}=\frac{x-6}{x-5}$

b. $\frac{4x^2-1}{2x^2+5x+2}=\frac{(2x+1)(2x-1)}{(2x+1)(x+2)}=\frac{2x-1}{x+2}$

c. $\frac{x^2+3x-4}{x^2-16}=\frac{(x+4)(x-1)}{(x+4)(x-4)}=\frac{x-1}{x-4}$

11. $\frac{6-x}{x-6}=-1$

13. 4 gal · 4 qts/1 gal = 16 qts; 3 to 16

15. Undefined when x – 1 = 0 or x + 3 = 0 or when x = 1 or x = –3

17. $\frac{4y^2}{9x^2}\cdot\frac{3x}{8y}=\frac{4yy}{3\cdot 3xx}\cdot\frac{3x}{2\cdot 4y}=\frac{y}{6x}$

19. $15x\left(\frac{1}{3x}+\frac{2}{5x}\right)=\frac{15x}{3x}+\frac{30x}{5x}=5+6=11$

21. $\frac{x^2+5x+4}{x^2-16}\cdot\frac{2x-8}{1-x^2}=\frac{(x+1)(x+4)}{(x-4)(x+4)}\cdot\frac{2(x-4)}{(1+x)(1-x)}=\frac{2}{1-x}$

23. $\frac{1}{x-3}+\frac{x}{x-3}=\frac{x+1}{x-3}$

25. $\frac{x}{x-1}-\frac{1}{1-x}=\frac{x}{x-1}-\frac{1}{-(x-1)}=\frac{x}{x-1}+\frac{1}{x-1}=\frac{x+1}{x-1}$

27. $\frac{x-2}{x^2-1}-\frac{2}{x-1}=\frac{x-2}{(x-1)(x+1)}-\frac{2(x+1)}{(x-1)(x+1)}=\frac{x-2-2(x+1)}{x^2-1}=\frac{x-2-2x-2}{x^2-1}=\frac{-x-4}{x^2-1}$

29. $\frac{1}{x^2-2x}-\frac{1}{3x^2-10x+8}=\frac{1}{x(x-2)}-\frac{1}{(x-2)(3x-4)}=\frac{3x-4}{x(x-2)(3x-4)}-\frac{x}{x(x-2)(3x-4)}$

$=\frac{3x-4-x}{x(x-2)(3x-4)}=\frac{2x-4}{x(x-2)(3x-4)}=\frac{2(x-2)}{x(x-2)(3x-4)}=\frac{2}{x(3x-4)}$

31. $\frac{A}{\frac{1}{2}h}=\frac{A}{1}\cdot\frac{2}{h}=\frac{2A}{h}$

33. $\frac{2500\ miles}{12.5\ miles\ per\ gal}=\frac{2500\ miles}{1}\cdot\frac{1\ gal}{12.5\ miles}=200\ gal$

35. $\frac{x+\frac{x}{3}}{3+\frac{2}{x}}=\frac{\frac{3x}{3}+\frac{x}{3}}{\frac{3x}{x}+\frac{2}{x}}=\frac{\frac{4x}{3}}{\frac{3x+2}{x}}=\frac{4x}{3}\cdot\frac{x}{3x+2}=\frac{4x^2}{9x+6}$

37. $\frac{\frac{x^2-x}{2}}{\frac{x^2-1}{x}}=\frac{x^2-x}{2}\cdot\frac{x}{x^2-1}=\frac{x(x-1)x}{2(x-1)(x+1)}=\frac{x^2}{2(x+1)}$

39. x ≠ 1;

$$\begin{array}{r} x^2 - 2x + 1 \\ x - 1 \overline{)\, x^3 - 3x^2 + 3x - 1} \\ \underline{-(x^3 - x^2)} \\ -2x^2 + 3x \\ \underline{-(-2x^2 + 2x)} \\ x - 1 \\ \underline{-(x - 1)} \\ 0 \end{array}$$

41. x ≠ −3;

$$\begin{array}{r} x^2 - 3x + 9 - \frac{26}{x+3} \\ x + 3 \overline{)\, x^3 + 0x^2 + 0x + 1} \\ \underline{-(x^3 + 3x^2)} \\ -3x^2 + 0x \\ \underline{-(-3x^2 - 9x)} \\ 9x + 1 \\ \underline{-(9x + 27)} \\ -26 \end{array}$$

43. x ≠ −1;

$$\begin{array}{r} x^3 - x^2 + x - 1 \\ x + 1 \overline{)\, x^4 + 0x^3 + 0x^2 + 0x - 1} \\ \underline{-(x^4 + x^3)} \\ -x^3 + 0x^2 \\ \underline{-(-x^3 - x^2)} \\ x^2 + 0x \\ \underline{-(x^2 + x)} \\ -x - 1 \\ \underline{-(-x - 1)} \\ 0 \end{array}$$

45. $\frac{6}{x} = \frac{15}{32}$

$32x\frac{6}{x} = \frac{15}{32}32x$

$192 = 15x$

$x = 12.8; \; x \neq 0$

47. $\frac{x}{2} = \frac{7}{x+5}$

$2(x+5)\frac{x}{2} = \frac{7}{x+5}2(x+5)$

$x^2 + 5x = 14$

$x^2 + 5x - 14 = 0$

$(x + 7)(x - 2) = 0$

$\{-7, 2\}; \; x \neq -5$

49. $\frac{x-2}{4}=\frac{x+1}{3}-3$

$(12)\frac{x-2}{4}=\frac{x+1}{3}(12)-3(12)$

$3x-6=4x+4-36$

$-6-4+36=4x-3x$

$26=\text{x}$

51. $\frac{x+3}{1}=\frac{-5}{2x}$

$2x\frac{x+3}{1}=\frac{-5}{2x}2x$

$2x^2+6x=-5$

$2x^2+6x+5=0$

$x=\frac{-6\pm\sqrt{6^2-4(2)(5)}}{2(2)}$

$x=\frac{-6\pm\sqrt{-4}}{4};\ x=\frac{-6\pm 2i}{4}$

$x=\frac{-3}{2}\pm\frac{1}{2}i;\ \{-1.5\pm 0.5i\}$

53. $\frac{1}{x}+\frac{5}{x+2}=\frac{13}{3x}$

$3x(x+2)\left(\frac{1}{x}+\frac{5}{x+2}\right)=\frac{13}{3x}(x+2)3x$

$3(x+2)+15x=13(x+2)$

$3x+6+15x=13x+26$

$18x-13x=26-6$

$5x=20$

$x=4;\ x\neq -2, 0$

55. $\frac{1}{x-1}+\frac{2}{x}=\frac{7}{6}$

$6x(x-1)\left(\frac{1}{x-1}+\frac{2}{x}\right)=\frac{7}{6}(x-1)6x$

$6x+12(x-1)=7x(x-1)$

$6x+12x-12=7x^2-7x$

$7x^2-7x-18x+12=0$

$7x^2-25x+12=0$

$(7x-4)(x-3)=0$

$\{\frac{4}{7}, 3\};\ x\neq 0, 1$

57. $\frac{x}{4}=\frac{6}{5};\ \frac{y}{4}=\frac{7}{5}$

$x=\frac{6\cdot 4}{5}=5.6;\ y=\frac{7\cdot 4}{5}=4.8$

59. $\frac{r}{7.5}=\frac{8}{6};\ s=\sqrt{6^2+8^2};\ t=\sqrt{7.5^2+r^2}$

$r=\frac{8\cdot 7.5}{6}=10;\ s=\sqrt{100}=10;$

$t=\sqrt{56.25+100}=\sqrt{156.25}=12.5$

61. $\frac{78\ ft}{149\ mph};\ 78\ ft\cdot\frac{1\ hr}{149\ mi}\cdot\frac{1\ mi}{5280\ ft}\cdot\frac{3600\text{ sec}}{1\ hr}\approx 0.357\text{ sec}$

63. $18\ yrs \cdot \frac{21\ in}{9\ mo} \cdot \frac{12\ mo}{1\ yr} = 504\ in$ (growth); 21 in. + 504 in = 525 in (total height)

$504\ in \cdot \frac{1\ ft}{12\ in} = 42\ ft$; $525\ in \cdot \frac{1\ ft}{12\ in} = 43.75\ ft$

65. Let x = MB and y = cost; $m = \frac{49.99 - 29.99}{128 - 64} = \frac{20}{64} = 0.3125$;

29.99 = 0.3125(64) + b; b = 29.99 – 20 = 9.99 (basic cost of stick manufacture);
y = 0.3125x + 9.99

67. Let x = gallons and y = cost; $m = \frac{186.55 - 114.80}{65 - 40} = \frac{71.75}{25} = 2.87$;

114.80 = 2.87(40) + b; b = 114.80 – 114.80 = 0; direct variation; y = 2.87x

69. Let D = distance; h = height; $D = k\sqrt{h}$

71. Let V = volume, l = length, w = width, h = height, k = 1; V = $klwh$

73. $d = ks^2$

75. $s = k\sqrt{df}$

77. cost = k(pounds); $\frac{k6}{k4} = 1.5$; The cost for 6 lbs is 1.5 times the cost of 4 lbs.

79. energy = k(height)^2; $\frac{k12^2}{k3^2} = \frac{144}{9} = 16$; The 12 ft wave has 16 times the energy of a 3 ft wave.

81. Divide total credits by number of credits per quarter: $y = \frac{90}{x}$; k = 90 credits

83. Divide total allocated by price per book; $y = \frac{100{,}000}{x}$; k = \$100,000

85. $\frac{1}{5.5} + \frac{1}{7} = \frac{1}{x}$; $38.5x\left(\frac{1}{5.5} + \frac{1}{7}\right) = \frac{1}{x}38.5x$: $7x + 5.5x = 38.5$; $12.5x = 38.5$; $x = 3.08$

Working together it takes 3.08 hours, longer than required by code.

87. a. The equation is the total distance divided by the total hours minus 3 hours not traveling.

b.

Total Trip Time (hours)	Rate (miles per hour)
4	$\frac{300}{4-3}=300$
5	$\frac{300}{5-3}=150$
6	$\frac{300}{6-3}=100$
7	$\frac{300}{7-3}=75$
8	$\frac{300}{8-3}=60$

c. Due to the 3 hour delay, we replace t with $t-3$ causing a shift of 3 units to the right.

d. $\frac{300}{24-3}=14.3$ mph

e. $\frac{300}{18-3}=20$ mph

87. f. $\frac{300}{10-3}=42.9$ mph

g. $\frac{300}{3.5-3}=600$ mph

h. $\frac{300}{3.1-3}=3000$ mph

i. $\frac{300}{3.01-3}=30{,}000$ mph

j. $\frac{300}{3-3}$; not possible, division by zero

89. The correct solution is x = 1.375; x = –1 makes the original equation undefined (as does x = 1); if the student had used the least common denominator of (x^2 – 1) only the correct answer would have been found.

91. The function varies inversely with the square of the input. The larger input will yield a smaller output. $f(a) > f(a+2)$

93. As x gets closer to 2, denominator gets closer to 0; equation of the asymptote is x = 2.

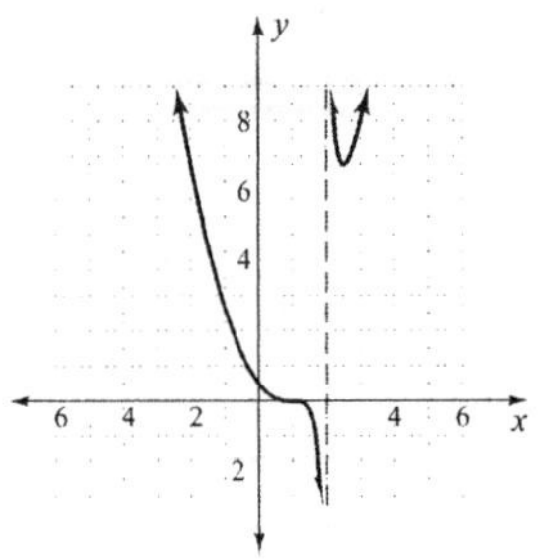

95. a. As x approaches 1 from the left ***y approaches*** $-\infty$.

b. As x approaches 1 from the right ***y approaches*** ∞.

c. As x approaches -3 from the right ***y approaches*** $-\infty$.

d. As x approaches -3 from the left ***y approaches*** ∞.

e. In connected mode the calculator connects across x = −3 and x = 1 with a nearly vertical line. This line should not be there, as shown in dot mode. (Think of the calculator forgetting to "lift its pencil")

f. The vertical asymptotes are at x = −3 and x = 1

Chapter 7 Test

1. $\dfrac{ab^2c}{a^2bc^2}=\dfrac{abbc}{aabcc}=\dfrac{b}{ac};\ a\neq 0,\ b\neq 0,\ c\neq 0$

2. equation is already simplified; b ≠ 3

3. $\dfrac{3ac}{15ac^2}=\dfrac{3ac}{5\cdot 3acc}=\dfrac{1}{5c};\ a\neq 0,\ c\neq 0$

4. $\dfrac{15b^2c^3}{10b^3c}=\dfrac{5\cdot 3bbccc}{5\cdot 2bbbc}=\dfrac{3c^2}{2b};\ b\neq 0,\ c\neq 0$

5. $\dfrac{x^2-25}{20+x-x^2}=\dfrac{(x+5)(x-5)}{(5-x)(4+x)}=\dfrac{(x+5)(x-5)}{-(x-5)(x+4)}=-\dfrac{x+5}{x+4};\ x\neq -4,\ 5$

6. $\dfrac{12-2a}{18+3a-a^2}=\dfrac{2(6-a)}{(6-a)(3+a)}=\dfrac{2}{a+3};\ a\neq -3,\ 6$

7. $\dfrac{3x^2-7x+2}{2x^2-5x+2}=\dfrac{(3x-1)(x-2)}{(2x-1)(x-2)}=\dfrac{3x-1}{2x-1};\ x\neq \tfrac{1}{2},\ 2$

8. $\dfrac{2x^2+7x+6}{x^2+4x+4}=\dfrac{(2x+3)(x+2)}{(x+2)(x+2)}=\dfrac{2x+3}{x+2};\ x\neq -2$

9. $\dfrac{12xy^2}{7y-y^2}\div\dfrac{6x^2}{7-y}=\dfrac{12xy^2}{y(7-y)}\cdot\dfrac{7-y}{6x^2}=\dfrac{2y}{x};\ y\neq 0,\ 7;\ x\neq 0$

10. $\dfrac{x^2-9}{x^2+4x+3}\cdot\dfrac{x-3}{x+1}=\dfrac{(x+3)(x-3)(x-3)}{(x+3)(x+1)(x+1)}=\dfrac{(x-3)^2}{(x+1)^2}$; $x\neq -3, -1$

11. $\dfrac{x^2+8x+16}{x-2}\cdot\dfrac{x^2-4}{x+4}=\dfrac{(x+4)(x+4)(x+2)(x-2)}{(x-2)(x+4)}=(x+4)(x+2)$; $x\neq -4, 2$

12. $x(x+1)\left(\dfrac{3}{x}+\dfrac{2}{x+1}\right)=3(x+1)+2x=3x+3+2x=5x+3$; $x\neq -1, 0$

13. $\dfrac{2-x}{x+2}+\dfrac{x+2}{x-2}=\dfrac{(x-2)(2-x)+(x+2)(x+2)}{(x+2)(x-2)}=\dfrac{-x^2+4x-4+x^2+4x+4}{(x+2)(x-2)}$

$=\dfrac{8x}{(x+2)(x-2)}$; $x\neq \pm 2$

14. $\dfrac{x-2}{x+2}-\dfrac{x}{x^2+2x}=\dfrac{x(x-2)-x}{x(x+2)}=\dfrac{x[(x-2)-1]}{x(x+2)}=\dfrac{x-3}{x+2}$; $x\neq -2, 0$

15. $\dfrac{10{,}000\ cm^3}{10\ cm}=1000\ \text{cm}^2$

16. $1{,}000{,}000\ \text{hr}\cdot\dfrac{1\ day}{24\ hr}\cdot\dfrac{1\ yr}{365\ days}\approx 114.2\ \text{yr}$

17. $\dfrac{30\frac{mi}{gal}}{60\frac{mi}{hr}}=\dfrac{1}{2}\dfrac{mi}{gal}\dfrac{hr}{mi}=\dfrac{1}{2}\dfrac{hr}{gal}$

18. $\dfrac{\frac{3}{x}-x}{x+\frac{2}{x}}=\dfrac{\frac{3-x^2}{x}}{\frac{x^2+2}{x}}=\dfrac{3-x^2}{x}\cdot\dfrac{x}{x^2+2}=\dfrac{3-x^2}{x^2+2}$; $x\neq 0$

19. $\dfrac{\frac{x^2-3x}{3}}{\frac{9-x^2}{x}}=\dfrac{x^2-3x}{3}\cdot\dfrac{x}{9-x^2}=\dfrac{x(x-3)x}{3(3-x)(3+x)}=\dfrac{x^2(x-3)}{-3(x-3)(x+3)}=-\dfrac{x^2}{3(x+3)}$; $x\neq -3, 0, 3$

20. Defined for $\Re$, $x \neq -2$;

$$\begin{array}{r}
x^2+4x+4 \\
x+2\overline{)\,x^3+6x^2+12x+8} \\
-(x^3+2x^2) \\
\hline
4x^2+12x \\
-(4x^2+8x) \\
\hline
4x+8 \\
-(4x+8) \\
\hline
0
\end{array}$$

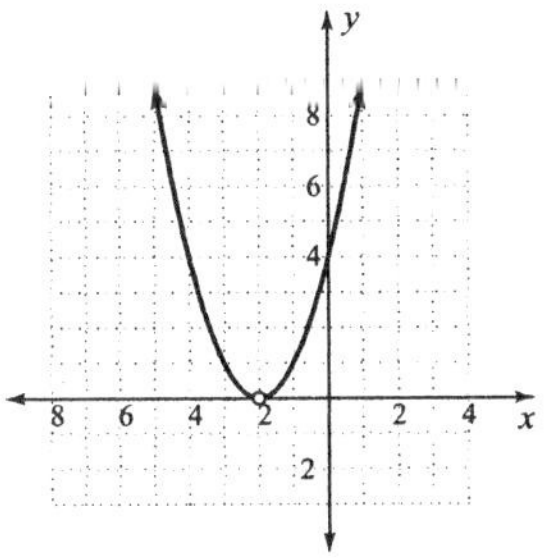

21. Defined for $\Re$, $x \neq -4$

$$\begin{array}{r}
x^2+2x+4-\frac{8}{x+4} \\
x+4\overline{)\,x^3+6x^2+12x+8} \\
-(x^3+4x^2) \\
\hline
2x^2+12x \\
-(2x^2+8x) \\
\hline
4x+8 \\
-(4x+16) \\
\hline
-8
\end{array}$$

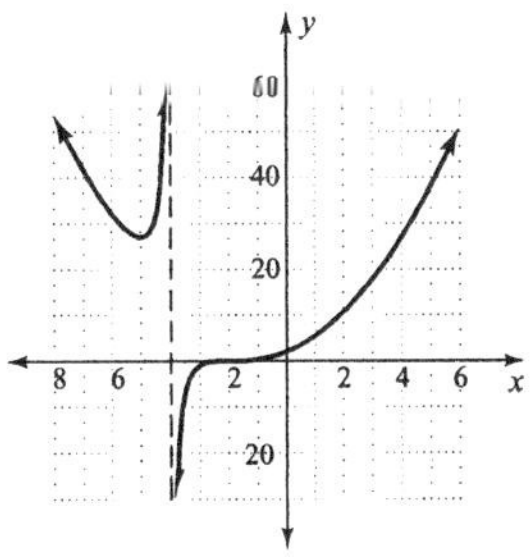

22. Defined for $\Re$, $x \neq 1$;

$$\begin{array}{r}
2x^3+2x^2+3x+3-\frac{1}{x-1} \\
x-1\overline{)\,2x^4+0x^3+x^2+0x-4} \\
-(2x^4-2x^3) \\
\hline
2x^3+x^2 \\
-(2x^3-2x^2) \\
\hline
3x^2+0x \\
-(3x^2-3x) \\
\hline
3x-4 \\
-(3x-3) \\
\hline
-1
\end{array}$$

23.

$$\frac{x-2}{10}=\frac{2}{x-1}$$

$$10(x-1)\frac{x-2}{10}=\frac{2}{x-1}10(x-1)$$

$$(x-1)(x-2)=20$$

$$x^2-3x+2=20$$

$$x^2-3x-18=0$$

$$(x-6)(x+3)=0$$

$$\{-3, 6\}\ \ x\neq 1$$

24.

$$\frac{1}{3}-\frac{1}{x}=\frac{2x}{3}$$

$$3x\left(\frac{1}{3}-\frac{1}{x}\right)=\frac{2x}{3}3x$$

$$x-3=2x^2$$

$$2x^2-x+3=0$$

$$x=\frac{-(-1)\pm\sqrt{(-1)^2-4(2)(3)}}{2(2)}$$

$$x=\frac{1\pm\sqrt{-23}}{4}$$

$$\left\{\frac{1}{4}\pm i\left(\frac{\sqrt{23}}{4}\right)\right\};\ x\neq 0$$

25.

$$\frac{1}{x-4}+\frac{1}{x}=\frac{10}{3x}$$

$$3x(x-4)\left(\frac{1}{x-4}+\frac{1}{x}\right)=\frac{10}{3x}3x(x-4)$$

$$3x+3(x-4)=10(x-4)$$

$$3x+3x-12=10x-40$$

$$-12+40=10x-6x$$

$$28=4x$$

$$x=7;\ x\neq 0, 4$$

26.

$$\frac{x-1}{9}=\frac{x+3}{10}$$

$$90\frac{x-1}{9}=\frac{x+3}{10}90$$

$$10(x-1)=9(x+3)$$

$$10x-10=9x+27$$

$$10x-9x=27+10$$

$$x=37$$

27.

$$\frac{3}{x}+\frac{2}{x+1}=4$$

$$x(x+1)\left(\frac{3}{x}+\frac{2}{x+1}\right)=4x(x+1)$$

$$3(x+1)+2x=4x^2+4x$$

$$3x+3+2x=4x^2+4x$$

$$4x^2+4x-5x-3=0$$

$$4x^2-x-3=0$$

$$(4x+3)(x-1)=0$$

$$\left\{-\frac{3}{4}, 1\right\};\ x\neq -1, 0$$

28.

$$\frac{7\ lb}{9\ mo}\cdot\frac{16\ oz}{1\ lb}\cdot\frac{1\ mo}{30\ day}\cdot\frac{1\ day}{24\ hr}\approx 0.0173\ {}^{oz}\!/_{hr}$$

29. $\dfrac{133\ lb}{18\ yr}\cdot\dfrac{16\ oz}{1\ lb}\cdot\dfrac{1\ yr}{365\ day}\cdot\dfrac{1\ day}{24\ hr}\approx 0.0135\ {}^{oz}/_{hr}$

30. $s=45, f=0.6;\ d=\dfrac{45^2}{30\cdot 0.6}=112.5\ \text{ft}$

31. $f=0.4, d=208;\ 208=\dfrac{s^2}{30\cdot 0.4};\ 208\cdot 12=s^2;\ s\approx 50\ \text{mph}$

32. $n=1350, l=1;\ 1350=\dfrac{k}{1^2};\ k=1350$

$n=\dfrac{1350}{4^2}=84\ \text{nails}$

33. Volume varies directly with the cube of the radius; constant of variation is $\frac{4}{3}\pi$.

34. a. $\dfrac{1}{5}+\dfrac{1}{4}=\dfrac{1}{x};\ 20x\left(\dfrac{1}{5}+\dfrac{1}{4}\right)=\dfrac{1}{x}20x;\ 4x+5x=20;\ 9x=20;\ x=2.2$ minutes

b. $\dfrac{1}{5}+\dfrac{1}{4}+\dfrac{1}{x}=\dfrac{1}{2};\ 20x\left(\dfrac{1}{5}+\dfrac{1}{4}+\dfrac{1}{x}\right)=\dfrac{1}{2}20x;\ 4x+5x+20=10x;\ 9x+20=10x;\ x=20$ min

35. $\text{AB}=\sqrt{6^2+15^2}=3\sqrt{29}\approx 16.155$

$\dfrac{EF}{4}=\dfrac{15}{6};\ EF=\dfrac{4\cdot 15}{6}=10$

$\text{ED}=\sqrt{4^2+10^2}=2\sqrt{29}\approx 10.770$

Exercises 8.1

1. a. $x^4x^5 = x^{4+5} = x^9$ b. $y^2y^a = y^{2+a}$
   c. $a^na^3 = a^{n+3}$

3. a. $(3x)^4 = 3^4x^4 = 81x^4$ b. $(ab)^3 = a^3b^3$
   c. $(4n)^2 = 4^2n^2 = 16n^2$

5. a. $(x^2)^4 = x^{2\cdot 4} = x^8$ b. $(2x^3)^5 = 2^5x^{3\cdot 5} = 32x^{15}$
   c. $(3a^4b^2)^3 = 3^3a^{4\cdot 3}b^{2\cdot 3} = 27a^{12}b^6$

7. a. $x^5/x^2 = x^{5-2} = x^3$ b. $y^3/y^5 = y^{3-5} = y^{-2} = 1/y^2$
   c. $a^5a^3/a^6 = a^{5+3-6} = a^2$

9. a. $(x^n)^4 = x^{4n}$ b. $xy^3/x^3y = x^{1-3}y^{3-1} = x^{-2}y^2 = y^2/x^2$
   c. $x^ky^1/xy^2 = x^{k-1}y^{1-2} = x^{k-1}y^{-1} = x^{k-1}/y$

11. a. $2x^{-1} = \dfrac{2}{x}$ b. $\left(\dfrac{y}{x}\right)^{-1} = \dfrac{y^{-1}}{x^{-1}} = y^{-1}x^{-(-1)} = \dfrac{x}{y}$

    c. $\left(\dfrac{s}{t}\right)^0 = 1$ d. $(x^2y^{-1})^{-1} = x^{2(-1)}y^{-1(-1)} = x^{-2}y^1 = \dfrac{y}{x^2}$

    e. $(x^{-1}y^3)(x^3y^{-2}) = x^{-1}x^3y^3y^{-2} = x^{-1+3}y^{3-2} = x^2y$

    f. $(x^3y^{-2})^{-1} = x^{3(-1)}y^{(-2)(-1)} = x^{-3}y^2 = \dfrac{y^2}{x^3}$

13. a. $(2y)^{-2} = \dfrac{1}{(2y)^2} = \dfrac{1}{2^2y^2} = \dfrac{1}{4y^2}$ b. $2x^{-4} = \dfrac{2}{x^4}$

    c. $\left(\dfrac{2x}{y}\right)^{-2} = \left(\dfrac{y}{2x}\right)^2 = \dfrac{y^2}{2^2x^2} = \dfrac{y^2}{4x^2}$ d. $\left(\dfrac{a}{2c}\right)^{-3} = \left(\dfrac{2c}{a}\right)^3 = \dfrac{2^3c^3}{a^3} = \dfrac{8c^3}{a^3}$

    e. $\dfrac{2x^3y^{-2}}{6x^{-1}y^2} = \dfrac{x^{3-(-1)}y^{(-2)-2}}{3} = \dfrac{x^4y^{-4}}{3} = \dfrac{x^4}{3y^4}$ f. $b^1b^n = b^{1+n}$

15. a. $\left(\frac{3a^2}{c}\right)^{-3} = \left(\frac{c}{3a^2}\right)^3 = \frac{c^3}{3^3 a^{2\cdot 3}} = \frac{c^3}{27a^6}$

b. $\left(\frac{-a}{c^2}\right)^{-2} = \left(\frac{c^2}{-a}\right)^2 = \frac{c^{2\cdot 2}}{(-1)^2 a^2} = \frac{c^4}{a^2}$

c. $\frac{-1}{c^{-3}} = -1 \cdot c^3 = -c^3$

d. $\frac{2}{a^{-2}} = 2 \cdot a^2 = 2a^2$

e. $\frac{12a^3b^{-2}}{4a^{-1}b^{-5}} = 3a^{3-(-1)}b^{(-2)-(-5)} = 3a^4b^3$

f. $\frac{3}{3^x} = 3 \cdot 3^{-x} = 3^{1-x}$

17. a. $5^{-1} = \frac{1}{5} = 0.2$

b. $(-2)^{-1} = \frac{1}{-2} = -\frac{1}{2} = -0.5$

c. $0.5^{-2} = \left(\frac{1}{2}\right)^{-2} = \left(\frac{2}{1}\right)^2 = 2^2 = 4$

d. $-0.25^{-2} = -\left(\frac{1}{4}\right)^{-2} = -\left(\frac{4}{1}\right)^2 = -4^2 = -16$

e. $0.2^{-1} = \left(\frac{1}{5}\right)^{-1} = \left(\frac{5}{1}\right)^1 = 5$

f. $(-3)^{-2} = \left(\frac{1}{(-3)}\right)^2 = \frac{1^2}{(-3)^2} = \frac{1}{9}$

19. a. $1.5^{-1} = \left(\frac{3}{2}\right)^{-1} = \left(\frac{2}{3}\right)^1 = \frac{2}{3}$

b. $-2^{-1} = -\frac{1}{2}$

c. $(-2)^{-1} = \frac{1}{(-2)} = -\frac{1}{2}$

d. $(-4)^{-2} = \left(\frac{1}{(-4)}\right)^2 = \frac{1}{(-4)^2} = \frac{1}{16}$

e. $-4^{-2} = -\left(\frac{1}{4}\right)^2 = -\frac{1}{4^2} = -\frac{1}{16}$

f. $(-0.5)^{-2} = \left(-\frac{1}{2}\right)^{-2} = \left(-\frac{2}{1}\right)^2 = (-2)^2 = 4$

21. a. Adding like terms: $y^3 + 2y^3 = y^3(1 + 2) = 3y^3$

b. Adding like terms: $x^{-1} + x^{-1} = x^{-1}(1 + 1) = 2x^{-1} = \frac{2}{x}$

c. Neither: already simplified

23. a. Neither: $x^2 + x^{-2} = x^2 + \frac{1}{x^2}$

b. Adding like terms: $y^{-3} + y^{-3} = 2y^{-3} = \frac{2}{y^3}$

c. Multiplying like bases: $y^{-4}y^2 = y^{-4+2} = y^{-2} = \frac{1}{y^2}$

25. a. Neither: already simplified

b. Multiplying like bases: $x^3x^{-1} = x^{3-1} = x^2$

c. Adding like terms: $x^2 + 2x^2 = 3x^2$

27. a. $\left(\frac{-3x}{2}\right)^{-3} = \frac{(-3x)^{-3}}{(2)^{-3}} = \frac{2^3}{(-3)^3x^3} = \frac{8}{-27x^3} = -\frac{8}{27x^3}$

b. $\left(\frac{4x^3}{y^2}\right)^{-2} = \frac{(4x^3)^{-2}}{(y^2)^{-2}} = \frac{(y^2)^2}{4^2(x^3)^2} = \frac{y^{2\cdot2}}{16x^{3\cdot2}} = \frac{y^4}{16x^6}$

c. $\frac{3x^{-2}y}{6x^2y^{-3}} = \frac{1}{2}x^{-2-2}y^{1-(-3)} = \frac{1}{2}x^{-4}y^4 = \frac{y^4}{2x^4}$

29. $\dfrac{\frac{4\pi(2r)^3}{3}}{\frac{4\pi r^3}{3}} = \frac{4\pi 2^3r^3}{3}\cdot\frac{3}{4\pi r^3} = 2^3 = 8$, 8 times larger

31. No output for 0^0; undefined at x = 0

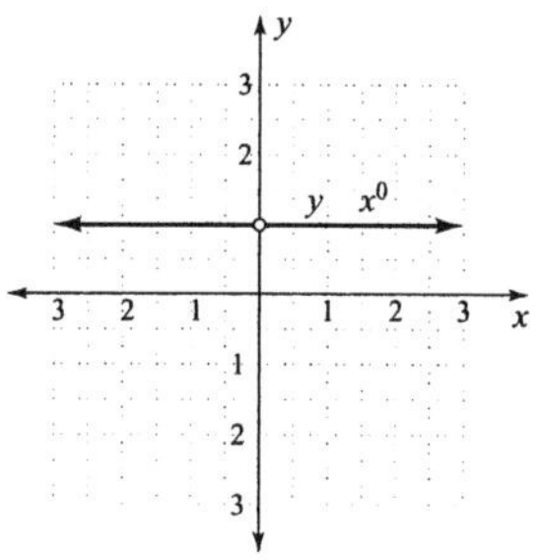

33. $-x^2 \neq (-x)^2$; In $-x^2$ the implied -1 is not included in the base.

35. The exponent does not apply to the 3. Answer is $\frac{3}{x^2}$.

37. $6\left(\frac{5}{6}\right)^4 = 6\cdot\frac{5}{6}\cdot\left(\frac{5}{6}\right)^3 = 5\left(\frac{5}{6}\right)^3$

Exercises 8.2

1. a. $0.001 = 10^{-3}$

b. $\frac{1}{100{,}000} = 0.000\,01 = 10^{-5}$

c. $10{,}000 = 10^4$

d. 1 hundredth $= 0.01 = 10^{-2}$

e. 1 million $= 1{,}000\,000 = 10^6$

f. 1 millionth $= 0.000\,001 = 10^{-6}$

3. a. $3000 = 3 \cdot 1000 = 3 \times 10^3$ **b.** $350 = 3.5 \cdot 100 = 3.5 \times 10^2$

c. $350,\bar{0}00 = 3.500 \cdot 100,000 = 3.500 \times 10^5$

d. $0.003\ 50 = 3.50 \cdot 0.001 = 3.50 \times 10^{-3}$

5. 3a - 1 digit; 3b - 2 digits; 3c - 4 digits; 3d - 3 digits

7. a. $2.9979 \times 10^{10} = 29{,}979{,}000{,}000$ cm/sec

b. $1.6726 \times 10^{-24} = 0.000\ 000\ 000\ 000\ 000\ 000\ 000\ 001\ 672\ 6$ g

c. $-1.6022 \times 10^{-19} = -0.000\ 000\ 000\ 000\ 000\ 000\ 160\ 22$ C

d. $1.6022 \times 10^{-19} = 0.000\ 000\ 000\ 000\ 000\ 000\ 160\ 22$ J

9. $4.0 \cdot 1.5 = 6;\ 10^{-2} \cdot 10^{6} = 10^4;\ 6 \times 10^4$

11. $6 \cdot 2.5 = 15 = 1.5 \times 10;\ 10^{-3} \cdot 10^{-5} \cdot 10^1 = 10^{-7};\ 1.5 \times 10^{-7}$

13. $7.5 \div 2.5 = 3.0;\ 10^{-2} \div 10^4 = 10^{-6};\ 3.0 \times 10^{-6}$

15. $2.4 \div 0.3 = 8;\ 10^6 \div 10^{-2} = 10^8;\ 8 \times 10^8$

17. a. $4\bar{0}0 \times 10^{-3} = 4.00 \times 10^2 \times 10^{-3} = 4.00 \times 10^{-1}$

b. $4\bar{0}0 \times 10^3 = 4.00 \times 10^2 \times 10^3 = 4.00 \times 10^5$

19. a. $0.05\bar{0} \times 10^{-4} = 5.0 \times 10^{-2} \times 10^{-4} = 5.0 \times 10^{-6}$

b. $0.05\bar{0} \times 10^5 = 5.0 \times 10^{-2} \times 10^5 = 5.0 \times 10^3$

21. a. 1 E 10; calculator reads one number **b.** 1 E 10; calculator reads two numbers

c. 1 E 11; calculator reads one number **d.** 1 E 11; calculator reads two numbers

e. 1 E 52; calculator reads one number **f.** 1 E 53; calculator reads one number

g. Tells the calculator to multiply the preceding number by 10 raised to the exponent that follows.

23. a. $5.33 \times 10^{11} - 2.6 \times 10^{10} = 5.33 \times 10^{11} - 0.26 \times 10^{11} = (5.33 - 0.26) \times 10^{11}$
$= 5.07 \times 10^{11}$

b. $2^n(2.6 \times 10^{10}) = 5.33 \times 10^{11};\ 2^n = \dfrac{5.33 \times 10^{11}}{2.6 \times 10^{10}} = 20.5$; Use guess and check to find n.

$n \approx 4.4$, National debt doubled over 4 times.

23. c. $1975 - 1918 = 57$ yrs; $\dfrac{57}{4.4} \approx 13$; $1918 + 13 = 1931$; $1931 + 13 = 1944$; $1944 + 13 = 1957$; $1957 + 13 = 1970$; $1970 + 13 = 1983$, $1983 + 13 = 1996$; $1996 + 13 = 2009$

d. $2.6 \times 10^{10} \cdot 2 = 5.2 \times 10^{10}$ (1931); $5.2 \times 10^{10} \cdot 2 = 1.04 \times 10^{11}$ (1944);
$1.04 \times 10^{11} \cdot 2 = 2.08 \times 10^{11}$ (1957); $2.08 \times 10^{11} \cdot 2 = 4.16 \times 10^{11}$ (1970);
$4.16 \times 10^{11} \cdot 2 = 8.32 \times 10^{11}$ (1983); $8.32 \times 10^{11} \cdot 2 = 1.66 \times 10^{12}$ (1996)
growth was faster than predicted

25. a. 2.8253×10^7 acre·feet · 4.3560×10^4 ft³/acre·ft $= 12.31 \times 10^{11} = 1.231 \times 10^{12}$ ft³

b. $1.231 \times 10^{12}\ ft^3 \cdot \left(\dfrac{1\ mi}{5280\ ft}\right)^2 \cdot \dfrac{Rhode\ Island}{1045\ mi^2} \approx 42\ ft$

27. a. $V_{circ} = \sqrt{\dfrac{(6.67 \times 10^{-11}\, N \cdot m^2 \cdot kg^{-2})(3.30 \times 10^{23}\, kg)}{3.24 \times 10^6\, m}}$;

To keep units correct set up two problems, one with numbers, the other with units.

$$\sqrt{\frac{(6.67 \times 10^{-11})(3.30 \times 10^{23})}{3.24 \times 10^6}} = \sqrt{\frac{6.67 \cdot 3.30 \times 10^{23-11-6}}{3.24}} \approx 2.61 \times 10^3$$

$$\sqrt{\frac{(Nm^2kg^{-2})(kg)}{m}} = \sqrt{\frac{(kg \cdot m \cdot \sec^{-2})(m^2 kg^{-2+1})}{m}} = \sqrt{kg^{1-1} m^{1+2-1} \sec^{-2}} = \sqrt{\frac{m^2}{\sec^2}} = \frac{m}{\sec}$$

$V_{circ} = 2.61 \times 10^3$ m/sec

b. Note: the units will be the same as in part a.

$$\sqrt{\frac{(6.67 \times 10^{-11})(1.90 \times 10^{27})}{8.14 \times 10^7}} = \sqrt{\frac{6.67 \cdot 1.90 \times 10^{27-11-7}}{8.14}} \approx 3.95 \times 10^4$$

$V_{circ} = 3.95 \times 10^4$ m/sec

c. Find V_{circ} first: units will be the same as in part a.

$$\sqrt{\frac{(6.67 \times 10^{-11})(5.98 \times 10^{24})}{7.18 \times 10^6}} = \sqrt{\frac{6.67 \cdot 5.98 \times 10^{24-11-6}}{7.18}} \approx 7.45 \times 10^3 \text{ m/sec}$$

$V_{esc} = \sqrt{2} \cdot 7.45 \times 10^3 \approx 1.05 \times 10^4$ m/sec

d. Find V_{circ} first:

$$\sqrt{\frac{(6.67 \times 10^{-11})(6.44 \times 10^{23})}{4.19 \times 10^6}} = \sqrt{\frac{6.67 \cdot 6.44 \times 10^{23-11-6}}{4.19}} \approx 3.20 \times 10^3 \text{ m/sec}$$

$V_{esc} = \sqrt{2} \cdot 3.19 \times 10^3 \approx 4.53 \times 10^3$ m/sec

e. If velocity drops below circular velocity the vehicle will be pulled back toward the planet.

f. If the velocity is between circular velocity and escape velocity the orbit becomes oval instead of circular.

Exercises 8.3

Guess and check is recommended for exercise 1, only final solutions are shown.

1. a. $27^x = 9$; $x = \frac{2}{3}$ **b.** $9^x = 27$; $x = \frac{3}{2}$ **c.** $81^x = 27$; $x = \frac{3}{4}$

3. $81 = 3^4 = 9^2$; $3 \cdot 3 \cdot 3 \cdot 3 = (3 \cdot 3) \cdot (3 \cdot 3)$

5. a. $4^x = (2 \cdot 2) \cdot (2 \cdot 2) \cdot (2 \cdot 2) \cdot (2 \cdot 2) \cdot (2 \cdot 2) = 1024$; $x = 5$

b. $8^x = (2 \cdot 2 \cdot 2) \cdot (2 \cdot 2 \cdot 2) \cdot (2 \cdot 2 \cdot 2) \cdot 2 = 1024$; $x = 3\frac{1}{3} = \frac{10}{3}$

c. $16^x = (2 \cdot 2 \cdot 2 \cdot 2) \cdot (2 \cdot 2 \cdot 2 \cdot 2) \cdot 2 \cdot 2 = 1024$; $x = 2\frac{1}{2} = \frac{5}{2}$

d. $32^x = (2 \cdot 2 \cdot 2 \cdot 2 \cdot 2) \cdot (2 \cdot 2 \cdot 2 \cdot 2 \cdot 2) = 1024$; $x = 2$

e. $64^x = (2 \cdot 2 \cdot 2 \cdot 2 \cdot 2 \cdot 2) \cdot 2 \cdot 2 \cdot 2 \cdot 2 = 1024$; $x = 1\frac{2}{3} = \frac{5}{3}$

f. $128^x = (2 \cdot 2 \cdot 2 \cdot 2 \cdot 2 \cdot 2 \cdot 2) \cdot 2 \cdot 2 \cdot 2 = 1024$; $x = 1\frac{3}{7} = \frac{10}{7}$

g. $256^x = (2 \cdot 2 \cdot 2 \cdot 2 \cdot 2 \cdot 2 \cdot 2 \cdot 2) \cdot 2 \cdot 2 = 1024$; $x = 1\frac{1}{4} = \frac{5}{4}$

7. a. $64 = (2 \cdot 2 \cdot 2) \cdot (2 \cdot 2 \cdot 2)$
$64^{1/2} = (2 \cdot 2 \cdot 2) = 8$

b. $16 = 2 \cdot 2 \cdot 2 \cdot 2$
$16^{1/4} = 2$

c. $32 = 2 \cdot 2 \cdot 2 \cdot 2 \cdot 2$
$32^{1/5} = 2$

9. a. $32 = 2 \cdot 2 \cdot 2 \cdot 2 \cdot 2$
$32^{2/5} = (32^{1/5})^2 = 2^2 = 4$

b. $16 = (2 \cdot 2) \cdot (2 \cdot 2)$
$16^{3/2} = (16^{1/2})^3 = 4^3 = 64$

c. $36 = 6 \cdot 6$
$36^{3/2} = (36^{1/2})^3 = 6^3 = 216$

d. $32 = 2 \cdot 2 \cdot 2 \cdot 2 \cdot 2$
$32^{6/5} = (32^{1/5})^6 = 2^6 = 64$

11. a. $16 = 2 \cdot 2 \cdot 2 \cdot 2$
$16^{5/4} = (16^{1/4})^5 = 2^5 = 32$

b. $64 = (2 \cdot 2 \cdot 2) \cdot (2 \cdot 2 \cdot 2)$
$64^{3/2} = (64^{1/2})^3 = 8^3 = 512$

c. $25 = 5 \cdot 5$
$25^{3/2} = (25^{1/2})^3 = 5^3 = 125$

d. $81 = 3 \cdot 3 \cdot 3 \cdot 3$
$81^{3/4} = (81^{1/4})^3 = 3^3 = 27$

13. Find the intersection of the y value and the graph; trace vertically to the x-axis and estimate the x value.

a. $x \approx 4.3$ **b.** $x \approx 4.7$ **c.** $x \approx 5$

15. a. $1.5 = \frac{3}{2}$; $25^{3/2} = (25^{1/2})^3 = 5^3 = 125$

b. $2.5 = \frac{5}{2}$; $4^{5/2} = (4^{1/2})^5 = 2^5 = 32$

c. $16^{5/2} = (16^{1/2})^5 = 4^5 = 1024$

d. $0.5 = \frac{1}{2}$; $0.09 = 0.3 \cdot 0.3$; $0.09^{1/2} = 0.3$

17. $\$6.00(1+0.05)^{10} = \$9.77$

19. $\$1000(1+0.07)^{5} = \$1402.55$

21. a. once a year = annual
b. twice a year = semiannual
c. 4 times a year = quarterly
d. 12 times a year = monthly
e. 52 times a year = weekly
f. 365 times a year = daily

23. a. 10,000 * (1 + 0.04 ÷ 12) ^ (12 * 3) = 11,272.72
b. 10,000 (1 + 0.04 ÷ 4) (4 * 3) = 11,268.25

25. $10,000 principal invested at 4% for 3 years; compounded monthly in **a** and quarterly in **b.**

27. $\$6.00(1+\frac{0.05}{2})^{2\cdot 10} = \$9.83$

29. $\$200(1+0.01)^{30} = \$269.57$

31. 8 months = $\frac{8}{12} = \frac{2}{3}$ yr.; $\$1350(1+\frac{0.16}{12})^{12\cdot\frac{2}{3}} = \$1500.91$

33. 2 months = $\frac{2}{12} = \frac{1}{6}$ yr; $\$650(1+\frac{0.16}{12})^{12\cdot\frac{1}{6}} = 667.45$

35. $1\frac{3}{4} = \frac{7}{4}$; $\$1000(1+\frac{0.08}{12})^{12\cdot\frac{7}{4}} \approx \$1,149.73$

37. $2\frac{1}{4} = \frac{9}{4}$; $\$1000(1+\frac{0.08}{12})^{12\cdot\frac{9}{4}} \approx \$1,196.50$

39. $5\frac{1}{3} = \frac{16}{3}$; $\$1000(1+\frac{0.07}{4})^{4\cdot\frac{16}{3}} \approx \$1,447.88$

41. $\$1000(1+\frac{0.07}{4})^{4\cdot 4.75} \approx \$1,390.45$

43. $\frac{r}{4} > \frac{r}{12}$; base is smaller

45. $1+\frac{r}{n}$

47. Answers could vary; 2 possible solutions are: $a = b = 2$; and $a = 2$, $b = 4$;

49. This is true for all b > 0.

51. fourth ≈ $8.84; fifth ≈ $9.37; Results could differ due to rounding errors.

53. Calculator tables are used to evaluate formulas; answers are given for reference only

a.

S (wind speed)	$W = 35.74 + 0.6215(25) - 35.75(S^{0.16}) + 0.4275(25)(S^{0.16})$ (wind chill temperature)
5	19
10	15
15	13
20	11
25	9
30	8

b.

S (wind speed)	$W = 35.74 + 0.6215(-5) - 35.75(S^{0.16}) + 0.4275(-5)(S^{0.16})$ (wind chill temperature)
5	−16
10	−22
15	−26
20	−29
25	−31
30	−33

c. The new model produces smaller changes and the temperatures are warmer.

Exercises 8.4

1. **a.** $256 = 2 \cdot 2 \cdot 2 \cdot 2 \cdot 2 \cdot 2 \cdot 2 \cdot 2$

b. $\sqrt[8]{256} = \sqrt[8]{2^8} = 2$

c. $\sqrt[4]{256} = \sqrt[4]{4^4} = 4$

d. $\sqrt{256} = 16$

e. $(256)^{5/8} = 2^5 = 32$

f. $(256)^{0.75} = (256)^{3/4} = 4^3 = 64$

g. $(256)^{7/8} = 2^7 = 128$

3. **a.** $\sqrt[3]{64} = \sqrt[3]{4^3} = 4$

b. $\sqrt[5]{32} = \sqrt[5]{2^5} = 2$

5. **a.** $\sqrt[3]{-27} = \sqrt[3]{(-3)^3} = -3$

b. $\sqrt[4]{625} = \sqrt[4]{5^4} = 5$

7. **a.** $\sqrt{-4}$, not a real number

b. $\sqrt[4]{-16}$, not a real number

9. **a.** $-\sqrt[4]{10{,}000} = -\sqrt[4]{10^4} = -10$

b. $\sqrt[3]{-1000} = \sqrt[3]{(-10)^3} = -10$

11. a. $-\sqrt[3]{64} = -\sqrt[3]{4^3} = -4$ **b.** $-\sqrt[3]{-1000} = -\sqrt[3]{(-10)^3} = -(-10) = 10$

13. a. $-\sqrt[4]{256} = -\sqrt[4]{4^4} = -4$ **b.** $\sqrt[4]{-625}$, not a real number

15. a. $x^{1/6} = \sqrt[6]{x}$ **b.** $\sqrt[3]{x} = x^{1/3}$ **c.** $x^{0.5} = x^{1/2} = \sqrt{x}$

17. a. $x^{3/2} = (\sqrt{x})^3$ *or* $\sqrt{x^3}$ **b.** $\sqrt[4]{x^3} = (x^3)^{1/4} = x^{3/4}$ **c.** $x^{0.8} = x^{4/5} = (\sqrt[5]{x})^4$ *or* $\sqrt[5]{x^4}$

19. a. $(\sqrt[3]{-8})^5 = (-8)^{5/3} = [(-2)^3]^{5/3} = (-2)^5 = -32$ **b.** $\sqrt[3]{27^2} = 27^{2/3} = (3^3)^{2/3} = 3^2 = 9$

21. a. $\sqrt[4]{16^2} = 16^{2/4} = 16^{1/2} = 4$ **b.** $(\sqrt[4]{-64})^2$, not a real number

23. a. $\sqrt{y^2} = |y|$ **b.** $\sqrt[4]{z^8} = z^{8/4} = z^2$ **c.** $\sqrt{x^6} = |x^3|$

25. a. $\sqrt[4]{x} \cdot \sqrt[3]{x^2} = x^{1/4} \cdot x^{2/3} = x^{1/4+2/3} = x^{3/12+8/12} = x^{11/12}$

b. $x^{2/3} \cdot x^{1/5} = x^{2/3 + 1/5} = x^{10/15 + 3/15} = x^{13/15}$

27. a. $x^{5/8} \div x^{1/2} = x^{5/8 - 1/2} = x^{5/8 - 4/8} = x^{1/8}$

b. $\sqrt[8]{x^3} \cdot \sqrt{x^4} = x^{3/8} \cdot x^{4/2} = x^{3/8+2} = x^{3/8+16/8} = x^{19/8}$

29. a. $\sqrt[4]{x^5} \div \sqrt[3]{x^2} = x^{5/4-2/3} = x^{15/12-8/12} = x^{7/12}$

b. $x^{4/5} \div x^{2/3} = x^{4/5 - 2/3} = x^{12/15 - 10/15} = x^{2/15}$

31. If $m > n$ in $\dfrac{a^n}{a^m}$ then subtract n from m and write $\dfrac{1}{a^{m-n}}$.

33. a. $(x^4)^{1/2} = x^{4/2} = x^2$ **b.** $(x^{2/3})^3 = x^{6/3} = x^2$ **c.** $(y^6)^{1/6} = y^{6/6} = |y|$

35. a. $\sqrt[2]{\sqrt[4]{x}} = (x^{1/4})^{1/2} = x^{1/8} = \sqrt[8]{x};\ x \geq 0$ **b.** $\sqrt[4]{\sqrt[3]{x^5}} = (x^{5/3})^{1/4} = x^{5/12} = \sqrt[12]{x^5}\ ;\ x \geq 0$

37. a. $\sqrt{a}\sqrt{a^3} = \sqrt{a^{1+3}} = \sqrt{a^4} = a^2;\ a \geq 0$ **b.** $\sqrt[3]{b}\sqrt[3]{b^2} = \sqrt[3]{b^{1+2}} = \sqrt[3]{b^3} = b$

c. $\dfrac{\sqrt{x^5}}{\sqrt{x^3}} = \sqrt{\dfrac{x^5}{x^3}} = \sqrt{x^{5-3}} = \sqrt{x^2} = x;\ x > 0$

39. a. $\sqrt{100n} = \sqrt{100}\sqrt{n} = 10\sqrt{n}$

b. $\sqrt{0.01n} = \sqrt{0.01}\sqrt{n} = 0.1\sqrt{n}$

c. $\sqrt{\dfrac{n}{100}} = \dfrac{\sqrt{n}}{\sqrt{100}} = \dfrac{\sqrt{n}}{10}$

d. $\sqrt[3]{\dfrac{x}{1000}} = \dfrac{\sqrt[3]{x}}{\sqrt[3]{1000}} = \dfrac{\sqrt[3]{x}}{10}$

e. $\sqrt[3]{8n} = \sqrt[3]{8}\sqrt[3]{n} = 2\sqrt[3]{n}$

41. a. $\sqrt{2}\sqrt{18} = \sqrt{2\cdot 18} = \sqrt{36} = 6$

b. $-\sqrt{3}\sqrt{27} = -\sqrt{3\cdot 27} = -\sqrt{81} = -9$

c. $\sqrt{2}\sqrt{8} = \sqrt{2\cdot 8} = \sqrt{16} = 4$

d. $\sqrt[2]{\sqrt[3]{64}} = \sqrt[6]{64} = 2$

43. a. $\dfrac{\sqrt[4]{243}}{\sqrt[4]{3}} = \sqrt[4]{\dfrac{243}{3}} = \sqrt[4]{81} = 3$

b. $\sqrt[2]{\dfrac{27}{3}} = \sqrt[2]{9} = 3$

c. $\sqrt[3]{16}\sqrt[3]{4} = \sqrt[3]{64} = 4$

d. $\dfrac{\sqrt[4]{64}}{\sqrt[4]{4}} = \sqrt[4]{\dfrac{64}{4}} = \sqrt[4]{16} = 2$

45. a. $x^{3/4}x^{1/3} = x^{3/4+1/3} = x^{9/12+4/12} = x^{13/12}$

b. $x^{3/4}x^{2/3} = x^{3/4+2/3} = x^{9/12+8/12} = x^{17/12}$

c. $x^{-1/3} = \dfrac{1}{x^{1/3}}$

d. $\sqrt[3]{x^2}\sqrt[4]{x} = x^{2/3}x^{1/4} = x^{2/3+1/4} = x^{8/12+3/12} = x^{11/12}$

47. a. $\sqrt[3]{x}\sqrt[2]{x^3} = x^{1/3}x^{3/2} = x^{1/3+3/2} = x^{2/6+9/6} = x^{11/6}$

b. $\dfrac{x^{3/4}}{x^{1/3}} = x^{3/4-1/3} = x^{9/12-4/12} = x^{5/12}$

c. $\left(x^{3/4}\right)^{2/3} = x^{3\cdot 2/(4\cdot 3)} = x^{6/12} = x^{1/2}$

d. $\sqrt[3]{\sqrt[2]{x^5}} = \sqrt[6]{x^5} = x^{5/6}$

49. a. $\sqrt[2]{\sqrt[3]{x^2}} = \sqrt[6]{x^2} = x^{2/6} = x^{1/3}$

b. $\left(\dfrac{16x^4}{y^8}\right)^{3/4} = \dfrac{16^{3/4}x^{12/4}}{y^{24/4}} = \dfrac{(16^{1/4})^3x^3}{y^6} = \dfrac{2^3x^3}{y^6} = \dfrac{8x^3}{y^6}$

c. $\dfrac{x^{-1/2}}{x^{3/2}} = x^{-1/2-3/2} = x^{-4/2} = x^{-2} = \dfrac{1}{x^2}$

d. $\dfrac{x^{-2/3}}{x} = x^{-2/3-1} = x^{-2/3-3/3} = x^{-5/3} = \dfrac{1}{x^{5/3}}$

51. a. $\sqrt[3]{16} = \sqrt[3]{2^4} = \sqrt[3]{2^3\cdot 2} = \sqrt[3]{2^3}\cdot\sqrt[3]{2} = 2\sqrt[3]{2}$

b. $\sqrt{75a^5} = \sqrt{3\cdot 25\cdot a^4\cdot a} = \sqrt{25}\cdot\sqrt{a^4}\cdot\sqrt{3a} = 5a^2\sqrt{3a}$

51. c. $\sqrt[4]{32x^9} = \sqrt[4]{16 \cdot 2 \cdot x^8 \cdot x} = \sqrt[4]{16} \cdot \sqrt[4]{x^8} \cdot \sqrt[4]{2x} = 2x^2\sqrt[4]{2x}$

d. $\sqrt[5]{-32a^4b^5c^6} = \sqrt[5]{-32} \cdot \sqrt[5]{b^5} \cdot \sqrt[5]{c^5} \cdot \sqrt[5]{a^4c} = -2bc\sqrt[5]{a^4c}$

e. $\sqrt[3]{x^4y^8z^9} = \sqrt[3]{x^3y^6z^9} \cdot \sqrt[3]{xy^2} = xy^2z^3\sqrt[3]{xy^2}$

53. $\sqrt[p]{x} \cdot \sqrt[q]{x} = x^{1/p}x^{1/q} = x^{1/p+1/q} = x^{q/pq+p/pq} = x^{(p+q)/pq}$

55. For the principal nth root, the radicand must be positive; for the odd nth root the radicand may be negative.

57. The even root of a negative number is undefined in the real numbers, so we exclude negative x for even roots from the definition.

59. A real number raised to an even power can yield a negative number. ***false***

61. The principal root is always positive. ***true***

63. yes, when $x < 0$, $-x > 0$

65. $5 \cdot 5^n = 5^1 \cdot 5^n = 5^{1+n}$; add exponents

67. x^2 is always positive

Mid-Chapter 8 Test

1. $x^{1/2} = \sqrt{x}\,;\, x \geq 0$

2. $\frac{1}{x} = x^{-1}; x \neq 0$

3. a. $\frac{a^{-1}b^2}{c^0} = \frac{b^2}{a^1(1)} = \frac{b^2}{a}$

b. $\frac{a^2b^{-1}}{c^{-2}} = \frac{a^2c^2}{b}$

4. a. $\frac{b^8}{b^5} = b^{8-5} = b^3$

b. $\frac{x^4y^5}{x^{-1}y^9} = \frac{x^{4-(-1)}}{y^{9-5}} = \frac{x^5}{y^4}$

5. a. $a \cdot a^x = a^{x+1}$

b. $(x^2)^n = x^{2n}$

6. a. $\left(\frac{2x}{3}\right)^{-2} = \left(\frac{3}{2x}\right)^2 = \frac{3^2}{2^2x^2} = \frac{9}{4x^2}$

b. $\frac{x}{y^{-2}} = xy^2$

7. **a.** $4.3 \times 10^{-3} = 0.0043$

b. $0.000\,123 = 1.23 \times 10^{-4}$

c. $\dfrac{3.6\times10^{-2}}{1.2\times10^{3}} = 3\times10^{-2-3} = 3\times10^{-5}$

d. 0.060 has two significant digits

e. Answers will vary, could include the zeros between a number and the decimal point, either to the left or right of the decimal.

8. **a.** $4^x = 32$
$(2^2)^x = 2^5$
$2^{2x} = 2^5$
$2x = 5$
$x = 2.5$

b. $9^x = 27$
$(3^2)^x = 3^3$
$3^{2x} = 3^3$
$2x = 3$
$x = 1.5$

c. $\left(\dfrac{1}{4}\right)^x = \dfrac{1}{8}$
$(2^{-2})^x = 2^{-3}$
$2^{-2x} = 2^{-3}$
$-2x = -3$
$x = 1.5$

d. $\pi^x = 1;\ x = 0$

9. **a.** $\$1000\left(1+\dfrac{0.07}{52}\right)^{5\cdot52} = \$1418.73$

b. $\$1000\left(1+\dfrac{0.07}{52}\right)^{2.75\cdot52} = \$1212.11$

10. $\left(1+\dfrac{0.04}{2}\right)^{2\cdot1} = 1.0404;\ (1+0.04)^1 = 1.040$; Receiving 2% twice a year is the same as getting a 4.04% increase once a year.

11. **a.** $\sqrt{4n} = 2\sqrt{n};\ n \ge 0$

b. $\sqrt{40n^2} = \sqrt{4n^2}\sqrt{10} = 4|n|\sqrt{10}$

c. $\sqrt{400n} = 20\sqrt{n};\ n \ge 0$

d. $\sqrt{4000n} = \sqrt{400}\sqrt{10n} = 20\sqrt{10n};\ n \ge 0$

12. **a.** $8^{2/3} = \left(\sqrt[3]{8}\right)^2 = (2)^2 = 4$

b. $125^{1/3} = \sqrt[3]{125} = 5$

13. **a.** $16^{3/4} = \left(\sqrt[4]{16}\right)^3 = (2)^3 = 8$

b. $(a^3)^{2/3} = \left(\sqrt[3]{a^3}\right)^2 = (a)^2 = a^2$

14. $\sqrt[3]{x}\cdot\sqrt[4]{x} = x^{1/3}x^{1/4} = x^{4/12+3/12} = x^{7/12}$

15. a. $\dfrac{x^{0.75}}{x^{2/3}} = x^{3/4-2/3} = x^{9/12-8/12} = x^{1/12}$ **b.** $\sqrt[3]{\sqrt[3]{x^6}} = \sqrt[3\cdot3]{x^6} = \sqrt[9]{x^6} = x^{6/9} = x^{2/3}$

16. a. $\sqrt[4]{-10{,}000x^4}$; not a real number **b.** $-\sqrt[4]{\dfrac{81x^8}{y^{-4}}} = -\sqrt[4]{81x^8y^4} = -3x^2y$

17. a. $\sqrt[3]{-64} = -4$ **b.** $\sqrt[4]{16y^4} = 2y$

18. a. $\sqrt{9x^5y^2} = \sqrt{9x^4y^2}\sqrt{x} = 3x^2y\sqrt{x}$ **b.** $\sqrt[3]{24x^5y^6} = \sqrt[3]{8x^3y^6}\sqrt[3]{3x^2} = 2xy^2\sqrt[3]{3x^2}$

19. Yes, because the 4th power is always positive.

20. $2|y|$

Exercises 8.5

1. $3\sqrt{2} - 4\sqrt{2} + 7\sqrt{2} = (3-4+7)\sqrt{2} = 6\sqrt{2}$

3. a. $\sqrt{20} + \sqrt{45} = 2\sqrt{5} + 3\sqrt{5} = (2+3)\sqrt{5} = 5\sqrt{5}$
b. $\sqrt{72} - \sqrt{18} = 6\sqrt{2} - 3\sqrt{2} = (6-3)\sqrt{2} = 3\sqrt{2}$

5. a. $\sqrt{75} + \sqrt{27} = 5\sqrt{3} + 3\sqrt{3} = (5+3)\sqrt{3} = 8\sqrt{3}$
b. $\sqrt{125} - \sqrt{80} = 5\sqrt{5} - 4\sqrt{5} = (5-4)\sqrt{5} = \sqrt{5}$

7. $\sqrt{9x} + \sqrt{16x} - \sqrt{x} = 3\sqrt{x} + 4\sqrt{x} - \sqrt{x} = (3+4-1)\sqrt{x} = 6\sqrt{x}$

9. $\sqrt{0.01x} + \sqrt{49x} - \sqrt{16x} = 0.1\sqrt{x} + 7\sqrt{x} - 4\sqrt{x} = (0.1+7-4)\sqrt{x} = 3.1\sqrt{x}$

11. $a\sqrt{b} + c\sqrt{d}$; not like terms

13. $\sqrt{ab} + 2\sqrt{ab} = (1+2)\sqrt{ab} = 3\sqrt{ab}$

15. $3\sqrt[3]{x} - \sqrt[3]{x} = (3-1)\sqrt[3]{x} = 2\sqrt[3]{x}$ **17.** $3\sqrt[4]{x} - 2\sqrt[2]{x}$; not like terms

19. $\sqrt[3]{8x}+\sqrt[3]{27x}=2\sqrt[3]{x}+3\sqrt[3]{x}=(2+3)\sqrt[3]{x}=5\sqrt[3]{x}$

21. $a\sqrt[4]{81ab}-\sqrt[4]{a^5b}=3a\sqrt[4]{ab}-a\sqrt[4]{ab}=(3a-a)\sqrt[4]{ab}=2a\sqrt[4]{ab}$

23. $\sqrt[4]{16x^4y}+x\sqrt[4]{y}=2x\sqrt[4]{y}+x\sqrt[4]{y}=(2x+x)\sqrt[4]{y}=3x\sqrt[4]{y}$

25. a. $(2-\sqrt{3})(2+\sqrt{3})=4-(\sqrt{3})^2=4-3=1$

b. $(5-\sqrt{2})(5-\sqrt{2})=5^2-10\sqrt{2}+(\sqrt{2})^2=25-10\sqrt{2}+2=27-10\sqrt{2}$

27. a. $(x-\sqrt{5})(x+\sqrt{5})=x^2-(\sqrt{5})^2=x^2-5$

b. $(x-\sqrt{7})(x-\sqrt{7})=x^2-2x\sqrt{7}+(\sqrt{7})^2=x^2-2x\sqrt{7}+7$

29. a. $(\sqrt{x}+3)^2=(\sqrt{x})^2+6\sqrt{x}+3^2=x+6\sqrt{x}+9$

b. $(\sqrt{a}-3)^2=(\sqrt{a})^2-6\sqrt{a}+3^2=a-6\sqrt{a}+9$

31. a. $(1-\sqrt{a})^2=1^2-2\sqrt{a}+(\sqrt{a})^2=1-2\sqrt{a}+a$

b. $(\sqrt{a}-\sqrt{b})(\sqrt{a}-\sqrt{b})=(\sqrt{a})^2-2\sqrt{a}\sqrt{b}+(\sqrt{b})^2=a-2\sqrt{ab}+b$

33. $(1+\sqrt{3})^2-2(1+\sqrt{3})-2=0$

$1+2\sqrt{3}+3-2-2\sqrt{3}-2=0$

$4-4+2\sqrt{3}-2\sqrt{3}=0$

$0=0$

35. $(-2-\sqrt{2})^2+4(-2-\sqrt{2})+2=0$

$4+4\sqrt{2}+2-8-4\sqrt{2}+2=0$

$6-6+4\sqrt{2}-4\sqrt{2}=0$

$0=0$

37. $2\left(\frac{3}{2}+\frac{\sqrt{3}}{2}\right)^2-6\left(\frac{3}{2}+\frac{\sqrt{3}}{2}\right)+3=0$

$2\left(\frac{9+6\sqrt{3}+3}{4}\right)-9-3\sqrt{3}+3=0$

$\frac{9}{2}+3\sqrt{3}+\frac{3}{2}-9-3\sqrt{3}+3=0$

$\frac{12}{2}-9+3+3\sqrt{3}-3\sqrt{3}=6-6=0$

39.

Multiply	x^2	$-x\sqrt[3]{2}$	$+\sqrt[3]{4}$
x	x^3	$-x^2\sqrt[3]{2}$	$x\sqrt[3]{4}$
$+\sqrt[3]{2}$	$+x^2\sqrt[3]{2}$	$-x\sqrt[3]{2}\,\sqrt[3]{2}=-x\sqrt[3]{4}$	$+\sqrt[3]{2}\,\sqrt[3]{4}=\sqrt[3]{8}=2$

$(x+\sqrt[3]{2})(x^2-x\sqrt[3]{2}+\sqrt[3]{4})=x^3+2$

41. a. $3-\sqrt{2}$ **b.** $3+\sqrt{a}$

c. $a+\sqrt{b}$ **d.** $\sqrt{2}+\sqrt{3}$

43. a. $\frac{4}{\sqrt{5}}=\frac{4\sqrt{5}}{\sqrt{5}\sqrt{5}}=\frac{4\sqrt{5}}{5}$ **b.** $\frac{8}{\sqrt{6}}=\frac{8\sqrt{6}}{\sqrt{6}\sqrt{6}}=\frac{8\sqrt{6}}{6}=\frac{4\sqrt{6}}{3}$

45. a. $\frac{a}{\sqrt{c}}=\frac{a\sqrt{c}}{\sqrt{c}\sqrt{c}}=\frac{a\sqrt{c}}{c}$ **b.** $\frac{a}{\sqrt{a}}=\frac{a\sqrt{a}}{\sqrt{a}\sqrt{a}}=\frac{a\sqrt{a}}{a}=\sqrt{a}$

47. $\frac{4}{7-\sqrt{5}}=\frac{4(7+\sqrt{5})}{(7-\sqrt{5})(7+\sqrt{5})}=\frac{4(7+\sqrt{5})}{49-5}=\frac{4(7+\sqrt{5})}{44}=\frac{7+\sqrt{5}}{11}$

49. $\frac{x}{x-\sqrt{y}}=\frac{x(x+\sqrt{y})}{(x-\sqrt{y})(x+\sqrt{y})}=\frac{x^2+x\sqrt{y}}{x^2-y}$

51. a. $\frac{1}{\sqrt[3]{2}}=\frac{\sqrt[3]{4}}{\sqrt[3]{2}\sqrt[3]{4}}=\frac{\sqrt[3]{4}}{\sqrt[3]{8}}=\frac{\sqrt[3]{4}}{2}$ **b.** $\frac{2}{\sqrt[3]{4}}=\frac{2\sqrt[3]{2}}{\sqrt[3]{4}\sqrt[3]{2}}=\frac{2\sqrt[3]{2}}{\sqrt[3]{8}}=\frac{2\sqrt[3]{2}}{2}=\sqrt[3]{2}$

c. $\frac{3}{\sqrt[3]{3}}=\frac{3\sqrt[3]{9}}{\sqrt[3]{3}\sqrt[3]{9}}=\frac{3\sqrt[3]{9}}{\sqrt[3]{27}}=\frac{3\sqrt[3]{9}}{3}=\sqrt[3]{9}$

53. Slope $\overline{BP}=\frac{y}{x+r}$

$$=\frac{\sqrt{r^2-x^2}}{x+r}=\frac{\sqrt{r^2-x^2}\cdot\sqrt{r^2-x^2}}{(x+r)\sqrt{r^2-x^2}}=\frac{r^2-x^2}{(x+r)\sqrt{r^2-x^2}}=\frac{(r+x)(r-x)}{(x+r)\sqrt{r^2-x^2}}$$

$$=\frac{r-x}{\sqrt{r^2-x^2}}=\frac{(-1)(x-r)}{\sqrt{r^2-x^2}}=\frac{(-1)(x-r)}{y}$$; which is the negative reciprocal of slope $\overline{AP}$

55. $d=\sqrt{(\Delta x)^2+(\Delta y)^2}=\sqrt{(x_2-x_1)^2+(y_2-y_1)^2}$

57. slope $= \dfrac{-3-3}{4-(-6)} = \dfrac{-6}{10} = \dfrac{-3}{5}$;

distance $= \sqrt{10^2 + (-6)^2} = \sqrt{100+36} = \sqrt{136} = 2\sqrt{34} \approx 11.662$

59. slope $= \dfrac{-4-7}{-3-6} = \dfrac{-11}{-9} = \dfrac{11}{9}$;

distance $= \sqrt{9^2 + 11^2} = \sqrt{81+121} = \sqrt{202} \approx 14.213$

61. slope $= \dfrac{b-\frac{b}{2}}{a-\frac{a}{2}} = \dfrac{\frac{b}{2}}{\frac{a}{2}} = \dfrac{b}{a}$; distance $= \sqrt{(\frac{a}{2})^2 + (\frac{b}{2})^2} = \sqrt{\frac{a^2}{4} + \frac{b^2}{4}} = \sqrt{\frac{a^2+b^2}{4}} = \dfrac{\sqrt{a^2+b^2}}{2}$

63. a. $\dfrac{(1+\sqrt{5})}{2} \cdot \dfrac{(1-\sqrt{5})}{2} = \dfrac{1-5}{4} = \dfrac{-4}{4} = -1$

b. $\dfrac{1+\sqrt{5}}{2} \approx 1.6180$; $\dfrac{1-\sqrt{5}}{2} \approx -0.6180$

c. reciprocal of $\dfrac{1+\sqrt{5}}{2}$ is $\dfrac{2}{1+\sqrt{5}} = \dfrac{2(1-\sqrt{5})}{(1+\sqrt{5})(1-\sqrt{5})} = \dfrac{2(1-\sqrt{5})}{1-5} = \dfrac{2(1-\sqrt{5})}{-4} = -\dfrac{1-\sqrt{5}}{2}$

reciprocal of $\dfrac{1-\sqrt{5}}{2}$ is $\dfrac{2}{1-\sqrt{5}} = \dfrac{2(1+\sqrt{5})}{(1-\sqrt{5})(1+\sqrt{5})} = \dfrac{2(1+\sqrt{5})}{1-5} = \dfrac{2(1+\sqrt{5})}{-4} = -\dfrac{1+\sqrt{5}}{2}$

the two equations appear to be negative reciprocals of each other

d. $\left(\dfrac{1+\sqrt{5}}{2}\right)^2 - \dfrac{1+\sqrt{5}}{2} - 1 = \dfrac{1+2\sqrt{5}+5}{4} - \dfrac{1+\sqrt{5}}{2} - 1 = \dfrac{6+2\sqrt{5}}{4} - \dfrac{1+\sqrt{5}}{2} - 1$

$= \dfrac{3+\sqrt{5}}{2} - \dfrac{1+\sqrt{5}}{2} - \dfrac{2}{2} = \dfrac{3-1-2+\sqrt{5}-\sqrt{5}}{2} = 0$

$\left(\dfrac{1-\sqrt{5}}{2}\right)^2 - \dfrac{1-\sqrt{5}}{2} - 1 = \dfrac{1-2\sqrt{5}+5}{4} - \dfrac{1-\sqrt{5}}{2} - 1 = \dfrac{6-2\sqrt{5}}{4} - \dfrac{1-\sqrt{5}}{2} - 1$

$= \dfrac{3-\sqrt{5}}{2} - \dfrac{1-\sqrt{5}}{2} - \dfrac{2}{2} = \dfrac{3-1-2-\sqrt{5}+\sqrt{5}}{2} = 0$

Exercises 8.6

1. $(\frac{1}{2}, 1), (\frac{1}{4}, 2), (\frac{1}{8}, 3), (\frac{1}{16}, 4)$

3. $(2, 1), (5, 2), (8, 3), (11, 4)$

5.

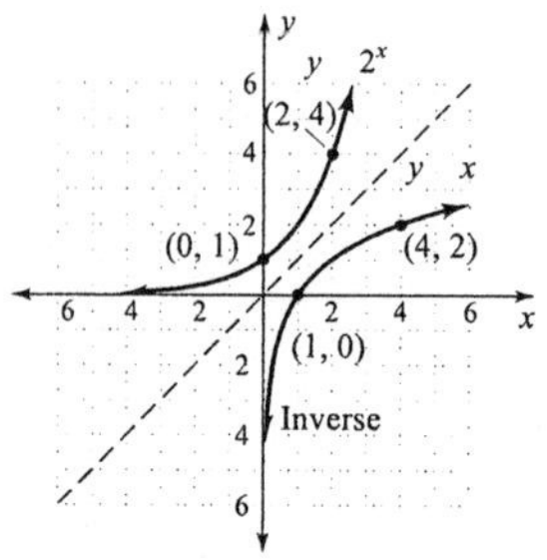

7.

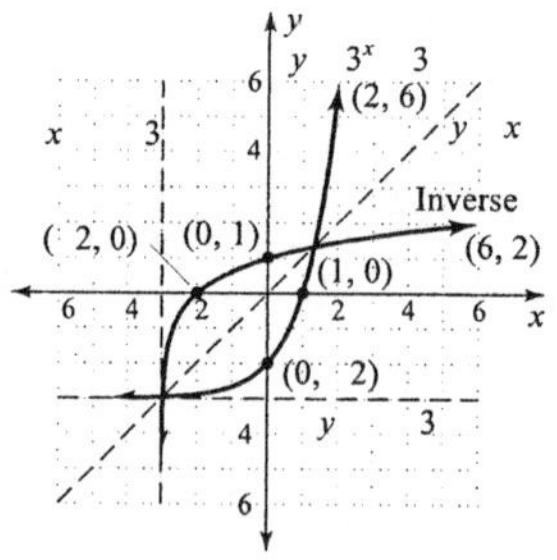

9.

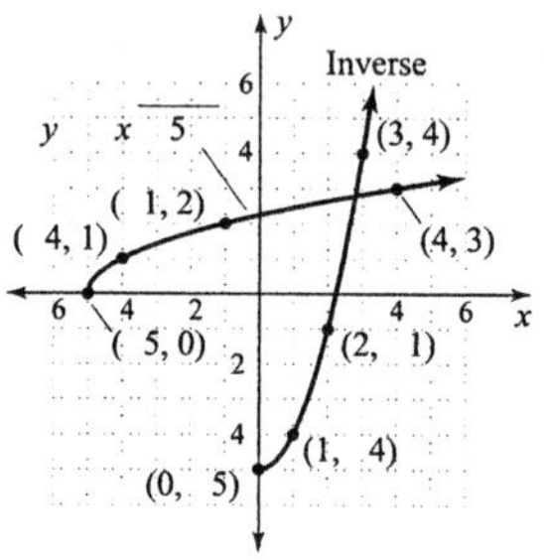

11.

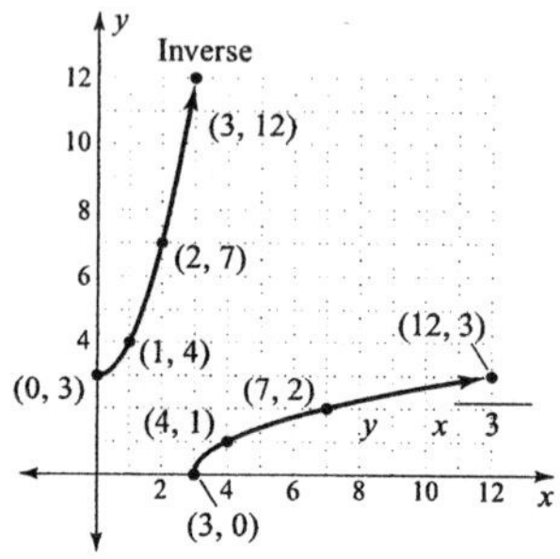

13.

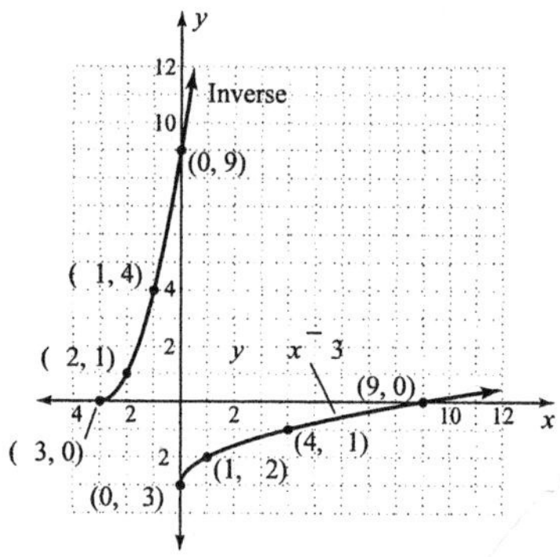

15.

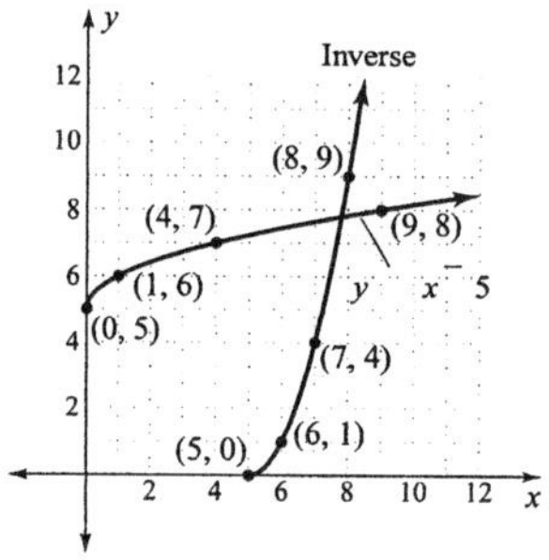

17. $x = 3y;$

$y = \dfrac{x}{2}$

19. $x = 4y - 2;\ x + 2 = 4y$

$\dfrac{x+2}{4} = y;\ y = \frac{1}{4}x + \frac{1}{2}$

21. $x = y + 3;\ y = x - 3$

23. $x = \sqrt{y};\ y = x^2$

25. $x = \sqrt[5]{y};\ y = x^5$

27. $x = y^4;\ y = \sqrt[4]{x}$

29. $x = y$

31. $x = -\frac{1}{y}$; $xy = -1$; $y = -\frac{1}{x}$

33. $x = 1 - y$; $x + y = 1$; $y = 1 - x$

35. The inverse is the same as the original.

37. $x = \sqrt{y+5}$; $x^2 = y + 5$
$y = x^2 - 5 = f^{-1}(x); y \geq -5, x \geq 0$

39. $x = \sqrt{y-3}$; $x^2 = y - 3$
$y = x^2 + 3 = f^{-1}(x)$; $y \geq 3$, $x \geq 0$

41. $x = \sqrt{y} - 3$; $x + 3 = \sqrt{y}$
$y = (x+3)^2 = f^{-1}(x)$; $y \geq 0$, $x \geq -3$

43. $x = \sqrt{y} + 5$; $x - 5 = \sqrt{y}$
$y = (x-5)^2 = f^{-1}(x)$; $y \geq 0$, $x \geq 5$

45. The even root of a negative number is not a real number.

47. Outputs are positive real numbers.

49. For any real number x, there is a corresponding real number y on the graph.

51. Principal square root is always positive.

53. a. Yes, S is a function.
b. Reversing the ordered pairs gives the same set
c. No, fails the vertical line test

55. With the condition, $x \geq 2$, no two inputs have the same output and an inverse function exists.

57. In order for a function to have an inverse function, there is exactly one ***input***, for each ***output***.

59. The $[x^{-1}]$ is the reciprocal key. The -1 is an exponent.

Exercises 8.7

1. from the graph, x = 14

3. from the graph, { } or ∅

Calculator graphs are not included for exercises 5 through 25.

5. defined when 2 – x ≥ 0 or x ≤ 2

$\sqrt{2-x}=2;\ \left(\sqrt{2-x}\right)^2=2^2$

$2-x=4;\ -x=2;\ x=-2$

$\sqrt{2-(-2)}=\sqrt{4}=2$

7. defined when x – 5 ≥ 0 or x ≥ 5

$\sqrt{x-5}=2;\ \left(\sqrt{x-5}\right)^2=2^2$

$x-5=4;\ x=9$

$\sqrt{9-5}=\sqrt{4}=2$

9. defined when x ≥ 5 (see #7)

$\sqrt{x-5}-6=-2;\ \sqrt{x-5}=4$

$\left(\sqrt{x-5}\right)^2=4^2;\ x-5=16;\ x=21$

$\sqrt{21-5}-6=\sqrt{16}-6=4-6=-2$

11. defined when 2x + 14 ≥ 0 or x ≥ –7

$\sqrt{2x+14}=0;\ \left(\sqrt{2x+14}\right)^2=0$

$2x+14=0;\ 2x=-14;\ x=-7$

$\sqrt{2(-7)+14}=\sqrt{-14+14}=0$

13. defined when x ≥ –7 (see #11)

$\sqrt{2x+14}=6;\ \left(\sqrt{2x+14}\right)^2=6^2$

$2x+14=36;\ 2x=22,\ x=11$

$\sqrt{2(11)+14}=\sqrt{22+14}=\sqrt{36}=6$

15. defined when 3x – 5 ≥ 0 or x ≥ $\frac{5}{3}$

$\sqrt{3x-5}=2;\ \left(\sqrt{3x-5}\right)^2=2^2$

$3x-5=4;\ 3x=9;\ x=3$

$\sqrt{3(3)-5}=\sqrt{9-5}=\sqrt{4}=2$

17. defined when x ≥ $\frac{5}{3}$ (see #15)

$\sqrt{3x-5}=-1$

Principal square root must be positive; { } or ∅

19. defined when 5x – 1 ≥ 0 or x ≥ $\frac{1}{5}$

$\sqrt{5x-1}=3;\ \left(\sqrt{5x-1}\right)^2=3^2$

$5x-1=9;\ 5x=10;\ x=2$

$\sqrt{5(2)-1}=\sqrt{10-1}=\sqrt{9}=3$

21. defined when 2x + 7 ≥ 0 or x ≥ –3.5 **and** x ≥ 0; use x ≥ 0

$\sqrt{2x+7}=\sqrt{x}+2;\ \left(\sqrt{2x+7}\right)^2=\left(\sqrt{x}+2\right)^2;\ 2x+7=x+4\sqrt{x}+4;\ 2x-x+7-4=4\sqrt{x}$

$x+3=4\sqrt{x};\ (x+3)^2=\left(4\sqrt{x}\right)^2;\ x^2+6x+9=16x;\ x^2-10x+9=0;\ (x-9)(x-1)=0$

$\{1, 9\}$

23. defined when x + 4 ≥ 0 or x ≥ –4 **and** 3x ≥ 0 or x ≥ 0; use x ≥ 0

$\sqrt{3x}-2=\sqrt{x+4};\ \left(\sqrt{3x}-2\right)^2=\left(\sqrt{x+4}\right)^2;\ 3x-4\sqrt{3x}+4=x+4;\ 2x=4\sqrt{3x}$

$x=2\sqrt{3x};\ (x)^2=\left(2\sqrt{3x}\right)^2=x^2=4(3x);x^2=12x;\ x=12$

25. defined when $3x - 3 \geq 0$ or $x \geq 1$ **and** $4x \geq 0$ or $x \geq 0$; use $x \geq 1$

$\sqrt{3x-3} = \sqrt{4x} - 1$; $\left(\sqrt{3x-3}\right)^2 = \left(\sqrt{4x}-1\right)^2$; $3x - 3 = 4x - 2\sqrt{4x} + 1$; $x + 4 = 2\sqrt{4x}$

$(x+4)^2 = \left(2\sqrt{4x}\right)^2$; $x^2 + 8x + 16 = 4(4x)$; $x^2 + 8x + 16 = 16x$; $x^2 - 8x + 16 = 0$;

$(x-4)^2 = 0$; $x = 4$

27. The extra number is an extraneous root.

29. a. $x^2 = 1$; $\sqrt{x^2} = \sqrt{1}$; $|x| = 1$; $x = \pm 1$

$x^2 - 1 = 0$; $(x + 1)(x - 1) = 0$; $x = \pm 1$

b. $x^4 = 1$; $\sqrt[4]{x^4} = \sqrt[4]{1}$; $|x| = 1$; $x = \pm 1$

$x^4 - 1 = 0$; $(x^2 + 1)(x^2 - 1) = 0$; $x^2 + 1 = 0$, $x^2 = -1$; $x = \pm i$

$x^2 - 1 = 0$; $(x+1)(x-1) = 0$; $x = \pm 1$

c. $x^3 = 1$; $\sqrt[3]{x^3} = \sqrt[3]{1}$; $x = 1$

$x^3 - 1 = 0$; $(x-1)(x^2 + x + 1) = 0$; $x = 1$ *or* $x = \dfrac{-1 \pm \sqrt{1^2 - 4(1)(1)}}{2(1)} = -\dfrac{1}{2} \pm \dfrac{i\sqrt{3}}{2}$

31. $4x^2 = 25$; $4x^2 - 25 = 0$; $(2x - 5)(2x + 5) = 0$; $x = \pm 2.5$; no roots with imaginary units

33. $16x^4 - 25 = 600$; $16x^4 = 625$; $(16x^4)^{1/4} = (625)^{1/4}$; $2|x| = 5$; $|x| = 2.5$;
$x = \pm 2.5$; 2 solutions with imaginary units (4th power has 4 solutions)

35. $100 = 40(1 + x)^5$; $2.5 = (1 + x)^5$; $\sqrt[5]{2.5} = \sqrt[5]{1+x}$; $1+x \approx 1.201$; $x \approx 0.201$

37. $30 = 10(1 + x)^{15}$; $3 = (1 + x)^{15}$; $\sqrt[15]{3} = \sqrt[15]{1+x}$; $1 + x \approx 1.076$; $x \approx 0.076$

39. $40 = 100(1 + x)^5$; $0.4 = (1 + x)^5$; $\sqrt[5]{0.4} = \sqrt[5]{1+x}$; $1 + x \approx 0.833$; $x \approx -0.167$

41. $10 = 30(1 + x)^{15}$; $\frac{1}{3} = (1 + x)^{15}$; $\sqrt[15]{\frac{1}{3}} = \sqrt[15]{1+x}$; $1 + x \approx 0.929$; $x \approx -0.071$

43. a. $r=\sqrt{24L};\ r^2=24L;\ L=\dfrac{r^2}{24}$

b. $r=\sqrt{12L};\ r^2=12L;\ L=\dfrac{r^2}{12}$

c. $r_{dry}=\sqrt{24(50)}\approx 34.64;\ r_{wet}=\sqrt{12(50)}\approx 24.49;$

$\dfrac{r_{wet}}{r_{dry}}=\dfrac{\sqrt{12L}}{\sqrt{24L}}=\sqrt{\dfrac{12L}{24L}}=\sqrt{\dfrac{1}{2}}=\dfrac{1}{\sqrt{2}}=\dfrac{\sqrt{2}}{2}$; slow down when the road is wet.

45. $v=\sqrt{2gs};\ v^2=2gs;\ s=\dfrac{v^2}{2g}$

47. $t=13;\ 119{,}000=55{,}000(1+r)^{13}$

$\dfrac{119{,}000}{55{,}000}=(1+r)^{13}$

$\left(\dfrac{119}{55}\right)^{1/13}=1+r$

$r=\left(\dfrac{119}{55}\right)^{1/13}-1\approx 6.1\%$

in 2010 $t=25;\ S=55{,}000(1+0.061)^{25}$

$S\approx \$243{,}000$

49. $t=25;\ 75{,}000=14{,}000(1+r)^{25}$

$\dfrac{75{,}000}{14{,}000}=(1+r)^{25}$

$\left(\dfrac{75}{14}\right)^{1/25}=1+r$

$r=\left(\dfrac{75}{14}\right)^{1/25}-1\approx 6.94\%$

in 2010 $t=40;\ S=14{,}000(1+0.0694)^{40}$

$S\approx \$205{,}315.25$

51. $t=61;\ 68{,}400=3000(1+r)^{61}$

$\dfrac{68{,}400}{3000}=(1+r)^{61}$

$\left(\dfrac{684}{30}\right)^{1/61}=1+r$

$r=\left(\dfrac{684}{30}\right)^{1/61}-1\approx 5.3\%$

in 2010 $t=73;\ S=3000(1+0.053)^{73}$

$S\approx \$127{,}000$

53. $t=8;\ 52.02=6.953(1+r)^{8}$

$\dfrac{52.02}{6.953}=(1+r)^{8}$

$\left(\dfrac{52.02}{6.953}\right)^{1/8}=1+r$

$r=\left(\dfrac{52.02}{6.953}\right)^{1/8}-1\approx 28.6\%$

55. $600=800(1+r)^3$

$0.75=(1+r)^3;\ (0.75)^{1/3}=1+r$

$r=(0.75)^{1/3}-1\approx -9.1\%$

57. $t=28;\ 5=96(1+r)^{28}$

$\dfrac{5}{96}=(1+r)^{28};\ \left(\dfrac{5}{96}\right)^{1/28}=1+r$

$r=\left(\dfrac{5}{96}\right)^{1/28}-1\approx -10\%$

59. See exercise for prices in 1994 and 1998. t = 8 from 1994

$$120 = 470(1 + r)^8;\ \frac{120}{470} = (1 + r)^8;\ \left(\frac{120}{470}\right)^{1/8} = 1 + r;\ r = \left(\frac{120}{470}\right)^{1/8} - 1 \approx -15.7\%$$

t = 4 from 1998

$$120 = 290(1 + r)^4;\ \frac{120}{290} = (1 + r)^4;\ \left(\frac{120}{290}\right)^{1/4} = 1 + r;\ r = \left(\frac{120}{290}\right)^{1/4} - 1 \approx -19.8\%$$

61. a. $100\pi = \frac{4\pi r^3}{3}$

$\frac{3(100\pi)}{4\pi} = r^3$

$r^3 = 75;\ r = 75^{1/3} \approx 4.22$ in

b. $200\pi = \frac{4\pi r^3}{3}$

$\frac{3(200\pi)}{4\pi} = r^3$

$r^3 = 150;\ r = 150^{1/3} \approx 5.31$ in

c. $400\pi = \frac{4\pi r^3}{3}$

$\frac{3(400\pi)}{4\pi} = r^3$

$r^3 = 300;\ r = 300^{1/3} \approx 6.69$ in

d. $800\pi = \frac{4\pi r^3}{3}$

$\frac{3(800\pi)}{4\pi} = r^3$

$r^3 = 600;\ r = 600^{1/3} \approx 8.43$ in

The volume of one sphere is 8 times the volume of another sphere. The radius of the first is ***two*** times the radius of the second.

63. $Q_b = \sigma A T^4$

$\frac{Q_b}{\sigma A} = T^4$

$T = \sqrt[4]{\frac{Q_b}{\sigma A}}$,

65. $K = \frac{mV^2}{2}$

$\frac{2K}{m} = V^2$

$V = \sqrt{\frac{2K}{m}}$

Chapter 8 Review Exercises

1. a. $\frac{a^0 b^{-1}}{c^2} = \frac{1}{bc^2}$

b. $\frac{a^3 b^{-2}}{c^{-1}} = \frac{a^3 c}{b^2}$

c. $\left(\frac{2x}{y^2}\right)^4 = \frac{2^4 x^4}{(y^2)^4} = \frac{16x^4}{y^8}$

3. a. $\left(x^3 y^{-4}\right)^2 = x^6 y^{-8} = \frac{x^6}{y^8}$

b. $\left(3x^3 y^2\right)^2 \left(4xy^{-3}\right) = 9x^6 y^4 (4xy^{-3}) = 36x^{6+1} y^{4-3} = 36x^7 y$

5. a. $3.45 \times 10^{-15} = 0.000\ 000\ 000\ 000\ 003\ 45$ **b.** $6.400 \times 10^{-3} = 0.006\ 400$
c. $4.005 \times 10^{5} = 400{,}500$ **d.** $4.7800 \times 10^{3} = 4780.0$

7. a. $64^{1/2} = 8$ **b.** $64^{1/6} = 2$ **c.** $64^{5/6} = (64^{1/6})^5 = 2^5 = 32$

9. a. $16^{0.75} = 16^{3/4} = (16^{1/4})^3 = 2^3 = 8$ **b.** $25^{3/2} = (25^{1/2})^3 = 5^3 = 125$
c. $27^{4/3} = (27^{1/3})^4 = 3^4 = 81$

11. $\$7(1 + 0.04)^{10} = \$10.36$; $\$7(1 + \frac{0.04}{2})^{2 \cdot 10} = \$10.40$

13. $\$1600(1 + \frac{0.05}{12})^{12 \cdot 4} = \$1953.43$ **15.** $\$1600(1 + \frac{0.05}{12})^{12 \cdot 3.75} = \$1929.21$

17. a. $(a^2)^{3/2} = (a^{2/2})^3 = a^3$ **b.** $x^{1/3} \cdot x^{2/3} = x^{1/3 + 2/3} = x^1 = x$
c. $a^{1/4}a^{3/2} = a^{1/4 + 3/2} = a^{1/4 + 6/4} = a^{7/4}$ **d.** $\dfrac{b^{2/3}}{b^{1/2}} = b^{2/3-1/2} = b^{4/6-3/6} = b^{1/6}$
e. $x \cdot x^n = x^{n+1}$ **f.** $(x^3y^4)^{1/2} = x^{3/2}y^{4/2} = x^{3/2}y^2$

19. a. $(a^{3/4} \cdot b^{1/4})^2 = a^{6/4}b^{2/4} = a^{3/2}b^{1/2}$ **b.** $(a^{4/3})^{3/4} = a^{12/12} = a^1 = a$
c. $\dfrac{x^n}{x^{n-1}} = x^{n-(n-1)} = x^{n-n+1} = x^1 = x$ **d.** $x \cdot x^{n-1} = x^{1+n-1} = x^n$
e. $b^{1/2} \cdot b = b^{1/2+1} = b^{3/2}$ **f.** $\left(\dfrac{x^3}{y^{1/2}}\right)^{2/3} = \dfrac{x^{6/3}}{y^{2/6}} = \dfrac{x^2}{y^{1/3}}$

21. a. $\sqrt[3]{\dfrac{27}{8}} = \left(\dfrac{27}{8}\right)^{1/3} = \dfrac{27^{1/3}}{8^{1/3}} = \dfrac{3}{2}$ **b.** $\sqrt[5]{-32} = (-32)^5 = -2$
c. $\sqrt{4^3} = 4^{3/2} = (4^{1/2})^3 = 2^3 = 8$

23. a. $125^{2/3} = \left(\sqrt[3]{125}\right)^2 = 5^2 = 25$ **b.** $64^{1/3} = \sqrt[3]{64} = 4$
c. $32^{2/5} = \left(\sqrt[5]{32}\right)^2 = 2^2 = 4$

25. a. $(-64)^{1/3} = \sqrt[3]{-64} = -4$ **b.** $-64^{1/2} = -\sqrt{64} = -8$
c. $(-16)^{1/4}$ is not a real number

27. a. $x^{2/3} = \left(\sqrt[3]{x}\right)^2$ **b.** $x^{1.5} = x^{3/2} = \left(\sqrt{x}\right)^3$ **c.** $a^{0.75} = a^{3/4} = \left(\sqrt[4]{a}\right)^3$

29. a. $\sqrt{9n} = 3\sqrt{n};\ n \ge 0$ **b.** $\sqrt{90n^2} = \sqrt{9n^2}\sqrt{10} = 3|n|\sqrt{10}$

c. $\sqrt{0.09x} = 0.3\sqrt{x};\ x \ge 0$ **d.** $\sqrt{900x^4} = 30x^2$

31. a. $\sqrt{z}\sqrt{z^5} = \sqrt{z^6} = z^3$ **b.** $\sqrt[3]{x^2}\cdot\sqrt[3]{x} = \sqrt[3]{x^3} = x$

c. $\sqrt[3]{x^5}\cdot\sqrt[3]{x^1} = \sqrt[3]{x^6} = x^2$

33. a. $\sqrt[3]{x}\cdot\sqrt[2]{x} = x^{1/3}x^{1/2} = x^{1/3+1/2} = x^{2/6+3/6} = x^{5/6}$

b. $\sqrt{x^3}\cdot\sqrt{x^2} = \sqrt{x^5} = x^{5/2}$ **c.** $\sqrt[2]{\sqrt[3]{x}} = \sqrt[2\cdot3]{x} = \sqrt[6]{x} = x^{1/6}$

35. a. $\dfrac{\sqrt{x}}{\sqrt{x^3}} = \sqrt{\dfrac{x}{x^3}} = \sqrt{\dfrac{1}{x^2}} = \dfrac{1}{x}$ **b.** $\dfrac{\sqrt[3]{x^5}}{\sqrt[3]{x^2}} = \sqrt[3]{\dfrac{x^5}{x^2}} = \sqrt[3]{x^3} = x$

c. $\dfrac{\sqrt[3]{x}}{\sqrt[4]{x}} = \dfrac{x^{1/3}}{x^{1/4}} = x^{1/3-1/4} = x^{4/12-3/12} = x^{1/12}$

37. a. $\sqrt{3} + 2\sqrt{12} = \sqrt{3} + 2\sqrt{4}\sqrt{3} = \sqrt{3} + 4\sqrt{3} = 5\sqrt{3}$

b. $\sqrt{9x} + \sqrt{x} = 3\sqrt{x} + \sqrt{x} = 4\sqrt{x},\ x \ge 0$

39. a. $(3 + \sqrt{6})(3 - \sqrt{6}) = 9 - 6 = 3$ **b.** $(3 - \sqrt{x})(3 - \sqrt{x}) = 9 - 6\sqrt{x} + \text{x}$

41. a. $\dfrac{1}{b-\sqrt{a}}\cdot\dfrac{b+\sqrt{a}}{b+\sqrt{a}} = \dfrac{b+\sqrt{a}}{b^2-a}$ **b.** $\dfrac{\sqrt{x}-3}{\sqrt{x}+4}\cdot\dfrac{\sqrt{x}-4}{\sqrt{x}-4} = \dfrac{x-7\sqrt{x}+12}{x-16}$

c. $\dfrac{1}{\sqrt{a}+\sqrt{b}}\cdot\dfrac{\sqrt{a}-\sqrt{b}}{\sqrt{a}-\sqrt{b}} = \dfrac{\sqrt{a}-\sqrt{b}}{a-b}$ **d.** $\dfrac{a}{\sqrt[4]{a^2b}}\cdot\dfrac{\sqrt[4]{a^2b^3}}{\sqrt[4]{a^2b^3}} = \dfrac{a\sqrt[4]{a^2b^3}}{ab} = \dfrac{\sqrt[4]{a^2b^3}}{b}$

e. $\dfrac{1}{\sqrt[3]{9x}}\cdot\dfrac{\sqrt[3]{3x^2}}{\sqrt[3]{3x^2}} = \dfrac{\sqrt[3]{3x^2}}{\sqrt[3]{27x^3}} = \dfrac{\sqrt[3]{3x^2}}{3x}$ **f.** $\dfrac{3x}{\sqrt[5]{x^3}}\cdot\dfrac{\sqrt[5]{x^2}}{\sqrt[5]{x^2}} = \dfrac{3x\sqrt[5]{x^2}}{\sqrt[5]{x^5}} = \dfrac{3x\sqrt[5]{x^2}}{x} = 3\sqrt[5]{x^2}$

43. $(1-\sqrt{2})^2 - 2(1-\sqrt{2}) - 1 = 1 - 2\sqrt{2} + 2 - 2 + 2\sqrt{2} - 1 = 0$

To find the other solution use quadratic formula:

$$x = \frac{-(-2) \pm \sqrt{(-2)^2 - 4(1)(-1)}}{2(1)} = \frac{2 \pm \sqrt{8}}{2} = \frac{2 \pm 2\sqrt{2}}{2} = 1 \pm \sqrt{2}$$

The other solution is $1 + \sqrt{2}$

45. $2\left(\frac{1+\sqrt{7}}{2}\right)^2 - 2\left(\frac{1+\sqrt{7}}{2}\right) - 3 = 2\left(\frac{1+2\sqrt{7}+7}{4}\right) - 1 - \sqrt{7} - 3$

$= \frac{8+2\sqrt{7}}{2} - 4 - \sqrt{7} = 4 + \sqrt{7} - 4 - \sqrt{7} = 0$

$2\left(\frac{1-\sqrt{7}}{2}\right)^2 - 2\left(\frac{1-\sqrt{7}}{2}\right) - 3 = 2\left(\frac{1-2\sqrt{7}+7}{4}\right) - 1 + \sqrt{7} - 3$

$= \frac{8-2\sqrt{7}}{2} - 4 + \sqrt{7} = 4 - \sqrt{7} - 4 + \sqrt{7} = 0$

47. a. (3, 3), (5, 4), (7, 5); yes it is a function

b. (–1, –2), (–2, 0), (–3, 2); yes it is a function

49. a. $x = 2y - 1$; $x + 1 = 2y$; $y = \frac{x}{2} + \frac{1}{2}$

b. $x = 3y$; $y = \frac{x}{3}$

c. $x = y^3$; $y = \sqrt[3]{x}$

d. $x = \sqrt{y}$; $y = x^2$, $x \geq 0$, $y \geq 0$

e. $x = -y$; $y = -x$

f. $x = \frac{1}{y}$; $y = \frac{1}{x}$; $x \neq 0$, $y \neq 0$

51. a. $x \geq 0$, $y \geq 1$ for function to have an inverse;

$x = \sqrt{y} + 1$; $x - 1 = \sqrt{y}$; $y = (x-1)^2$; $y \geq 0$; $x \geq 1$

b. $x \geq 1$, $y \geq 0$ for function to have an inverse;

$x = \sqrt{y-1}$; $x^2 = y - 1$; $y = x^2 + 1$; $x \geq 0$, $y \geq 1$

53. $8.00 = 5.50(1 + r)^{10}$

$\frac{8.00}{5.50} = (1 + r)^{10}$

$\left(\frac{8.00}{5.50}\right)^{1/10} - 1 = r$

$r \approx 3.82\%$

55. $17{,}000 = 7000(1 + r)^{22}$

$\frac{17{,}000}{7{,}000} = (1 + r)^{22}$

$\left(\frac{17}{7}\right)^{1/22} - 1 = r$

$r \approx 7.67\%$

57. $r = \sqrt[3]{\frac{3V}{4\pi}}$; $r^3 = \frac{3V}{4\pi}$; $V = \frac{4\pi r^3}{3}$

59. a. $x^3 + y^3 = a$

$y^3 = a - x^3$

$y = \sqrt[3]{a - x^3}$

b. $x^2 = \dfrac{8}{27k}(y-k)^3$

$\dfrac{27kx^2}{8} = (y-k)^3$

$y - k = \sqrt[3]{\dfrac{27kx^2}{8}}$

$y = \dfrac{3}{2}\sqrt[3]{kx^2} + k$

61. $F = k\dfrac{M_1M_2}{r^2}$

$M_1 = \dfrac{Fr^2}{kM_2}$

63. $x^4 = a^2(x^2 - y^2)$

$\dfrac{x^4}{a^2} = x^2 - y^2$

$y^2 = x^2 - \dfrac{x^4}{a^2} = x^2(1 - \dfrac{x^2}{a^2})$

$y = \pm\sqrt{x^2\dfrac{(a^2 - x^2)}{a^2}} = \pm\dfrac{x}{a}\sqrt{a^2 - x^2}$

65. a. From the graph x = 12

b. $\sqrt{6x-8} = 4;\ 6x - 8 = 16;\ 6x = 24;\ x = 4$

c. $\sqrt{3x} + 2 = 2;\ \sqrt{3x} = 0;\ x = 0$

d. $\sqrt{3x} + 2 = \sqrt{6x-8};\ \left(\sqrt{3x} + 2\right)^2 = \left(\sqrt{6x-8}\right)^2;\ 3x + 4\sqrt{3x} + 4 = 6x - 8$

$4\sqrt{3x} = 3x - 12;\ \left(4\sqrt{3x}\right)^2 = (3x-12)^2;\ 16(3x) = 9x^2 - 72x + 144;\ 48x = 9x^2 - 72x + 144$

$9x^2 - 120x + 144 = 0;\ (9x-12)(x-12) = 0;\ x = 12;\ x = \frac{4}{3}$ is an extraneous solution

67. a. $\sqrt[3]{8000} = \sqrt[3]{8\times10^3} = 20$ **b.** $\sqrt[3]{800} = \sqrt[3]{8\times10^2} \approx 9.283$ **c.** $\sqrt[3]{80} = \sqrt[3]{8\times10^1} \approx 4.309$

d. $\sqrt[3]{8} = \sqrt[3]{8\times10^0} = 2$ **e.** $\sqrt[3]{0.8} = \sqrt[3]{8\times10^{-1}} \approx 0.9283$ **d.** $\sqrt[3]{0.08} = \sqrt[3]{8\times10^{-2}} \approx 0.4309$

g. $\sqrt[3]{27000} = \sqrt[3]{27\times10^3} = 30$ **h.** $\sqrt[3]{2700} = \sqrt[3]{27\times10^2} \approx 13.93$ **i.** $\sqrt[3]{270} = \sqrt[3]{27\times10^1} \approx 6.463$

j. $\sqrt[3]{27} = \sqrt[3]{27\times10^0} = 3$ **k.** $\sqrt[3]{2.7} = \sqrt[3]{27\times10^{-1}} \approx 1.393$ **l.** $\sqrt[3]{0.27} = \sqrt[3]{27\times10^{-2}} \approx 0.6463$

Chapter 8 Test

1. a. $\dfrac{a^0b^2}{b^{-1}} = 1\cdot b^{2-(-1)} = b^3$

b. $\dfrac{a^{-2}b^3}{(bc)^2} = \dfrac{b^3}{a^2b^2c^2} = \dfrac{b}{a^2c^2}$

c. $\left(\dfrac{x^2y}{3x}\right)^{-2} = \left(\dfrac{3x}{x^2y}\right)^2 = \dfrac{9x^2}{x^4y^2} = \dfrac{9}{x^2y^2}$

2. a. $(x^{-1}y^3)^2 = x^{-2}y^6 = \dfrac{y^6}{x^2}$

b. $a^6 \cdot a^3 = a^{6+3} = a^9$

c. $\dfrac{x}{x^n} = x^{1-n}$

3. a. $\dfrac{b^{3/4}}{b^2} = \dfrac{1}{b^{2-0.75}} = \dfrac{1}{b^{1.25}}$

b. $\left(\dfrac{a^{1/3}}{b^{1/2}}\right)^6 = \dfrac{a^{6/3}}{b^{6/2}} = \dfrac{a^2}{b^3}$

c. $\left(x^{3/4}y^{2/3}\right)^6 = x^{18/4}y^{12/3} = x^{4.5}y^4$

4. a. $(b^2)^{3/2} = b^{6/2} = b^3$

b. $b^{2/3}b^{1/3} = b^{3/3} = b$

c. $3 \cdot 3^n = 3^{1+n}$

5. $(2x)^2 = 2^2x^2 = 4x^2$; in $2x^2$ the 2 is not part of the base

6. $a\left(\dfrac{b}{a}\right)^4 = a\dfrac{b^4}{a^4} = \dfrac{b^4}{a^3}$; $b\left(\dfrac{b}{a}\right)^3 = b\dfrac{b^3}{a^3} = \dfrac{b^4}{a^3}$; the expressions are equal

7. a. $3.450 \times 10^{-5} = 0.000\ 034\ 50$

b. $\dfrac{6.3\times10^{-2}}{7\times10^4} = 0.9\times10^{-2-4} = 0.9\times10^{-6} = 9\times10^{-7}$ or 0.000 000 9

8. $\sqrt{4x^2} = \sqrt{4}\sqrt{x^2} = 2|x|$; x can be any real number while square root remains positive.

in $\sqrt{4x^2} = 2x$; x must be ≥ 0 to satisfy principal square root being positive.

9. a. $\sqrt{36x} = 6\sqrt{x};\ x \geq 0$

b. $\sqrt{0.36x^2} = 0.6|x|$

c. $\sqrt{360x^4} = \sqrt{36x^4}\sqrt{10} = 6x^2\sqrt{10}$

10. a. $64^{2/3} = \left(\sqrt[3]{64}\right)^2 = 4^2 = 16$ **b.** $32^{4/5} = \left(\sqrt[5]{32}\right)^4 = 2^4 = 16$

c. $81^{3/4} = \left(\sqrt[4]{81}\right)^3 = 3^3 = 27$

11. a. $\sqrt[4]{\frac{16}{81}} = \frac{\sqrt[4]{16}}{\sqrt[4]{81}} = \frac{2}{3}$ **b.** $\sqrt[3]{-27} = -3$

c. $-\sqrt[3]{-64} = -(-4) = 4$

12. a. $\sqrt[3]{\sqrt[5]{x^{10}}} = \sqrt[15]{x^{10}} = x^{10/15} = x^{2/3}$ **b.** $\sqrt[3]{x^2}\sqrt[3]{x^4} = \sqrt[3]{x^{2+4}} = \sqrt[3]{x^6} = x^{6/3} = x^2$

c. $\frac{\sqrt[3]{54x^2}}{\sqrt[3]{128x^{-1}}} = \sqrt[3]{\frac{54x^{2-(-1)}}{128}} = \sqrt[3]{\frac{27x^3}{64}} = \frac{\sqrt[3]{27x^3}}{\sqrt[3]{64}} = \frac{3x}{4}$

13. a. $(2 - \sqrt{x})(2 + \sqrt{x}) = 4 - x$

b. $(8 - \sqrt{3})(3 + \sqrt{8}) = 24 + 8\sqrt{8} - 3\sqrt{3} - \sqrt{3}\sqrt{8} = 24 + 16\sqrt{2} - 3\sqrt{3} - 2\sqrt{6}$

14. a. $\frac{1}{x+\sqrt{y}} \cdot \frac{x-\sqrt{y}}{x-\sqrt{y}} = \frac{x-\sqrt{y}}{x^2-y}$ **b.** $\frac{1}{\sqrt[3]{x}} \cdot \frac{\sqrt[3]{x^2}}{\sqrt[3]{x^2}} = \frac{\sqrt[3]{x^2}}{\sqrt[3]{x^3}} = \frac{\sqrt[3]{x^2}}{x}$

15. $(4-\sqrt{10})^2 - 8(4-\sqrt{10}) + 6 = 16 - 8\sqrt{10} + 10 - 32 + 8\sqrt{10} + 6 = 0$

16. $2000\left(1 + \frac{0.08}{4}\right)^{4\cdot 3} = \$2536.48$

17. $18{,}000 = 12{,}000(1 + \text{r})^5$

$\left(\frac{18{,}000}{12{,}000}\right) = (1+r)^5$

$\left(\frac{18{,}000}{12{,}000}\right)^{1/5} = 1 + r$

$r = \left(\frac{18{,}000}{12{,}000}\right)^{1/5} - 1 \approx 8.45\%$

18. $M = \frac{mgl}{\pi r^2 s}$

$r^2 = \frac{mgl}{M\pi s}$

$r = \sqrt{\frac{mgl}{M\pi s}}$

19. $d = \frac{Wl^3}{3EI}$; $l^3 = \frac{3EId}{W}$; $l = \sqrt[3]{\frac{3EId}{W}}$

20. a. $(-1,-1), (-3,-2), (-5,-3)$; yes it is a function

b. $(-1,-1), (-1,-2), (-1,-3)$; no, it is not a function (same input has several outputs)

21.

x	$y = (\frac{1}{2})^x$
0	1
1	0.5
2	0.25
3	0.125

y	$x = (\frac{1}{2})^y$
0	1
1	0.5
2	0.25
3	0.125

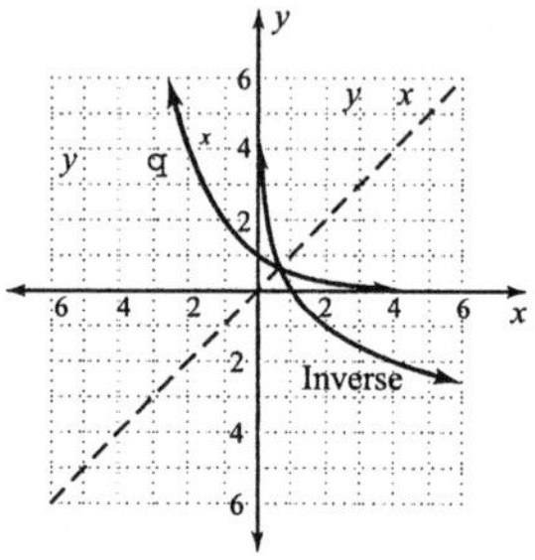

22. $3y - 2x = 6$; or $y = \frac{2}{3}x + 2$

23. a. from the graph $x = 8$

b. $\sqrt{x+1} = 2;\ \left(\sqrt{x+1}\right)^2 = 2^2;\ x+1 = 4;\ x = 3$

c. $\sqrt{2x} - 1 = 1; \sqrt{2x} = 2;\ \left(\sqrt{2x}\right)^2 = 2^2;\ 2x = 4;\ x = 2$

d. $\sqrt{x+1} = \sqrt{2x} - 1;\ \left(\sqrt{x+1}\right)^2 = \left(\sqrt{2x} - 1\right)^2;\ x+1 = 2x - 2\sqrt{2x} + 1;\ 2\sqrt{2x} = x$

$\left(2\sqrt{2x}\right)^2 = x^2;\ 4(2x) = x^2;\ x^2 - 8x = 0;\ x(x-8) = 0;\ x = 8;\ x = 0$ is an extraneous solution

Cumulative Review Chapters 1 to 8

1. $5 - 3(x + 2) = 32$

$5 - 3x - 6 = 32$

$-3x = 33$

$x = -11$

3. $A = \dfrac{a+b+c}{3}$

$3A = a + b + c$

$b = 3A - a - c$

5. Let y = total cost in \$; x = kilowatt hours used; assume a linear equation.

$m = \dfrac{83.33 - 68.46}{1410 - 1120} \approx 0.051$; $83.33 = 0.051(1410) + b$; $b = 11.42$; $y \approx 0.051x + 11.42$

If calculator linear regression is used equation is: $y \approx 0.051x + 11.03$, round off error

7. a. $\dfrac{-2-1}{4-(-1)} = -\dfrac{3}{5}$ **b.** negative reciprocal; $\dfrac{5}{3}$

9. inputs, $\Re$; outputs, 4

11. $50(2.72) + x(4.00) = (50 + x)(3.00)$

$136 + 4x = 150 + 3x$

$x = 14$ credits of A

13. $3.5^2 + 12^2 = 12.25 + 144 = 156.25 = 12.5^2$, right triangle

15. a. $\{-4, 1\}$ **b.** $\{-3, 0\}$ **c.** no real number solution

17. $x < -3$ or $x > 0$

19.

Multiply	x^3	$-3x^2$	$+3x$	-1
x	x^4	$-3x^3$	$+3x^2$	$-x$
−1	$-x^3$	$+3x^2$	$-3x$	$+1$

$(x^3 - 3x^2 + 3x - 1)(x - 1) = x^4 - 4x^3 + 6x^2 - 4x + 1$

21. a. $(8 - 2i)(8 + 2i) = 8^2 - (2i)^2 = 64 - 4(-1) = 64 + 4 = 68$

b. $(5 - 4i)(3 + 2i) = 15 - 2i - 8i^2 = 15 - 2i + 8 = 23 - 2i$

c. $\dfrac{1}{2+i} \cdot \dfrac{2-i}{2-i} = \dfrac{2-i}{4-(-1)} = \dfrac{2-i}{5}$

23. Shift $y = x^2$ *a* units horizontally.

25. $\dfrac{\$8.99}{package} \cdot \dfrac{1\ package}{56\ diapers} \cdot \dfrac{10\ diapers}{1\ day} \cdot 365\ days \approx \$585.96$

27. $\dfrac{a+b}{ab} = \dfrac{a}{ab} + \dfrac{b}{ab} = \dfrac{1}{b} + \dfrac{1}{a}$, **e**

29. $\dfrac{a}{1-a} = \dfrac{a}{-(a-1)} = \dfrac{-a}{a-1}$, **f**

31. $\dfrac{x-y}{y-x} = \dfrac{x}{y-x} - \dfrac{y}{y-x} = \dfrac{x}{y-x} - \dfrac{y}{-(x-y)} = \dfrac{x}{y-x} + \dfrac{y}{x-y}$; **d**

33.

$$\begin{array}{r} x^2 + x + 1 \\ x-1\overline{)x^3 + 0x^2 + 0x - 1} \\ \underline{-(x^3 - x^2)} \\ x^2 + 0x \\ \underline{-(x^2 - x)} \\ x - 1 \\ \underline{-(x-1)} \\ 0 \end{array}$$

35. a. $3.80 \times 10^{-16} = 0.000\ 000\ 000\ 000\ 000\ 380$

b. $4.23 \times 10^{15} = 4{,}230{,}000{,}000{,}000{,}000$

37. a. $32^{4/5} = (32^{1/5})^4 = 2^4 = 16$

b. $\sqrt[3]{125^2} = 125^{2/3} = (125^{1/3})^2 = 5^2 = 25$

c. $\sqrt[4]{-81}$, no real number solution

d. $-\sqrt{2}\sqrt{8} = -\sqrt{16} = -4$

e. $x^{3/5} \cdot x^{1/3} = x^{3/5 + 1/3} = x^{9/15 + 5/15} = x^{14/15}$

f. $\dfrac{x^{1/2}}{x^{-3/2}} = x^{1/2-(-3/2)} = x^{4/2} = x^2$

g. $\sqrt{18a^3} = \sqrt{9a^2}\sqrt{2a} = 3a\sqrt{2a}$

h. $\sqrt[3]{16x^3y^4z^5} = \sqrt[3]{8x^3y^3z^3}\sqrt[3]{2yz^2} = 2xyz\sqrt[3]{2yz^2}$

i. $(x^{2/3})^{1/2} = x^{2/6} = x^{1/3}$

j. $\sqrt{8x} + \sqrt{2x} + \sqrt{50x} = 2\sqrt{2x} + \sqrt{2x} + 5\sqrt{2x} = 8\sqrt{2x}$

k. $(2-\sqrt{2})(2+\sqrt{2}) = 4 - 2 = 2$

39.

$2500 = 1500(1 + r)^6$

$\left(\frac{2500}{1500}\right) = (1 + r)^6$

$\left(\frac{2500}{1500}\right)^{1/6} = 1 + r$

$\left(\frac{2500}{1500}\right)^{1/6} - 1 = r$

$r \approx 8.89\%$

$2500 = 1500\,(1 + \tfrac{r}{12})^{12 \cdot 6}$

$\left(\frac{2500}{1500}\right) = (1 + \tfrac{r}{12})^{72}$

$\left(\frac{2500}{1500}\right)^{1/72} = 1 + \tfrac{r}{12}$

$\left(\frac{2500}{1500}\right)^{1/72} - 1 = \tfrac{r}{12}$

$12\left[\left(\frac{2500}{1500}\right)^{1/72} - 1\right] = r$

$r \approx 8.54\%$

41.

$(-2, 2)$ & $(4, 1)$ $d = \sqrt{(4-(-2))^2 + (1-2)^2} = \sqrt{36+1} = \sqrt{37}$

$(-2, 2)$ & $(3, -5)$ $d = \sqrt{(3-(-2))^2 + (-5-2)^2} = \sqrt{25+49} = \sqrt{74}$

$(4, 1)$ & $(3, -5)$ $d = \sqrt{(3-4)^2 + (-5-1)^2} = \sqrt{1+36} = \sqrt{37}$

$\left(\sqrt{37}\right)^2 + \left(\sqrt{37}\right)^2 = \left(\sqrt{74}\right)^2$; figure is an isosceles right triangle

Exercises 9.1

1. a. a^x; base is a, exponent is x **b.** $4x^3$; base is x, exponent is 3
c. 2^{-x}; base is 2, exponent is $-x$ **d.** x^a; base is x, exponent is a

3. a. π^x; base is π, exponent is x **b.** $2x^4$; base is x, exponent is 4
c. a_1r^{n-1}; base is r, exponent is $n - 1$ **d.** $100(1.06)^t$; base is 1.06, exponent is t

5. a. $13.5 \div 4.5 = 3$; $4.5 \div 1.5 = 3$; $1.5 \div 0.5 = 3$; common ratio is 3
b. $a_1 = 0.5, r = 3; a_n = 0.5 \cdot 3^{n-1}$ **c.** $y = 0.167 \cdot 3^x$
d. $0.5 \cdot 3^{n-1} = 0.5 \cdot 3^n \cdot 3^{-1} = \frac{0.5}{3} \cdot 3^n = 0.167 \cdot 3^n$

7. a. $80 \div 40 = 2$; $40 \div 20 = 2$; $20 \div 10 = 2$; common ratio is 2
b. $a_1 = 10, r = 2; a_n = 10 \cdot 2^{n-1}$ **c.** $y = 5 \cdot 2^x$
d. $10 \cdot 2^{n-1} = 10 \cdot 2^n \cdot 2^{-1} = \frac{10}{2} \cdot 2^n = 5 \cdot 2^n$

9. a. $4 \div 2 = 2$; $2 \div 1 = 2$, $1 \div \frac{1}{2} = 2$; $\frac{1}{2} \div \frac{1}{4} = 2$; common ratio is 2
b. $a_1 = \frac{1}{4}, r = 2; a_n = \frac{1}{4} \cdot 2^{n-1}$ **c.** $y = 0.125 \cdot 2^x$
d. $\frac{1}{4} \cdot 2^{n-1} = \frac{1}{4} \cdot 2^n \cdot 2^{-1} = \frac{1}{4} \div 2 \cdot 2^n = \frac{1}{8} \cdot 2^n = 0.125 \cdot 2^n$

11. a. $16 \div 8 = 2$; $8 \div 4 = 2$; $4 \div 2 = 2$; $2 \div 1 = 2$; common ratio is 2
b. $a_1 = 1, r = 2; a_n = 1 \cdot 2^{n-1}$ **c.** $y = 0.5 \cdot 2^x$
d. $1 \cdot 2^{n-1} = 1 \cdot 2^n \cdot 2^{-1} = 1 \cdot \frac{1}{2} \cdot 2^n = 0.5 \cdot 2^n$

13. a. $2 \div 4 = \frac{1}{2}$; $4 \div 8 = \frac{1}{2}$; $8 \div 16 = \frac{1}{2}$; $16 \div 32 = \frac{1}{2}$; common ratio is $\frac{1}{2}$
b. $a_1 = 32, r = \frac{1}{2}; a_n = 32 \cdot (\frac{1}{2})^{n-1}$ **c.** $y = 64 \cdot 0.5^x$
d. $32 \cdot (\frac{1}{2})^{n-1} = 32 \cdot (\frac{1}{2})^n \cdot (\frac{1}{2})^{-1} = 32 \cdot 2 \cdot (\frac{1}{2})^n = 64 \cdot (\frac{1}{2})^n = 64 \cdot 0.5^n$

15. a. $\frac{1}{27} \div \frac{1}{9} = \frac{1}{27} \cdot \frac{9}{1} = \frac{1}{3}$; $\frac{1}{9} \div \frac{1}{3} = \frac{1}{3}$; $\frac{1}{3} \div 1 = \frac{1}{3}$; $1 \div 3 = \frac{1}{3}$; common ratio is $\frac{1}{3}$
b. $a_1 = 3, r = \frac{1}{3}; a_n = 3 \cdot (\frac{1}{3})^{n-1}$ **c.** $y = 9 \cdot 0.333^x$
d. $3 \cdot (\frac{1}{3})^{n-1} = 3 \cdot (\frac{1}{3})^n \cdot (\frac{1}{3})^{-1} = 3 \cdot 3 \cdot (\frac{1}{3})^n = 9 \cdot (\frac{1}{3})^n = 9 \cdot 0.333^n$

17. no common ratio; 1st differences = 6, 10, 14; 2nd differences = 4, 4; quadratic function
$2a = 4, a = 2$; $c = 2 - (6 - 4) = 0$; $2(1)^2 + b(1) = 2, b = 0$; $y = 2x^2$

19. common ratio is 3; exponential function; $a_n = 0.5 \cdot 3^{n-1}$; $y = 0.167 \cdot 3^x$

21. no common ratio; 1st differences = 4, 5, 6; 2nd differences = 1, 1; quadratic function
$2a = 1$, $a = 0.5$; $c = 3 - (4 - 1) = 0$; $0.5(1)^2 + b(1) = 3$, $b = 2.5$; $y = 0.5x^2 + 2.5x$

23. no common ration; 1st differences = 8, 8, 8, 8; linear function
$m = 8$; $30 = 8(1) + b$, $b = 22$; $y = 8x + 22$

25. common ratio is 2; exponential function; $a_n = 5 \cdot 2^{n-1}$; $y = 2.5 \cdot 2^x$

27. $2^{x+2} = 2^x \cdot 2^2 = 2^x \cdot 4 = 4 \cdot 2^x$

29. $2^{x-1} = 2^x \cdot 2^{-1} = 2^x \cdot \frac{1}{2} = \frac{2^x}{2}$

31. $(\frac{1}{2})^{-x} = (\frac{2}{1})^x = 2^x$

33. $\frac{2^n}{2} = 2^n \cdot 2^{-1} = 2^{n-1}$

35. $4 \cdot 2^n = 2^2 \cdot 2^n = 2^{n+2}$

37. $\frac{3^n}{3} = 3^n \cdot 3^{-1} = 3^{n-1}$

39. $9 \cdot 3^n = 3^2 \cdot 3^n = 3^{n-1}$

41. a. common ratio is 2; $a_n = \frac{1}{8} \cdot 2^{n-1}$; $\frac{1}{8} = 2^{-3}$; $f(x) = 2^{-3} \cdot 2^{x-1} = 2^{x-4}$

b. common ratio is 2; $a_n = 4 \cdot 2^{n-1}$; $4 = 2^2$; $f(x) = 2^2 \cdot 2^{x-1} = 2^{x+1}$

43. a. common ratio is $\frac{1}{2}$; $a_n = 4 \cdot (\frac{1}{2})^{n-1}$; $4 = (\frac{1}{2})^{-2}$; $f(x) = (\frac{1}{2})^{-2} \cdot (\frac{1}{2})^{x-1} = (\frac{1}{2})^{x-3}$

b. common ratio is $\frac{1}{2}$; $a_n = 8 \cdot (\frac{1}{2})^{n-1}$; $8 = (\frac{1}{2})^{-3}$; $f(x) = (\frac{1}{2})^{-3} \cdot (\frac{1}{2})^{x-1} = (\frac{1}{2})^{x-4}$

45. $-6 \div 2 = -3$; $18 \div -6 = -3$; $r = -3$; $a_n = 2 \cdot (-3)^{n-1}$; Base is negative, not an exponential function.

47. $-32 \div 64 = -\frac{1}{2}$; $16 \div -32 = -\frac{1}{2}$; $r = -\frac{1}{2}$; $a_n = 64 \cdot (-\frac{1}{2})^{n-1}$; base is negative, not an exponential function

49. a. $y \approx 24{,}778(1.049)^x$; rate of growth = 4.9%; $24{,}778(1.049)^{70} \approx 705{,}000$; actual: 448,607

b. $y \approx 6657(1.043)^x$; rate of growth = 4.3%; $6657(1.043)^{70} \approx 127{,}000$; actual: 73,344

c. $y \approx 42{,}738(0.998)^x$; rate of decline = −0.2%; $42{,}738(0.998)^{70} \approx 37{,}000$; actual: 40,407

d. $y \approx 38{,}587(1.005)^x$; rate of growth = 0.5%; $38{,}587(1.005)^{70} \approx 55{,}000$; actual: 62,916

51. a. common ratio = $\frac{1}{2}$; $a_n = 108(\frac{1}{2})^{n-1}$; $y = 216(\frac{1}{2})^x$; $216(\frac{1}{2})^{10} \approx 0.211$ in.
changing from $\frac{2}{3}$ to $\frac{1}{2}$ changes the base in the equation
b. common ratio is $\frac{2}{3}$; $a_n = 135(\frac{2}{3})^{n-1}$; $y = 202.5(\frac{2}{3})^x$

Exercises 9.2

1. a. $b > 1$, increasing **b.** $b < 1$, decreasing

3. a. $b > 1$, increasing **b.** $b < 1$, decreasing

5. a. $b > 1$, increasing **b.** $b > 1$, increasing

7. Use $x = 1$ to identify graphs; $a = 10^x$; $b = 2.72^x$; $c = 2^x$

9. Graph will never be below the x-axis. { } or ∅

11. $3 \cdot 2^0 = 3 \cdot 1 = 3$; $4 \cdot 2^0 = 4 \cdot 1 = 4$; $\frac{1}{2} \cdot 2^0 = \frac{1}{2} \cdot 1 = \frac{1}{2}$; y-intercept of $y = a \cdot 2^x$ is a.

13. a is the y-intercept value

15. Let $x = 0$ and simplify; $y = 2^{0+h}$, $y = 2^h$

17. When $x = 1$, $b^x = b$.

19. One possible answer is $y = \dfrac{1}{x^2}$

21. a. y-intercept is 16; $y = 2^4 \cdot 2^x$; $y = 2^{x+4}$
b. y-intercept is 8; $y = 2^3 \cdot 2^x$; $y = 2^{x+3}$

23. a. y-intercept is 27; $y = 3^3 \cdot 3^x$; $y = 3^{x+3}$
b. y-intercept is $\frac{1}{27}$; $y = 3^{-3} \cdot 3^x$; $y = 3^{x-3}$

25. a. $y = 2^x \cdot 2^{-4}$; $y = \frac{1}{16} \cdot 2^x$; y-intercept is $\frac{1}{16}$
b. $y = 2^x \cdot 2^1$; $y = 2 \cdot 2^x$; y-intercept is 2

27.

x	$y = 2^{x-1}$	$y = 2^x$	$y = 2^{x+1}$
−2	$2^{-2-1} = 2^{-3} = \frac{1}{8}$	$2^{-2} = \frac{1}{4}$	$2^{-2+1} = 2^{-1} = \frac{1}{2}$
−1	$2^{-1-1} = 2^{-2} = \frac{1}{4}$	$2^{-1} = \frac{1}{2}$	$2^{-1+1} = 2^0 = 1$
0	$2^{0-1} = 2^{-1} = \frac{1}{2}$	$2^0 = 1$	$2^{0+1} = 2^1 = 2$
1	$2^{1-1} = 2^0 = 1$	$2^1 = 2$	$2^{1+1} = 2^2 = 4$
2	$2^{2-1} = 2^1 = 2$	$2^2 = 4$	$2^{2+1} = 2^3 = 8$
graph	c	b	a

Adding 1 shift the graph 1 unit to the left, subtracting 1 shifts graph 1 unit to the right

29.

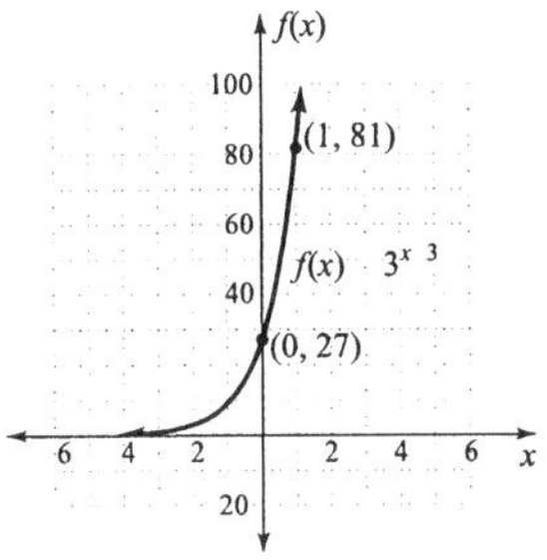

31.

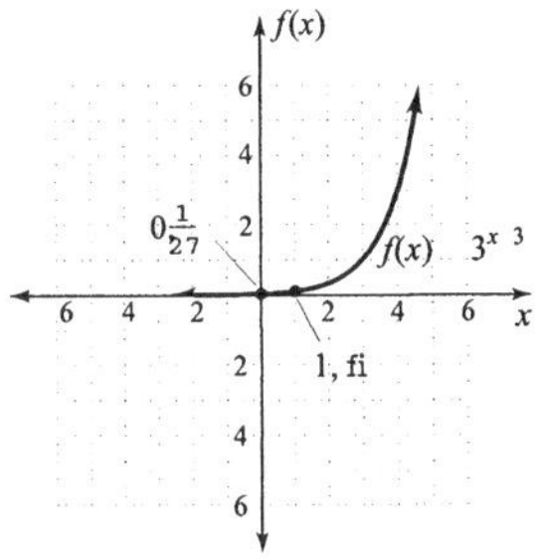

33. They do not intersect. $3^x < 3^{x+1}$ for all x.

35. a. $2^x = 256$
$2^x = 2^8$
$x = 8$

b. $16^x = 256$
$(2^4)^x = 2^8$
$2^{4x} = 2^8$
$4x = 8$
$x = 2$

37. a. $4^n = 8$
$(2^2)^n = 2^3$
$2^{2n} = 2^3$
$2n = 3$
$n = \frac{3}{2}$

b. $8^n = 16$
$(2^3)^n = 2^4$
$2^{3n} = 2^4$
$3n = 4$
$n = \frac{4}{3}$

39. a. $16^n = 4$

$(2^4)^n = 2^2$

$2^{4n} = 2^2$

$4n = 2$

$n = \frac{1}{2}$

b. $100^x = 1000$

$(10^2)^x = 10^3$

$10^{2x} = 10^3$

$2x = 3$

$x = \frac{3}{2}$

41. a. $64^x = \frac{1}{8}$

$(2^6)^x = 2^{-3}$

$2^{6x} = 2^{-3}$

$6x = -3$

$x = -\frac{1}{2}$

b. $125^x = \frac{1}{5}$

$(5^3)^x = 5^{-1}$

$5^{3x} = 5^{-1}$

$3x = -1$

$x = -\frac{1}{3}$

43. a. $8^n = \frac{1}{2}$

$(2^3)^n = 2^{-1}$

$2^{3n} = 2^{-1}$

$3n = -1$

$n = -\frac{1}{3}$

b. $81^x = \frac{1}{3}$

$(3^4)^x = 3^{-1}$

$3^{4x} = 3^{-1}$

$4x = -1$

$x = -\frac{1}{4}$

45. a. $(\frac{1}{2})^{x+5} = 32$

$(2^{-1})^{x+5} = 2^5$

$2^{-x-5} = 2^5$

$-x - 5 = 5$

$-x = 10$

$x = -10$

b. $(\frac{1}{8})^{x+3} = 8$

$(2^{-3})^{x+3} = 2^3$

$2^{-3x-9} = 2^3$

$-3x - 9 = 3$

$-3x = 12$

$x = -4$

47. a. $2^{x-1} = 2^{2x+5}$

$x - 1 = 2x + 5$

$-6 = x$

b. $3^{2x-1} = 3^{x+3}$

$2x - 1 = x + 3$

$x = 4$

49. a. $2^{x+5} = 4^{x-1}$

$2^{x+5} = (2^2)^{x-1}$

$2^{x+5} = 2^{2x-2}$

$x + 5 = 2x - 2$

$7 = x$

b. $5^{x+1} = 125^{x-3}$

$5^{x+1} = (5^3)^{x-3}$

$5^{x+1} = 5^{3x-9}$

$x + 1 = 3x - 9$

$10 = 2x$

$x = 5$

51. No; in the example $x^2 = 3^2$, x could be -3, which is a different base.

53. $4^{6t} = (2^2)^{6t} = 2^{12t}$; $8^{4t} = (2^3)^{4t} = 2^{12t}$ rules are the same

55. a. $x \approx 2.5$ **b.** $x \approx 4.8$ **c.** $y \approx 1.8$
d. $y \approx 4.5$ **e.** $a \approx 1.34$

57. a. $y = 1$ **b.** graph on calculator **c.** horizontal line at $y = 1$
d. yes **e.** $y = 1$

59. a. graph on calculator **b.** points only **c.** no
d. no, negative base

61. a. y-intercepts $= -1$ **b.** $y = -2^x$ and $y = -2^{-x}$ or $y = -(\frac{1}{2})^x$
c. A ***negative*** coefficient causes the exponential graph to reflect over the x-axis.
d. $y = -x^2$

Exercises 9.3

1. a.

x	$f(x) = 3^x$
-2	$3^{-2} = \frac{1}{9}$
-1	$3^{-1} = \frac{1}{3}$
0	$3^0 = 1$
1	$3^1 = 3$
2	$3^2 = 9$
3	$3^3 = 27$
4	$3^4 = 81$

b.

x	y
$\frac{1}{9}$	-2
$\frac{1}{3}$	-1
1	0
3	1
9	2
27	3
81	4

c.

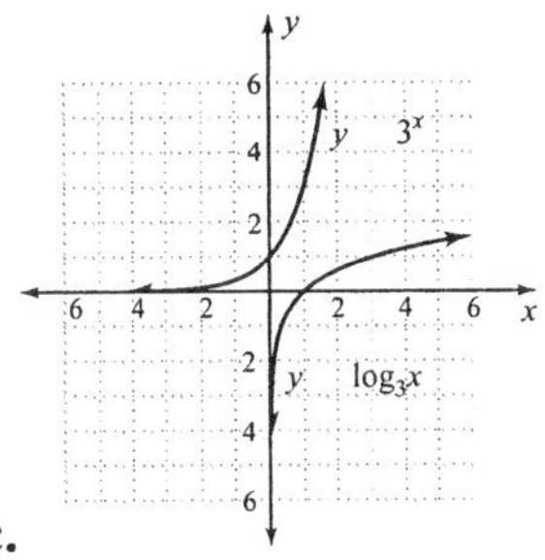

inverse is a function

d. $y = 3^x$; $x = 3^y$, $y = \log_3 x$

3.

Exponential Equation	Logarithmic Equation
$3^3 = 27$	$\log_3 27 = 3$
$2^3 = 8$	$\log_2 8 = 3$
$10^1 = 10$	$\log_{10} 10 = 1$
$5^3 = 125$	$\log_5 125 = 3$
$3^{-2} = \frac{1}{9}$	$\log_3\left(\frac{1}{9}\right) = -2$

5. **a.** base = 10; exponent = 3; $10^3 = 1000$ **b.** base = 10, exponent = 0; $10^0 = 1$
c. base = 3; exponent = 4; $3^4 = 81$ **d.** base = 10; exponent = 2; $10^2 = 100$

7. **a.** base = 10; exponent = −3; $10^{-3} = 0.001$ **b.** base = 4; exponent = 0; $4^0 = 1$
c. base = m; exponent = k; $m^k = n$ **d.** base = 5; exponent = −2; $5^{-2} = \frac{1}{25}$

9. **a.** $\log_2 32 = 5$ **b.** $\log_2 2 = 1$ **c.** $\log_2 1 = 0$ **d.** $\log_{10} 100 = 2$

11. **a.** $\log_{10} 0.001 = -3$ **b.** $\log_f g = d$ **c.** $\log_3\left(\frac{1}{9}\right) = -2$ **d.** $\log_4 1 = 0$

13. **a.** $\log_{10} 1 = 0$ **b.** $\log_{10} 6 = 0.77815$ **c.** $\log_{10} 10 = 1$
d. $\log_{10} 60 = 1.77815$ **e.** $\log_{10} 100 = 2$ **f.** $\log_{10} 600 = 2.77815$

15. **a.** $\log_{10} 0.6 = -0.22185$ **b.** $\log_{10} 0.1 = -1$ **c.** $\log_{10} 0.01 = -2$
d. $\log_{10} 0.06 = -1.22185$ **e.** $\log_{10} 0.006 = -2.22185$ **f.** $\log_{10} 0.000\,006 = -5.22185$

17. **a.** The decimal portion is the same, all are positive.
b. The decimal portion is the same, all are negative.

19. **a.** $\log_{10} 17 = x$; $x \approx 1.23045$ **b.** $\log_{10} 125 = x$; $x \approx 2.09691$
c. $\log_{10} 400 = x$; $x \approx 2.60206$ **d.** $\log_{10} 0.05 = x$; $x \approx -1.30103$

21.

Exponential Equation	Logarithmic Equation	Solve for x
$5^0 = x$	$\log_5 x = 0$	$x = 1$
$x^3 = 64$	$\log_x 64 = 3$	$64 = 4^3$; $x = 4$
$10^x = -10$	$\log_{10}(-10) = x$	{ } or $\varnothing$
$10^x = 0$	$\log_{10} 0 = x$	{ } or $\varnothing$
$10^{-1} = x$	$\log_{10} x = -1$	$x = \frac{1}{10}$

23. a. $\log_7 x = 2$

$7^2 = x$

$x = 49$

b. $\log_3 3 = x$

$3^x = 3$

$x = 1$

c. $\log_{10} 0.01 = x$

$10^x = 0.01$

$x = -2$

d. $\log_{10} x = 0$

$10^0 = x$

$x = 1$

e. $\log_x 64 = 3$

$x^3 = 64$

$x^3 = 4^3$

$x = 4$

f. $\log_a x = 1$

$a^1 = x$

$a = x$

25. a. $\log_2 x = 8$

$2^8 = x$

$x = 256$

b. $\log_{10} x = -2$

$10^{-2} = x$

$x = \frac{1}{100} = 0.01$

c. $\log_2\left(\frac{1}{4}\right) = x$

$2^x = \frac{1}{4}$

$x = -2$

d. $\log_{10} 1 = x$

$10^x = 1$

$x = 0$

e. $\log_x 100 = 2$

$x^2 = 100$

$x = 10$

f. $\log_a a = x$

$a^x = a$

$x = 1$

27. Line of symmetry is $y = x$.

29. The set of outputs from a function is its ***range.***

31. Because we swap x and y, the domain of a function is the ***range*** of its inverse.

33. If the ordered pair (a, b) belongs to a function, then the ordered pair ***(b, a)*** belongs to its inverse.

35. The logarithm is the ***exponent*** to which we raise a base to obtain a given number.

37. The ***base*** of a logarithmic equation must be a positive number not equal to 1.

39. The set of outputs (range) to a logarithmic equation, $y = \log_b x$, $\mathfrak{R}$.

41. The set of outputs (range) to an exponential equation, $y = b^x$, is $\boldsymbol{y > 0}$.

43. No, log is not a unit.

Mid-Chapter 9 Test

1. a. common ratio is 3; exponential; $a_n = 3 \cdot 3^{n-1}$; $f(x) = 3^x$

b. no common ratio; 1st differences = 6, 6, 6, 6; linear; $m = 6$; $-1 = 6(1) + b$, $b = -7$
$y = 6x - 7$

c. common ratio is $\frac{1}{3}$; exponential; $a_n = 81 \cdot (\frac{1}{3})^{n-1}$; $f(x) = 243 \cdot (\frac{1}{3})^x$

d. no common ratio; 1st differences = 5, 9, 13, 17; 2nd differences = 4, 4, 4; quadratic
$2a = 4$, $a = 2$; $c = 1 - (5 - 4) = 0$; $1 = 2(1)^2 + b(1)$, $b = -1$; $y = 2x^2 - x$

e. common ratio is 2; exponential; $a_n = \frac{1}{8} \cdot 2^{n-1}$; $f(x) = 0.0625 \cdot 2^x$

2.

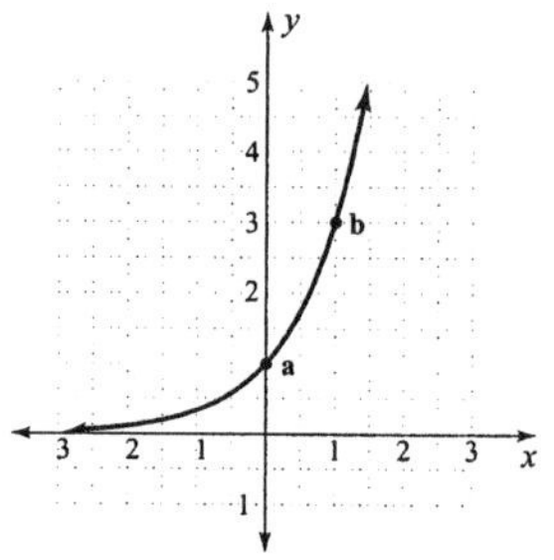

3. a. 2^{x-1} is 2^x shifted right 1 unit

b. no intersection, $2^x > 2^{x-1}$ for all x

4. a. $2^x = 10$, $x \approx 3.3$

b. $\log_2 10 = x$

5. $y = \frac{1}{2} \cdot 3^x$ increases rapidly as x gets large; $y = 2 \cdot (\frac{1}{3})^x$ approaches zero as x gets large

a. y-intercepts are the constants $\frac{1}{2}$ and 2

b. $b > 1$ indicates an increasing function; $0 < b < 1$ indicates a decreasing function

6. a. $2^x = 32$

$2^x = 2^5$

b. $(\frac{1}{2})^x = 0$

{ } or ∅

c. $(\frac{1}{2})^x = \frac{1}{16}$

$(2^{-1})^x = 2^{-4}$

$2^{-x} = 2^{-4}$

$-x = -4; x = 4$

d. $5^x = \frac{1}{125}$

$5^x = 5^{-3}$

$x = -3$

e. $10^x = -10$

{ } or ∅

f. $10^x = 0.01$

$10^x = 10^{-2}$

$x = -2$

7. a. $\log_4 x = -1$

$4^{-1} = x$

$x = \frac{1}{4}$

b. $\log_{10} 0.0001 = x$

$10^x = 0.0001$

$10^x = 10^{-4};\ x = -4$

c. $\log_3 x = 3$

$3^3 = x$

$x = 27$

d. $\log_2 x = 0$

$2^0 = x$

$x = 1$

e. $\log_4 2 = x$

$4^x = 2$

$(2^2)^x = 2^1$

$2^{2x} = 2^1$

$2x = 1; x = \frac{1}{2}$

f. $\log_4 (-2) = x$

$4^x = -2$

{ } or ∅

8. a.

x	$y = 4^x$
−2	$4^{-2} = \frac{1}{16}$
−1	$4^{-1} = \frac{1}{4}$
0	$4^0 = 1$
1	$4^1 = 4$
2	$4^2 = 16$
3	$4^3 = 64$

b.

x	y
$\frac{1}{16}$	−2
$\frac{1}{4}$	−1
1	0
4	1
16	2
64	3

8. c. 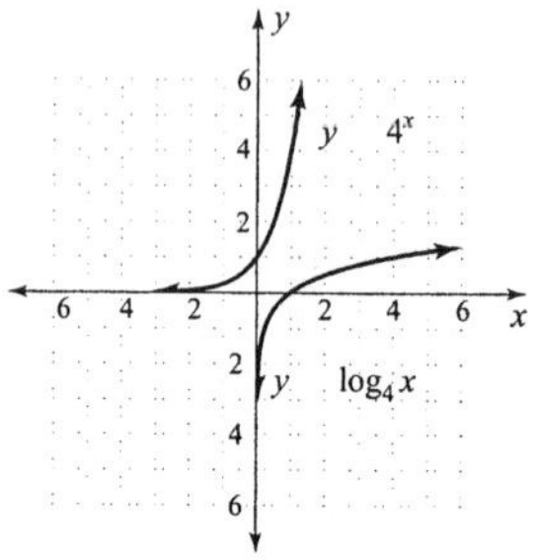

d. $x = 4^y$; $y = \log_4 x$

9.

Exponential Eqn.	Logarithmic Eqn.	Solve for x
$10^x = 15$	$\log_{10} 15 = x$	$x \approx 1.17609$
$10^x = 13$	$\log_{10} 13 = x$	$x \approx 1.11394$
$3^x = 81$	$\log_3 81 = x$	$3^x = 3^4$; $x = 4$
$3^9 = x$	$\log_3 x = 9$	$x = 19{,}683$
$4^{x+1} = 64$	$\log_4 64 = x + 1$	$4^{x+1} = 4^3$; $x + 1 = 3$; $x = 2$

Exercises 9.4

1. **a.** base = 3 **b.** base = 10 **c.** base = 4
   d. base = 4 **e.** base = 10 **f.** base = m

3. **a.** $2^x = 1$

   $2^x = 2^0$

   $x = 0$

   b. $2^x = 3$

   $\log_2 3 = x$

   $x = \dfrac{\log 3}{\log 2} \approx 1.58$

   c. $2^x = 4$

   $2^x = 2^2$

   $x = 2$

   d. $2^x = 5$

   $\log_2 5 = x$

   $x = \dfrac{\log 5}{\log 2} \approx 2.32$

   e. $2^x = 6$

   $\log_2 6 = x$

   $x = \dfrac{\log 6}{\log 2} \approx 2.58$

   f. $2^x = 10$

   $\log_2 10 = x$

   $x = \dfrac{\log 10}{\log 2} \approx 3.32$

   g. $2^x = 12$

   $\log_2 12 = x$

   $x = \dfrac{\log 12}{\log 2} \approx 3.58$

   h. $2^x = 20$

   $\log_2 20 = 2$

   $x = \dfrac{\log 20}{\log 2} = 4.32$

   i. $2^x = 24$

   $\log_2 24 = x$

   $x = \dfrac{\log 24}{\log 2} \approx 4.58$

5. **a.** $5^x = 2$

$\log_5 2 = x$

$x = \dfrac{\log 2}{\log 5} \approx 0.43$

b. $5^x = 5$

$5^x = 5^1$

$x = 1$

c. $5^x = 6$

$\log_5 6 = x$

$x = \dfrac{\log 6}{\log 5} \approx 1.11$

d. $5^x = 10$

$\log_5 10 = x$

$x = \dfrac{\log 10}{\log 5} \approx 1.43$

e. $5^x = 25$

$5^x = 5^2$

$x = 2$

f. $5^x = 30$

$\log_5 30 = x$

$x = \dfrac{\log 30}{\log 5} \approx 2.11$

7. **a.** $3^x = 10$

$\log_3 10 = x$

$x = \dfrac{\log 10}{\log 3} \approx 2.09590$

b. $x = \log_4 6$

$x = \dfrac{\log 6}{\log 4} \approx 1.29248$

c. $20 = 10(1.07)^x$

$2 = 1.07^x$

$\log_{1.07} 2 = x$

$x = \dfrac{\log 2}{\log 1.07} \approx 10.24477$

d. $15 = 30(0.95)^x$

$0.5 = 0.95^x$

$\log_{0.95} 0.5 = x$

$x = \dfrac{\log 0.5}{\log 0.95} \approx 13.51341$

e. $30 = 60(0.5)^x$

$0.5 = 0.5^x$

$0.5^1 = 0.5^x$

$x = 1$

9. x-intercept = 1; $f(1) = 0$

11. x-intercept = 1 for both graphs; the graph of $\log_6 x$ is above that of $\log_3 x$ to the left of the x-intercept and is below it to the right of the x-intercept.

13. False; if $b < 10$ then $\log_b x < \log_{10} x$ for $0 < x < 1$

15. False; this would imply $b^b = 1$

17. Graphs mirror each other across $y = x$. $2^0 = 1$ is the same fact as $\log_2 1 = 0$; $2^1 = 2$ is the same fact as $\log_2 2 = 1$. (2, 4) is on Y2, (4, 2) is on Y3.

19. a. $pH = -\log_{10}(1.0 \times 10^{-14}) = 14$ **b.** $pH = -\log_{10} 0.20 = 0.7$

c. $pH = -\log_{10}(5.01 \times 10^{-3}) = 2.3$ **d.** $pH = -\log_{10}(1.0 \times 10^{-10}) = 10$

e. $pH = -\log_{10}(1.26 \times 10^{-3}) = 2.9$

21. a. base **b.** acid **c.** acid

d. base **e.** acid

23. a.
$4.1 = -\log_{10}[H^+]$
$-4.1 = \log_{10}[H^+]$
$10^{-4.1} = [H^+]$
$[H^+] \approx 7.94 \times 10^{-5}$ M

b.
$-0.5 = -\log_{10}[H^+]$
$0.5 = \log_{10}[H^+]$
$10^{0.5} = [H^+]$
$[H^+] \approx 3.16$ M

c.
$12 = -\log_{10}[H^+]$
$-12 = \log_{10}[H^+]$
$10^{-12} = [H^+]$
$[H^+] = 1.0 \times 10^{-12}$ M

d.
$3.5 = -\log_{10}[H^+]$
$-3.5 = \log_{10}[H^+]$
$10^{-3.5} = [H^+]$
$[H^+] \approx 3.16 \times 10^{-4}$ M

25. pH is negative when $[H^+] > 1$; pH is positive when $[H^+] < 1$

27. a.
$I = 3.1623 \times 10^7$
$R = \log 3.1623 \times 10^7 = 7.5$

b.
$I = 1.2589 \times 10^8$
$R = \log 1.2589 \times 10^8 = 8.1$

29. a.
$8.6 = \log I$
$I = 10^{8.6} \approx 398{,}100{,}000$

b.
$7.9 = \log I$
$I = 10^{7.9} \approx 79{,}430{,}000$

31. $6.9 = \log I(sf)$; $I(sf) = 10^{6.9} \approx 7{,}943{,}000$; $6.8 = \log I(n)$; $I(n) = 10^{6.8} \approx 6{,}310{,}000$

$\frac{7{,}943{,}000}{6{,}310{,}000} \approx 1.3$ times a strong

33.
$1{,}000{,}000 = 1{,}000(1 + 0.02)^t$
$\frac{1{,}000{,}000}{1{,}000} = (1.02)^t$
$\log_{1.02} 1{,}000 = t$
$t = \frac{\log 1{,}000}{\log 1.02} \approx 349$ yr.

35.
$1{,}000{,}000 = 1{,}000(1 + 0.10)^t$
$\frac{1{,}000{,}000}{1{,}000} = (1.10)^t$
$\log_{1.10} 1{,}000 = t$
$t = \frac{\log 1{,}000}{\log 1.10} \approx 73$ yr

37. a. $r = 8\%$

b. $r = 6\%$

39. a. $r = -3\%$

b. $1 + r = 0.95;\ r = -0.05 = -5\%$

41.
$$2 = (1.06)^t$$
$$\log_{1.06} 2 = t$$
$$t = \frac{\log 2}{\log 1.06} \approx 11.9 \text{ yr.}$$

43.
$$0.5 = (1 - 0.10)^t$$
$$0.5 = (0.90)^t$$
$$\log_{0.9} 0.5 = t$$
$$t = \frac{\log 0.5}{\log 0.9} \approx 6.58 \text{ yr}$$

45. a. $108 \div 2 = 54;\ 54 = 108(\tfrac{2}{3})^{n-1}$
$$0.5 = (\tfrac{2}{3})^{n-1}$$
$$\log_{\frac{2}{3}} 0.5 = n - 1$$
$$n - 1 = \frac{\log 0.5}{\log \frac{2}{3}}$$
$$n = \frac{\log 0.5}{\log \frac{2}{3}} + 1 \approx 2.7$$

b. $32 = 108(\tfrac{2}{3})^{n-1}$
$$\tfrac{32}{108} = (\tfrac{2}{3})^{n-1}$$
$$\log_{\frac{2}{3}}\left(\tfrac{32}{108}\right) = n - 1$$
$$n - 1 = \frac{\log \frac{32}{108}}{\log \frac{2}{3}}$$
$$n = \frac{\log \frac{32}{108}}{\log \frac{2}{3}} + 1 = 4$$

c. $1 = 108(\tfrac{2}{3})^{n-1}$
$$\tfrac{1}{108} = (\tfrac{2}{3})^{n-1}$$
$$\log_{\frac{2}{3}}\left(\tfrac{1}{108}\right) = n - 1$$
$$n - 1 = \frac{\log \frac{1}{108}}{\log \frac{2}{3}}$$
$$n = \frac{\log \frac{1}{108}}{\log \frac{2}{3}} + 1 \approx 12.5$$

47.
$$2P = P(1 + r)^t$$
$$2 = (1 + r)^t$$
$$\log_{(1+r)} 2 = t$$
$$t = \frac{\log 2}{\log(1 + r)}$$

P does not appear in the formula;

doubling time is not dependent on P

49. a. $\dfrac{1\ half-life}{26.3\ hr} \cdot \dfrac{24\ hr}{1\ day} \cdot \dfrac{365\ day}{1\ yr} \approx 333$ half-lives

b. $\dfrac{1\ half-life}{2.4\ day} \cdot \dfrac{365\ day}{1\ yr} \approx 152$ half-lives

c. $\dfrac{1\ half-life}{14.3\ day} \cdot \dfrac{365\ day}{1\ yr} \approx 26$ half-lives

d. S = 1.7 curies; half-life = 24,000 yrs; $\frac{1}{2}$ S = 0.85 curies; $y = 1.7(0.99997)^x$; x in years

49. e. S = 12,000,000 curies; half-life = 15 hrs; $\frac{1}{2}$ S = 6,000,000 curies;
y = 12,000,000(0.955)^x; x in hrs.

f. S = 490,000 curies; half-life = 245 days; $\frac{1}{2}$ S = 245,000 curies;
y = 490,000(0.997)^x; x in days

51. a. convert units: $\frac{r}{day} \cdot \frac{365\ day}{yr} = 365r$ per year; $t\ day \cdot \frac{1\ yr}{365\ day} = \frac{t}{365}$ years

b. $S = P\left(1 + \frac{365r}{365}\right)^{365 \cdot \frac{t}{365}} = P(1 + r)^t$

c. $\$100(1 + 0.015)^{20} = \$134.69$

d. $\$100(1 + 0.015)^{30} = \$156.31$

53. b. Let x be # of years since 1986, y = cost; (0, 0.50), (9, 0.99), (15, 2)
y = 0.0964x + 0.392; slope is increase per year, 9.64 cents

c. y = 0.4808(1.095)^x; y-intercept is ≈ \$0.50 cost in 1986

d. r ≈ 9.5%

e. linear model assumes a constant price change each year; exponential model assumes a constant percent increase per year.

f. 2005 is 19 years after 1986; linear: 0.0964(19) + 0.392 ≈ \$2.22;
exponential: 0.4808(1.095)^{19} ≈ \$2.70

Exercises 9.5

1. a. $\log 2 + \log x = \log 2x$

b. $\log x^2 = 2\log x$

c. $\log\left(\frac{2}{x}\right) = \log 2 - \log x$

d. $\log 2^x = x\log 2$

e. $\log x - \log 2 = \log\left(\frac{x}{2}\right)$

3. log of a product property; $\log 2 + \log 3 = \log 6$

5. log of a power property: $\log 3^2 = 2\log 3$

7. a. $\log_b xz^3 = \log_b x + \log_b z^3 = \log_b x + 3\log_b z$

b. $\log_b\left(\frac{x^2}{y}\right) = \log_b x^2 - \log_b y = 2\log_b x - \log_b y$

9. a. $\log_b (xy)^2 = 2\log_b xy = 2(\log_b x + \log_b y)$

b. $\log_b\left(\frac{5}{x^2}\right) = \log_b 5 - \log_b x^2 = \log_b 5 - 2\log_b x$

11. a. $\log_b\left(\frac{\sqrt{x}}{y}\right) = \log_b \sqrt{x} - \log_b y = \log_b x^{\frac{1}{2}} - \log_b y = \frac{1}{2}\log_b x - \log_b y$

b. $\log_b (x-z)^2 = 2\log_b (x-z)$

13. a. $\log_b\left(\frac{xz^3}{y^2}\right) = \log_b (xz^3) - \log_b y^2 = \log_b x + \log_b z^3 - 2\log_b y$

$= \log_b x + 3\log_b z - 2\log_b y$

b. $\log_b\left(\frac{y^3z^2}{x}\right) = \log_b (y^3z^2) - \log_b x = \log_b y^3 + \log_b z^2 - \log_b x$

$= 3\log_b y + 2\log_b z - \log_b x$

15. a. $\log_b x + 2\log_b y = \log_b x + \log_b y^2 = \log(xy^2)$

b. $\log_b y - 2\log_b x = \log_b y - \log_b x^2 = \log_b\left(\frac{y}{x^2}\right)$

17. a. $\frac{1}{2}(\log_b y + 3\log_b z) = \frac{1}{2}(\log y + \log_b z^3) = \frac{1}{2}\log_b (yz^3) = \log_b (yz^3)^{\frac{1}{2}} = \log_b \sqrt{yz^3}$

b. $\frac{1}{2}\log_b x - \log_b 3 = \log_b x^{\frac{1}{2}} - \log_b 3 = \log_b \sqrt{x} - \log_b 3 = \log_b\left(\frac{\sqrt{x}}{3}\right)$

19. a. $\log(x+1) + \log(x-1) = \log[(x+1)(x-1)] = \log(x^2-1)$

b. $\log(x^2-1) - \log(x-1) = \log\left(\frac{x^2-1}{x-1}\right) = \log\left(\frac{(x-1)(x+1)}{(x-1)}\right) = \log(x+1)$

21. a. $\log(x^2+3x+2) - \log(x+1) = \log\left(\frac{x^2+3x+2}{x+1}\right) = \log\left(\frac{(x+2)(x+1)}{(x+1)}\right) = \log(x+2)$

b. $\log(x+3) + \log(x-2) = \log[(x+3)(x-2)] = \log(x^2+x-6)$

23. $\log\sqrt{x} = \log x^{\frac{1}{2}} = \frac{1}{2}\log x$

25. $\log\left(\frac{1}{x}\right) = \log x^{-1} = -1\log x = -\log x$

27. $x = \log_b a$

$b^x = a$

$\log b^x = \log a$

$x \log b = \log a$

$x = \dfrac{\log a}{\log b}$

$\log_b a = \dfrac{\log a}{\log b}$

29. $\log_2 x^2 + \log_2 x = \log_2 x^3 = 6$

$3\log_2 x = 6$

$\log_2 x = 2$

$\text{x} = 2^2 = 4$

31. $\log x + \log x = \log x^2 = 2$

$2\log x = 2$

$\log x = 1$

$\text{x} = 10^1 = 10$

33. $\log(x^2 + x) - \log(x+1) = \log\left(\dfrac{x^2 + x}{x+1}\right) = \log\left(\dfrac{x(x+1)}{(x+1)}\right) = \log x = 1; \text{x} = 10^1 = 10$

35. $\log_3 x^3 - \log_3 x = \log_3\left(\dfrac{x^3}{x}\right) = \log_3 x^2 = 2\log_3 x = 4;\ \log_3 x = 2;\ \text{x} = 3^2 = 9$

37. a. $\log 5^x = \log 20$

$x \log 5 = \log 20$

$x = \dfrac{\log 20}{\log 5} \approx 1.8614$

b. $\log_5 20 = x$

$x = \dfrac{\log 20}{\log 5}$

$x \approx 1.8614$

39. a. $\log 3^{x+2} = \log 48$

$(x+2)\log 3 = \log 48$

$x + 2 = \dfrac{\log 48}{\log 3}$

$x = \dfrac{\log 48}{\log 3} - 2 \approx 1.5237$

b. $\log_3 48 = x + 2$

$x + 2 = \dfrac{\log 48}{\log 3}$

$x = \dfrac{\log 48}{\log 3} - 2 \approx 1.5237$

41. a. $\log 4^{x-1} = \log 28$

$(x-1)\log 4 = \log 28$

$x-1 = \frac{\log 28}{\log 4}$

$x = \frac{\log 28}{\log 4} + 1 \approx 3.4037$

b. $\log_4 28 = x-1$

$x-1 = \frac{\log 28}{\log 4}$

$x = \frac{\log 28}{\log 4} + 1 \approx 3.4037$

43. a. $\log 9^{x+1} = \log 42$

$(x+1)\log 9 = \log 42$

$x+1 = \frac{\log 42}{\log 9}$

$x = \frac{\log 42}{\log 9} - 1 \approx 0.7011$

b. $\log_9 42 = x+1$

$x+1 = \frac{\log 42}{\log 9}$

$x = \frac{\log 42}{\log 9} - 1 \approx 0.7011$

45. a. $\log(1000(1.08)^t) = \log 2000$

$\log 1000 + \log 1.08^t = \log 2000$

$t \log 1.08 = \log 2000 - \log 1000$

$t \log 1.08 = \log\left(\frac{2000}{1000}\right)$

$t \log 1.08 = \log 2$

$t = \frac{\log 2}{\log 1.08} \approx 9.0065$

Exercise 45a. demonstrates that it is not necessary to divide by the constant before taking the log of each side however, it does make the solution less complicated.

b. $1.08^t = 2;\ \log_{1.08} 2 = t$

$t = \frac{\log 2}{\log 1.08} \approx 9.0065$

47. $15(1.07)^t = 45;\ 1.07^t = 3$

a. $\log 1.07^t = \log 3$

$t \log 1.07 = \log 3$

$t = \frac{\log 3}{\log 1.07} \approx 16.2376$

b. $\log_{1.07} 3 = t$

$t = \frac{\log 3}{\log 1.07} \approx 16.2376$

49. $1000(1-0.08)^t = 500;\ (0.92)^t = 0.5$

a. $\log 0.92^t = \log 0.5$

$t \log 0.92 = \log 0.5$

$t = \dfrac{\log 0.5}{\log 0.92} \approx 8.3130$

b. $\log_{0.92} 0.5 = t$

$t = \dfrac{\log 0.5}{\log 0.92} \approx 8.3130$

51. $45(1-0.07)^t = 15;\ (0.93)^t = \frac{1}{3}$

a. $\log 0.93^t = \log \frac{1}{3}$

$t \log 0.93 = \log \frac{1}{3}$

$t = \dfrac{\log \frac{1}{3}}{\log 0.93} \approx 15.1385$

b. $\log_{0.93}\left(\frac{1}{3}\right) = t$

$t = \dfrac{\log \frac{1}{3}}{\log 0.93} \approx 15.1385$

53.

x	y=3^x
−1	$3^{-1} = \frac{1}{3}$
0	$3^0 = 1$
1	$3^1 = 3$
2	$3^2 = 9$
3	$3^3 = 27$
4	$3^4 = 81$
5	$3^5 = 243$
6	$3^6 = 729$

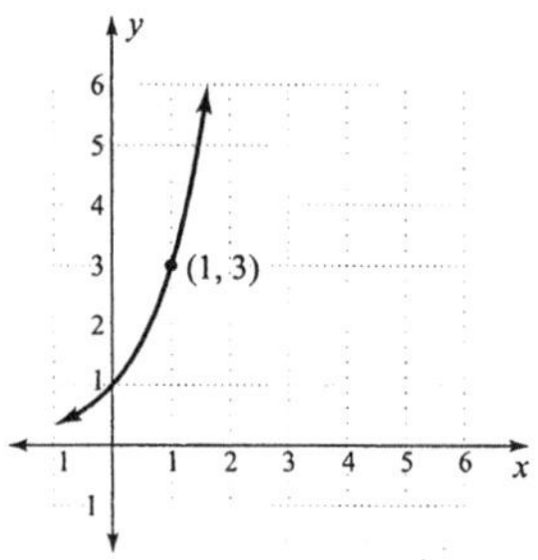

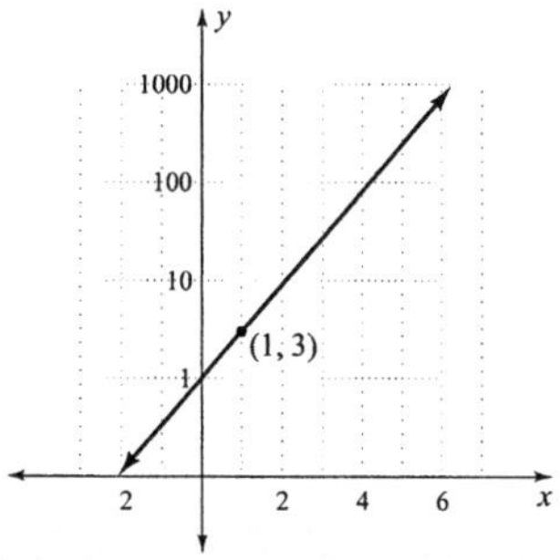

55. straight line

57. a. approximately 198 − 176 = 22 ft

b. 198 − 66 = 132 ft; ≈ 12 yrs

c. 132 − 66 = 66; 60 to 70 yrs

d. over 200 yrs

57. e. height loss is more rapid between year 2 and year 12; possible due to wind breakage

f. $\approx$ 16 ft.

59. a. - c.

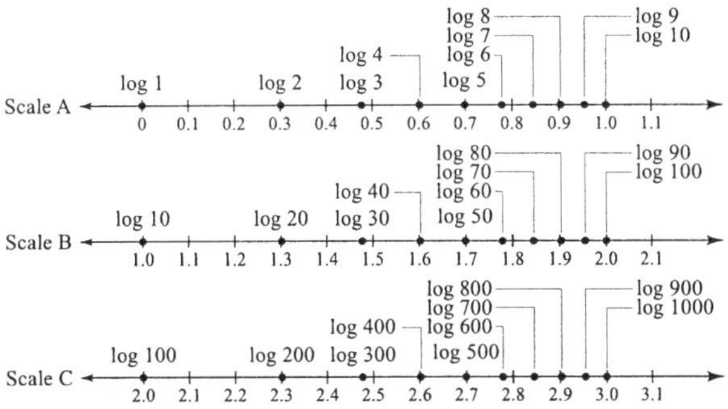

d. Dots are in the same relative positions because log 20 = 1 + log 2, etc.

Exercises 9.6

1. $\$1000(1 + 0.04)^3 \approx \$1{,}124.86$ $\qquad \$1000(1 + \frac{0.04}{12})^{3 \cdot 12} \approx \$1{,}127.27$

$\$1000(1 + \frac{0.04}{365})^{3 \cdot 365} \approx \$1{,}127.48$

$\$1{,}127.27 - \$1{,}124.86 = \$2.41$ $\qquad \$1{,}127.27 - \$1{,}127.48 = \$2.62$

Answers to exercises 3 and 5 are given for reference only.

3. a. $e^2 \approx 7.39$ **b.** $e^\pi \approx 23.14$ **c.** $2\sqrt{e} \approx 3.30$

5. a. $3^e \approx 19.81$ **b.** $e^{0.4 \cdot 2} \approx 2.23$ **c.** $e^{\pi/2} \approx 4.81$

7. e^(e^(1))

9. $\$1000e^{0.08 \cdot 1} = \$1083.28$

11. $\$1000e^{0.10 \cdot 1} = \$1105.17$

13. $\$1000e^{0.045 \cdot 1} = \$1046.02$

15. $\$500{,}000 = Pe^{0.10 \cdot 12}$

$$P = \frac{\$500{,}000}{e^{1.2}} \approx \$150{,}597.11$$

17. $\$500{,}000 = Pe^{0.10 \cdot 20}$

$$P = \frac{\$500{,}000}{e^{2.0}} \approx \$67{,}667.65$$

19.

Logarithmic Equation	Show base e in logarithm	Exponential Equation	y = ?
$\ln y = 1$	$\log_e y = 1$	$y = e^1$	$y = e$
$\ln y = 0$	$\log_e y = 0$	$y = e^0$	$y = 1$
$y = \ln(-1)$	$y = \log_e(-1)$	$e^y = -1$	$\{\,\}$ or $\varnothing$
$y = \ln e^2$	$y = \log_e e^2$	$e^y = e^2$	$y = 2$
$y = \ln e^e$	$y = \log_e e^e$	$e^y = e^e$	$y = e$

21. a. $e^x = 4;\ x = \ln 4 = 1.3863$ **b.** $\ln x = -2;\ x = e^{-2} = 0.1353$

23. a. $\ln x = 1.5;\ x = e^{1.5} = 4.4817$ **b.** $e^x = 2;\ x = \ln 2 = 0.6931$

25.
$$2 = e^{8r}$$
$$\ln 2 = 8r$$
$$r = \frac{\ln 2}{8} \approx 8.66\%$$

27.
$$2 = e^{rt}$$
$$\ln 2 = rt$$
$$r = \frac{\ln 2}{t}$$

29.
$$3 = e^{0.06t}$$
$$\ln 3 = 0.06t$$
$$t = \frac{\ln 3}{0.06} \approx 18.3 \text{ yr}$$

31.
$$0.5 = e^{0.08t}$$
$$\ln 0.5 = 0.08t$$
$$t = \frac{\ln 0.5}{0.08} \approx 8.7 \text{ yr}$$

33. a. $69 \div r = 8;\ r = 69 \div 8 = 8.625\%$ **b.** $r = 69 \div 11 \approx 6.27\%$

c. $t = \dfrac{69}{r};\ r = \dfrac{69}{t}$

d. Ex. 25 = 8.66%; Ex. 26 = 6.30%; rule of 69 was accurate to within 0.1%.

35. $2P = Pe^{rt};\ 2 = e^{rt};\ \ln 2 = rt;\ t = \dfrac{\ln 2}{r} \approx \dfrac{0.693}{r};\ t \approx \dfrac{0.693(100)}{r(100)} = \dfrac{69.3}{100r}$

37. a. $\dfrac{\log 8}{\log 6} \approx 1.1606;\ \dfrac{\ln 8}{\ln 6} \approx 1.1606$

b. $\log 8 \approx 0.903;\ \log 6 \approx 0.778;\ \ln 8 \approx 2.079;\ \ln 6 \approx 1.792$

c. ratios are equal, logarithms are not.

39. $K = e^{\left(-\frac{Ea}{RT}+C\right)}$

41. $\ln[H^+] = -\dfrac{EnF}{RT}$; $[H^+] = e^{\left(-\frac{EnF}{RT}\right)}$

43. a.

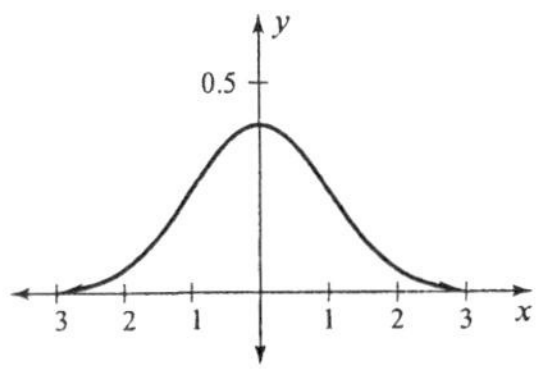

b. $y \approx 0.4$ at $x = 0$

c. y - intercept $= \dfrac{1}{\sqrt{2\pi}}$

d. Answers may vary; ≈ 0.65

45. a. $t = 0$; $1{,}849{,}568e^{-0.0147(0)} \approx 1{,}850{,}000$

b. $1{,}849{,}568e^{-0.0147(20)} \approx 1{,}380{,}000$

c. $1{,}849{,}568e^{-0.0147(40)} \approx 1{,}030{,}000$

d. The sign on the exponent

e. $0.5 = e^{-0.0147(t)}$; $t = \dfrac{\ln 0.5}{-0.0147} \approx 47$ yrs or 1997

f. $t = 50$; $1{,}849{,}568e^{-0.0147(50)} \approx 887{,}000$; decline in population is less than expected

47. a. $4! = 4 \cdot 3 \cdot 2 \cdot 1 = 24$

b. $10! = 10 \cdot 9 \cdot 8 \cdot 7 \cdot 6 \cdot 5 \cdot 4 \cdot 3 \cdot 2 \cdot 1 = 3{,}628{,}800$

c. $6! = 6 \cdot 5 \cdot 4 \cdot 3 \cdot 2 \cdot 1 = 720$

d. $9! = 9 \cdot 8 \cdot 7 \cdot 6 \cdot 5 \cdot 4 \cdot 3 \cdot 2 \cdot 1 = 362,880$

e. 9 terms shown ≈ 7.3873; $e^2 \approx 7.3891$

Chapter 9 Review Exercises

1. common ratio is 3; next number = $243 \cdot 3 = 729$; geometric sequence;
$y = 9 \cdot 3^{x-1}$ or $y = 3 \cdot 3^x$

3. no common ratio; first differences = 2, 4, 6, 8; second differences = 2, 2, 2;
next number is $21 + (8 + 2) = 31$; quadratic sequence; $2a = 2$, $a = 1$; $c = 1 - (2 - 2) = 1$;
$1 = 1(1)^2 + b(1) + 1$, $b = -1$; $y = x^2 - x + 1$

5. no common ratio; first differences = 3, 3, 3, 3; next number is $17 + 3 = 20$; arithmetic sequence; $m = 3$; $5 = 3(1) + b$, $b = 2$; $y = 3x + 2$

7. common ratio = 2; $a_n = \frac{1}{4} \cdot 2^{n-1}$; $f(x) = 0.125 \cdot 2^x$;
$\frac{1}{4} \cdot 2^{n-1} = 2^{-2} \cdot 2^{-1} \cdot 2^n = 2^{-3} \cdot 2^n = \frac{1}{8} \cdot 2^n = 0.125 \cdot 2^n$

9. common ratio = 2; $a_n = \frac{1}{16} \cdot 2^{n-1}$; $f(x) = 0.03125 \cdot 2^x$;

$\frac{1}{16} \cdot 2^{n-1} = 2^{-4} \cdot 2^{-1} \cdot 2^n = 2^{-5} \cdot 2^n = \frac{1}{32} \cdot 2^n = 0.03125 \cdot 2^n$

11. a. $2^{x-3} = 2^{-3} \cdot 2^x = \frac{1}{8} \cdot 2^x$; y-intercept = $\frac{1}{8}$ **b.** $3^{x+1} = 3^1 \cdot 3^x = 3 \cdot 3^x$; y-intercept = 3

c. $3^{x-3} = 3^{-3} \cdot 3^x = \frac{1}{27} \cdot 3^x$; y-intercept = $\frac{1}{27}$

13. a. y-intercept = 2; $2 \cdot 2^x = 2^1 \cdot 2^x = 2^{x+1}$ **b.** y-intercept = $\frac{1}{9}$; $\frac{1}{9} \cdot 3^x = 3^{-2} \cdot 3^x = 3^{x-2}$

15. a. $4^x = 64$

$4^x = 4^3$

$x = 3$

b. $2^x = 2$

$2^x = 2^1$

$x = 1$

c. $a^x = \frac{1}{a}$

$a^x = a^{-1}$

$x = -1$

d. $b^n = 1$

$b^n = b^0$

$n = 0$

17. a. $25^n = 125$

$(5^2)^n = 5^3$

$5^{2n} = 5^3$

$2n = 3$

$n = \frac{3}{2}$

b. $27^n = \frac{1}{3}$

$(3^3)^n = 3^{-1}$

$3^{3n} = 3^{-1}$

$3n = -1$

$n = -\frac{1}{3}$

c. $\left(\frac{1}{25}\right)^n = 125$

$(5^{-2})^n = 5^3$

$5^{-2n} = 5^3$

$-2n = 3$

$n = -\frac{3}{2}$

d. $\left(\frac{1}{16}\right)^n = 16$

$(4^{-2})^n = 4^2$

$4^{-2n} = 4^2$

$-2n = 2$

$n = -1$

19. a. $\left(\frac{1}{100}\right)^n = 10$

$(10^{-2})^n = 10^1$

$10^{-2n} = 10^1$

$-2n = 1$; $n = -\frac{1}{2}$

b. $100^n = 10$

$(10^2)^n = 10^1$

$10^{2n} = 10^1$

$2n = 1$; $n = \frac{1}{2}$

c. $4^{x-4} = 64$

$4^{x-4} = 4^3$

$x - 4 = 3$

$x = 7$

d. $27^{x+1} = 81$

$(3^3)^{x+1} = 3^4$

$3^{3x+3} = 3^4$

$3x + 3 = 4$; $3x = 1$; $x = \frac{1}{3}$

21.

Exponential Equation	Logarithmic Equation	Solve for x
$2^x = 16$	$\log_2 16 = x$	$2^x = 2^4$; $x = 4$
$x^2 = 25$	$\log_x 25 = 2$	$x^2 = 5^2$; $x = 5$
$3^x = 81$	$\log_3 81 = x$	$3^x = 3^4$; $x = 4$
$10^{1/2} = x$	$\log_{10} x = \frac{1}{2}$	$x = 10^{1/2} = \sqrt{10} \approx 3.162$
$10^x = 19$	$\log_{10} 19 = x$	$x \approx 1.2788$
$4^0 = x$	$\log_4 x = 0$	$x = 1$

23.
$10^x = 0.1$
$10^x = 10^{-1}$
$x = -1$

25.
$36 = 10^x$
$\log 36 = x$
$x = 1.556$

27.
$10^x = 0.75$
$\log 0.75 = x$
$x = -0.125$

29.
$4^{x+1} = 32$
$(2^2)^{x+1} = 2^5$
$2^{2x+2} = 2^5$
$2x + 2 = 5$
$2x = 3$
$x = \frac{3}{2}$

31.
$3^{2x} = 6$
$\log_3 6 = 2x$
$$2x = \frac{\log 6}{\log 3}$$
$$x = \frac{\log 6}{2\log 3} = 0.815$$

33.
$\log_{10} x = -1$
$10^{-1} = x$
$x = 0.1$

35.
$\log_2 x = 1$
$2^1 = x$
$x = 2$

37.
$\log_2 2 = x$
$2^x = 2$
$2^x = 2^1$
$x = 1$

39. $\log 10{,}000 = x$
$10^x = 10{,}000$
$10^x = 10^4$
$x = 4$

41. $\log 0.01 = x$
$10^x = 0.01$
$10^x = 10^{-2}$
$x = -2$

43. $\log_3 9 = x$
$3^x = 9$
$3^x = 3^2$
$x = 2$

45. $\log_{27} 9 = x$
$27^x = 9$
$(3^3)^x = 3^2$
$3^{3x} = 3^2$
$3x = 2$
$x = \frac{2}{3}$

47. $\log_b xyz = \log_b x + \log_b y + \log_b z$

49. $\log_b \dfrac{x}{y^2} = \log_b x - \log_b y^2 = \log_b x - 2\log_b y$

51. $\log_b \sqrt{\dfrac{z}{x}} = \log_b \dfrac{\sqrt{z}}{\sqrt{x}} = \log_b \sqrt{z} - \log_b \sqrt{x} = \log_b z^{1/2} - \log_b x^{1/2} = \frac{1}{2}\log_b z - \frac{1}{2}\log_b x$

53. $\log_b x + \frac{1}{2}\log_b y = \log_b x + \log_b y^{1/2} = \log_b xy^{1/2} = \log_b x\sqrt{y}$

55. $\log(x - 1) + \log(x - 2) = \log[(x - 1)(x - 2)] = \log(x^2 - 3x + 2)$

57. $\log(x^2 + x) - \log x = \log[x(x + 1)] - \log x = \log\left(\dfrac{x(x+1)}{x}\right) = \log(x + 1)$

59. $\log_3 x^2 - \log_3 x = 2$
$\log_3\left(\dfrac{x^2}{x}\right) = 2$
$\log_3 x = 2$
$3^2 = x;\ \ x = 9$

61. a. $e^e \approx 15.2$ **b.** $3^e \approx 19.8$ **c.** $e^3 \approx 20.1$

63. $e^x = \dfrac{1}{e^2}$

$e^x = e^{-2}$

$x = -2$

65. $\left(\dfrac{1}{e}\right)^x = e$

$(e^{-1})^x = e^1$

$-1x = 1;\ x = -1$

67. $\ln x = 3$

$e^3 = x$

69. $\ln x = \frac{1}{e}$

$e^{1/e} = x$

71. $\left(\frac{1}{2}\right)^x = \left(2^{-1}\right)^x = 2^{-x}$

73.

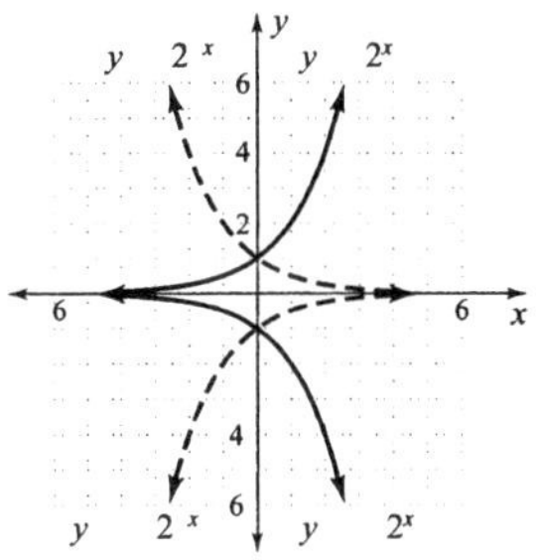

75.

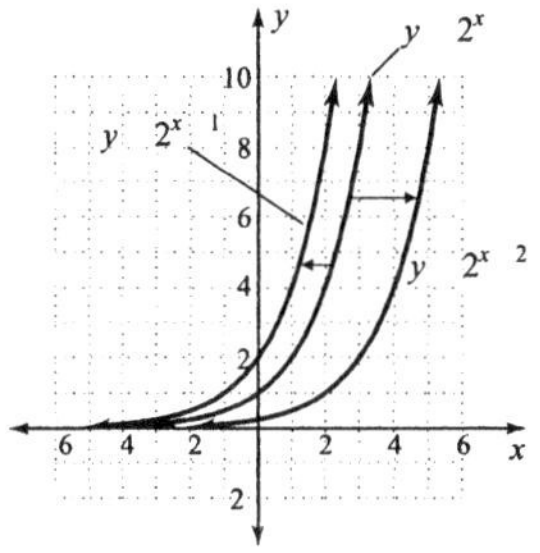

a. shift 1 unit left; **b.** shift 1 unit right

77. a. $3^0 = 1$ and $\log_3 1 = 0$ **b.** $3^1 = 3$ and $\log_3 3 = 1$

c. graphs mirror each other across y = x

79. straight line

81. $\log(5 \cdot 100) = \log 5 + \log 100 = \log 5 + 2$

83. All are 2 + a decimal

85. log (1000x) is log x + 3

87. inverse of $y = 2^x$ is $x = 2^y$ which is $\log_2 x = y$

89. $\log(8x^3) = \log(2^3x^3) = \log(2x)^3 = 3\log(2x)$

91. $\frac{1}{3}\log x = \log x^{1/3} = \log\sqrt[3]{x}$

93. $1000(1 + \frac{0.06}{12})^{12 \cdot 2} \approx \$1127.15;\ 2 = (1 + \frac{0.06}{12})^{12t};\ t = \dfrac{\log 2}{12\log(1 + \frac{0.06}{12})} \approx 11.58$ yr

95. $72 \div 6 = 12$ yr; $72 \div 8 - 9$ yr; $72 \div 12 = 6$ yr

97. a. $1000 = Pe^{0.06 \cdot 5}$ **b.** $1000 = Pe^{0.08 \cdot 5}$ **c.** $1000 = Pe^{0.12 \cdot 5}$

$P = \frac{1000}{e^{0.06 \cdot 5}} \approx \$740.82$ $P = \frac{1000}{e^{0.08 \cdot 5}} \approx \$670.32$ $P = \frac{1000}{e^{0.12 \cdot 5}} \approx \$548.81$

99. $69 \div 5 \approx 13.8\%$; $2 = e^{5r}$, $\ln 2 = 5r$, $r = \ln 2 \div 5 \approx 0.1386$

101. Use exponential regression: $y \approx 49.3(0.626)^x$, with x = 0 for 1989
$49.3(0.626)^{16} \approx \$0.027$ million

103. $P = 59{,}509(0.9936)^x$; rate of decay = 0.9936 – 1 = –0.64%;
$59{,}509(0.9936)^{30} \approx 49{,}011$ in 2010 which seems reasonable

105.a.

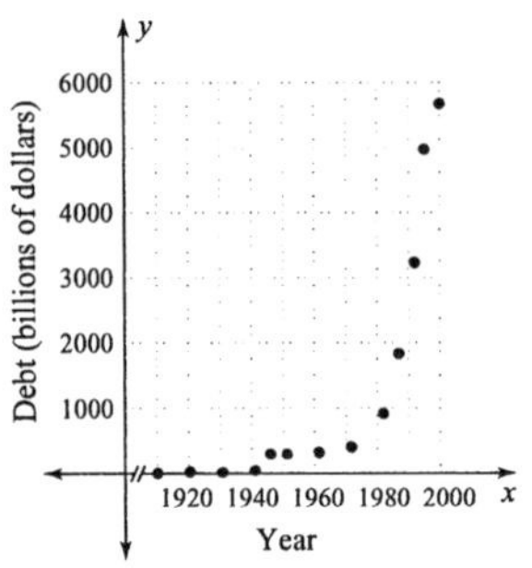

b. $y \approx (4.340 \times 10^9)(1.086)^x$

c. Answers may vary; could include wars; excess government spending

e.

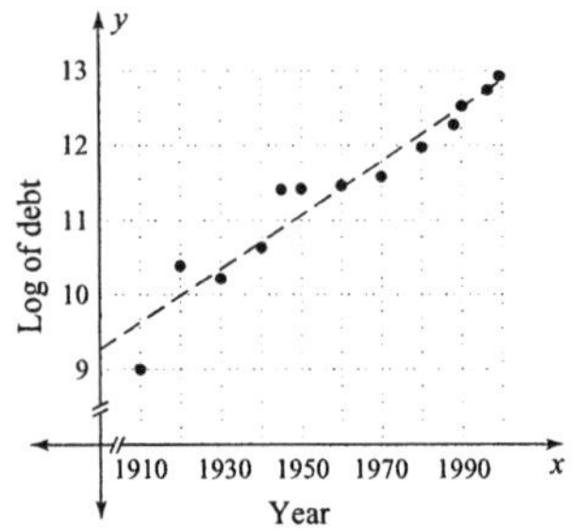

graph becomes somewhat linear

Chapter 9 Test

1. a. common ratio = 2; geometric sequence
b. $a_n = \frac{1}{8} \cdot 2^{n-1}$; $f(x) = 0.0625 \cdot 2^x$; $2^{-3} \cdot 2^{-1} \cdot 2^n = 2^{-4} \cdot 2^n = \frac{1}{16} \cdot 2^n = 0.0625 \cdot 2^n$

2. a. no common ratio; 1st differences = 9, 15, 21, 27; 2nd differences = 6, 6, 6; quadratic
c. $f(x) = 3x^2$

3. a. $3^x = 27$
$3^x = 3^3$
$x = 3$

b. $\left(\frac{1}{16}\right)^n = 64$
$(4^{-2})^n = 4^3$
$-2n = 3$
$n = -\frac{3}{2}$

c. $\frac{1}{1000} = 0.1^x$
$10^{-3} = (10^{-1})^x$
$-3 = -x$
$x = 3$

d. $4^{0.5x} = 16$
$4^{0.5x} = 4^2$
$0.5x = 2$
$x = 4$

e. $4^x = \frac{1}{64}$
$4^x = 4^{-3}$
$x = -3$

f. $100^n = 10$
$(10^2)^n = 10^1$
$2n = 1$
$n = \frac{1}{2}$

g. $27^x = 9$
$(3^3)^x = x^2$
$3x = 2$
$x = \frac{2}{3}$

h. $4^{x+1} = 8$
$(2^2)^{x+1} = 2^3$
$2x + 2 = 3$
$2x = 1;\ x = \frac{1}{2}$

4. a. $10^x = 3$
$\log 3 = x$
$x \approx 0.477$

b. $5^{2x} = 0.25$
$\log_5 0.25 = 2x$
$x = \dfrac{\log 0.25}{2\log 5} \approx -0.431$

c. $10^x = -1$
$\{\ \}$ or $\varnothing$

d. $e^x = \frac{1}{e}$
$e^x = e^{-1}$
$x = -1$

e. $5^{x-2} = \frac{1}{125}$
$5^{x-2} = 5^{-3}$
$x - 2 = -3$
$x = -1$

f. $\log_2 x = -2$
$2^{-2} = x$
$x = \frac{1}{4}$

g. $\log_5 x = 3$
$5^3 = x$
$x = 125$

h. $\log_x 32 = 5$
$x^5 = 32$
$x^5 = 2^5$
$x = 2$

i. $\log x = 1$
$10^1 = x$
$x = 10$

j. $\ln x = 1.5$
$e^{1.5} = x$
$x \approx 4.482$

5. a. $\ln e^2 = 2\ln e = 2(1) = 2$

b. $\log 1 = 0$

c. $\log_2 x + \log_2(x+1) = \log_2[x(x+1)] = \log_2(x^2+x)$

d. $\log_3(x^2-9) - \log_3(x+3) = \log_3\left(\dfrac{x^2-9}{x+3}\right) = \log_3\left(\dfrac{(x+3)(x-3)}{(x+3)}\right) = \log_3(x-3)$

6. a. $\log_b\left(\dfrac{z^2}{y}\right) = \log_b z^2 - \log_b y = 2\log_b z - \log_b y$

b. $\log_b\left(\dfrac{\sqrt{x}}{y}\right) = \log_b \sqrt{x} - \log_b y = \log_b x^{1/2} - \log_b y = \frac{1}{2}\log_b x - \log_b y$

7. $y = b^{x+2} = b^2 b^x$; y-intercept $= b^2$

8. $1360 = 1000(1 + 0.075)^t$

$1.360 = (1.075)^t$

$\log 1.360 = \log 1.075^t$

$\log 1.360 = t\log 1.075$

$$t = \frac{\log 1.360}{\log 1.075} \approx 4.25 \text{ yr}$$

9. $1906 = 1000e^{0.075t}$

$1.906 = e^{0.075t}$

$\ln 1.906 = 0.075t$

$$t = \frac{\ln 1.906}{0.075} \approx 8.6 \text{ yr}$$

10.

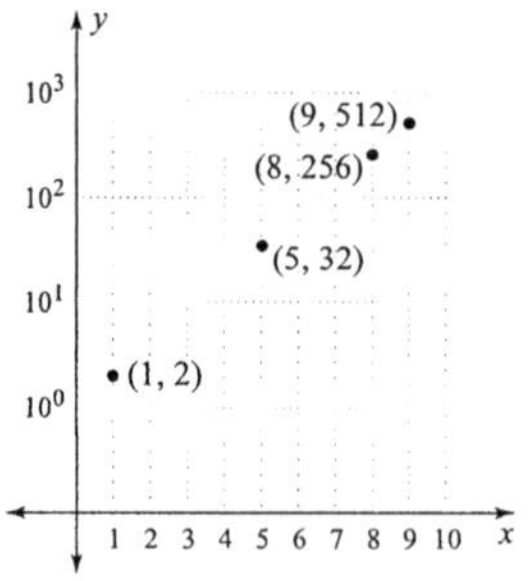

11. $f(x) = 3^x$

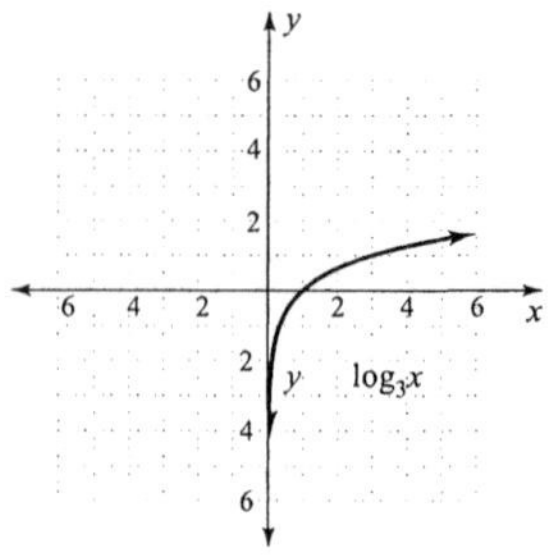

12. $f(x) = \log_2 x$

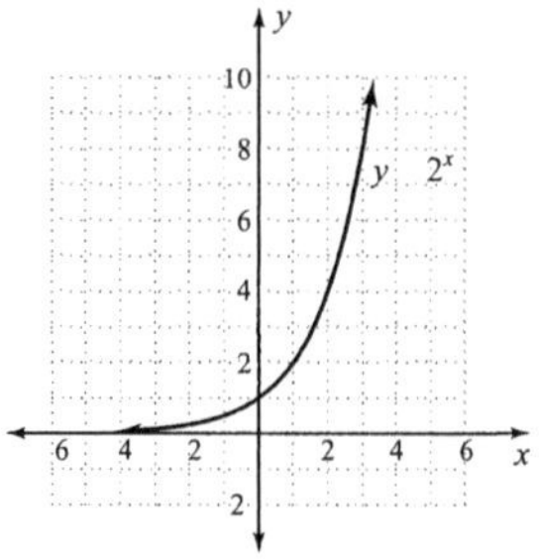

13.

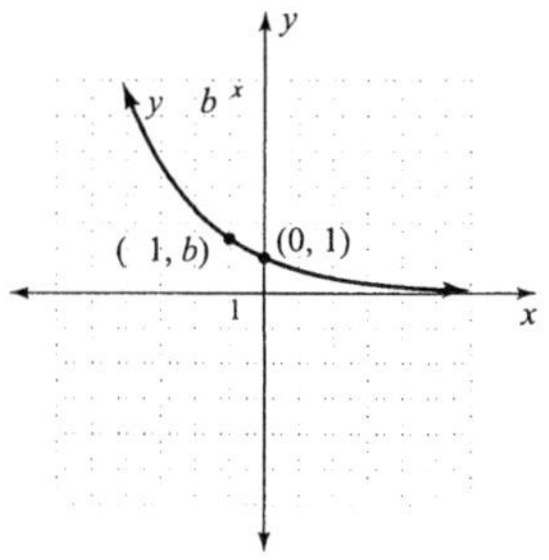

14. possible answers: (0, 1), (1, 0); (1, 10), (10, 1);
if $y = 10^x$, $x = \log y$; $y = \log x$ will have reversed coordinates

15. $1 + r = 0.93$; $r = 0.93 - 1 = -0.07$ (to find % multiply by 100)

16. $\log (100x) = \log 100 + \log x = 2 + \log x$

17. a. $\text{pH} = -\log (5.01 \times 10^{-9}) = 8.3$

b. $11.9 = -\log [H^+]$; $-11.9 = \log[H^+]$; $10^{-11.9} = [H^+] \approx 1.26 \times 10^{-12}$ M

18. Extend the logarithm of a product property to *n* factors of x. Write out the *n* factors using the logarithm of a product:

$$\log_b x \cdot x \cdot x \cdot .... \cdot x \text{ } (n \text{ factors of x}) = \log_b x + \log_b x + \log_b x + ... + \log_b x = n\log_b x$$

19. a.

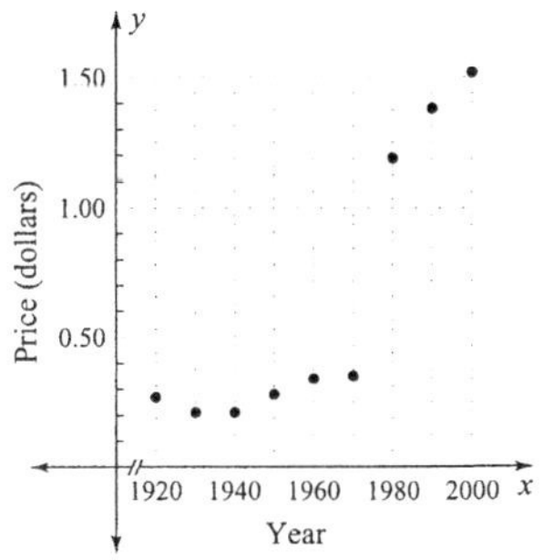

b. possible ans: exponential; data start off low and increases rapidly

c. $f(x) \approx 0.161(1.0263)^x$

20. a. $y = 15{,}649(1.0021)^x$, growth rate $\approx$ 0.21%, $15{,}649(1.0021)^{80} \approx 18{,}516$ in 2010

b. $y = 15{,}642(1.00695)^x$; growth rate $\approx$ 0.7%; $15{,}642(1.00695)^{40} \approx 20{,}637$ in 2010

c. Ask how much the college grew during the years and how much it is expected to grow.

Exercises 10.1

1. (±3, 0), (0, ±3)

For exercises 3 to 19 refer to text for sample equations.

3. $x^2 + y^2 = 5^2$; $x^2 + y^2 = 25$

5. $\dfrac{x^2}{5^2} + \dfrac{y^2}{2^2} = 1$; $\dfrac{x^2}{25} + \dfrac{y^2}{4} = 1$

7. no x-intercept indicates y^2 term comes first; $\dfrac{y^2}{4^2} - \dfrac{x^2}{3^2} = 1$; $\dfrac{y^2}{16} - \dfrac{x^2}{9} = 1$

9. opening to the right indicates squared term is y^2 and a is positive
$3 = a(-1)^2$; $a = 3$; $x = 3y^2$

11. opening downward indicates squared term is x^2 and a is negative
$-1 = a(3)^2$; $a = -\frac{1}{9}$; $y = -\frac{1}{9}x^2$

13. $x^2 + y^2 = 4^2$; $x^2 + y^2 = 16$

15. $\dfrac{x^2}{1^2} + \dfrac{y^2}{5^2} = 1$; $x^2 + \dfrac{y^2}{25} = 1$

17. no x-intercept indicates y^2 term comes first: $\dfrac{y^2}{3^2} - \dfrac{x^2}{5^2} = 1$; $\dfrac{y^2}{9} - \dfrac{x^2}{25} = 1$

19. opening upward indicates squared term is x^2 and a is positive
$1 = a(-2)^2$; $a = \frac{1}{4}$; $y = \frac{1}{4}x^2$

21. r < 0 has no meaning; circle becomes a point

For exercises 23 to 43, solving for y detail is not shown, graphs are combined.

23. parabola; y-intercept (0, 2)

25. circle; intercepts (±3, 0), (0 ±3)
for calculator graph: $y = \pm\sqrt{9 - x^2}$

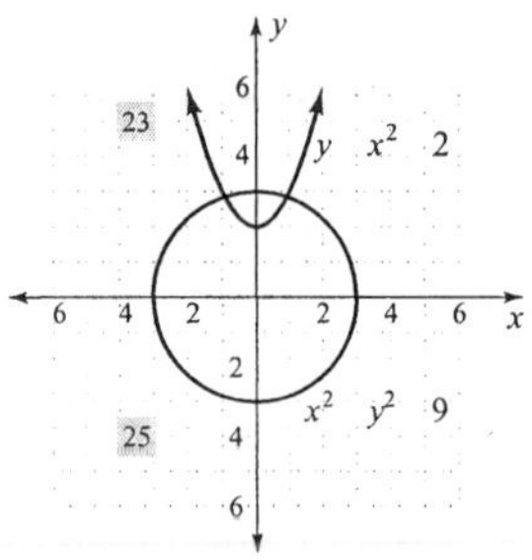

27. hyperbola; x-intercepts (±2, 0)

calculator: $y = \pm\sqrt{x^2 - 4}$

29. ellipse: intercepts (±2, 0), (0, ±1)

calculator: $y = \pm\sqrt{1 - \frac{x^2}{4}}$

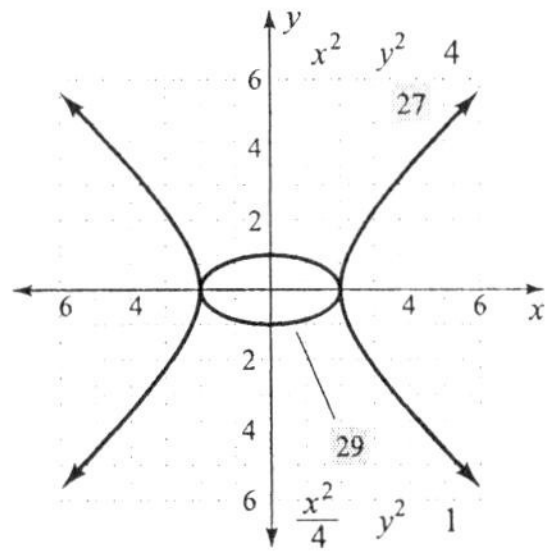

31. divide both sides by 100:

$\frac{x^2}{25} + \frac{y^2}{100} = 1$; ellipse,

intercepts: (±5, 0), (0, ±10)

calculator: y = $\pm\sqrt{100 - 4x^2}$

33. add 8 to both sides then divide by 2:

$x^2 + y^2 = 4$; circle

intercepts: (±2, 0), (0, ±2)

calculator: y = $\pm\sqrt{4 - x^2}$

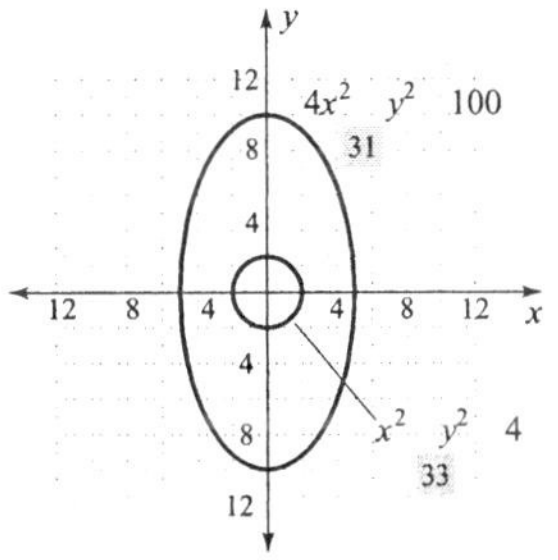

35. y = x; straight line

37. hyperbola; x-intercepts (±1, 0);

calculator: y = $\pm\frac{1}{2}\sqrt{x^2 - 1}$

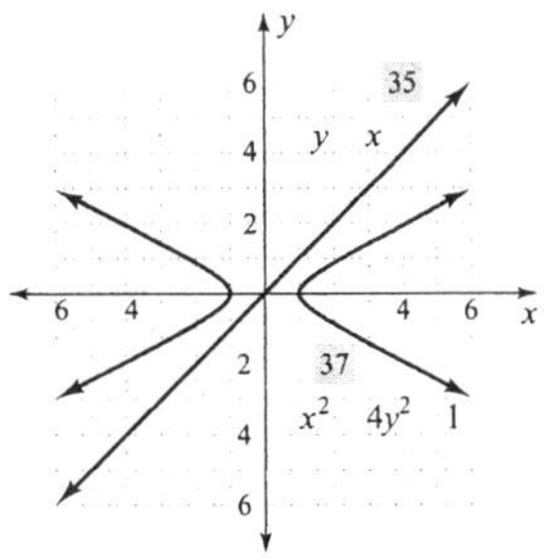

39. parabola; vertex at (0, 0)

calculator: $y = \pm\frac{1}{2}\sqrt{x}$

41. hyperbola; y-intercepts; $(0, \pm 1)$

calculator: $y = \pm\sqrt{1 + \dfrac{x^2}{4}}$

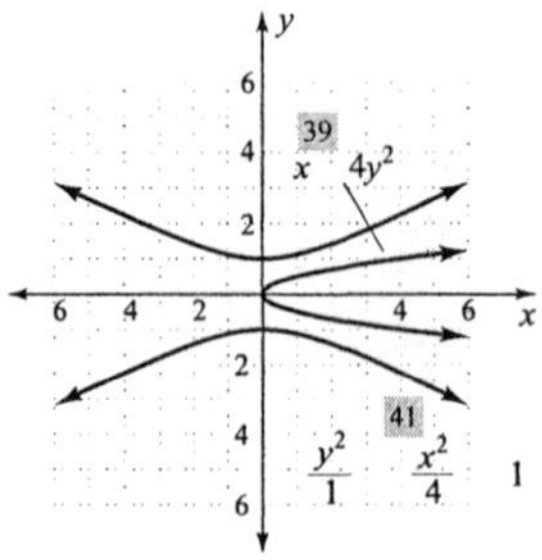

43. parabola; vertex at (0, 0)

calculator: $y = \pm\sqrt{-x}$

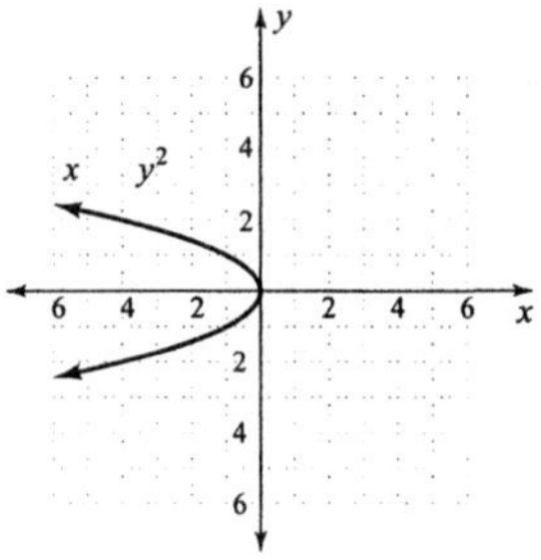

45. Branches become flatter, x-intercepts move to $(\pm a, 0)$;

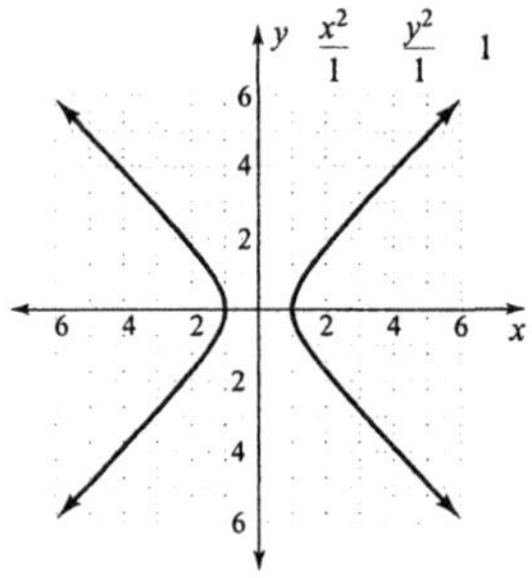

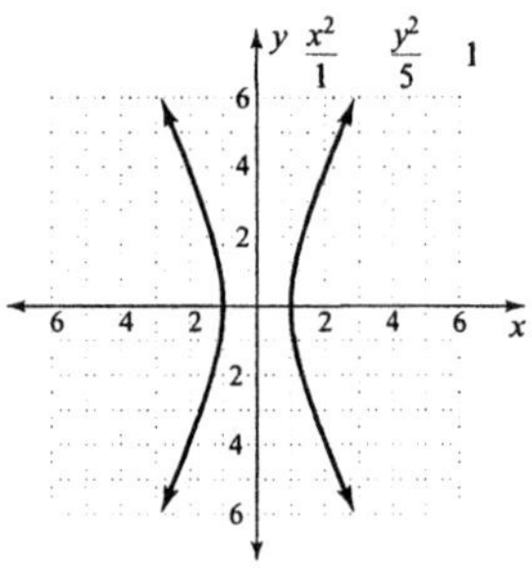

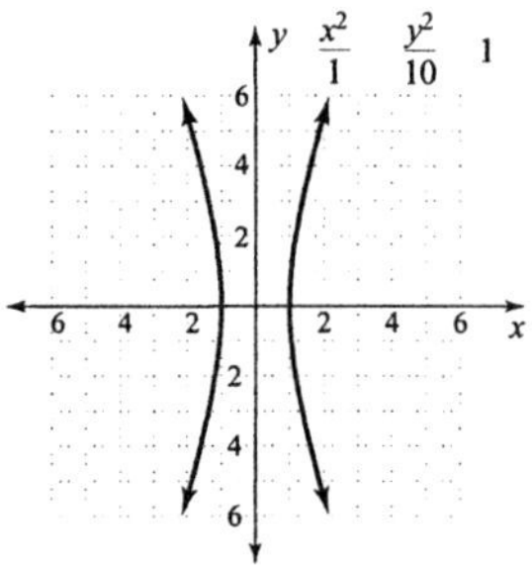

47. a.

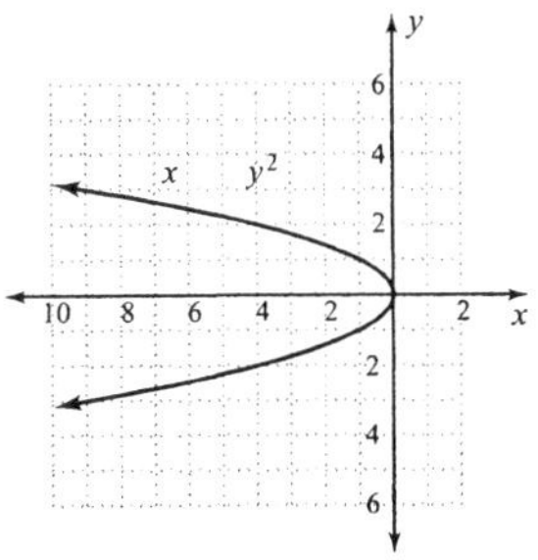

x	$x = -y^2$
-9	$-9 = -(\pm 3)^2$
-4	$-4 = -(\pm 2)^2$
-1	$-1 = -(\pm 1)^2$
0	$0 = -(0)^2$
1	$1 = -(i)^2$
4	$4 = -(2i)^2$
9	$9 = -(3i)^2$

b. Opposite of x makes the radicand appear negative.

c. $y = -\sqrt{-x}$ has meaning when x is a negative number.

Exercises 10.2

1. parabola and a line, set equal to each other and solve.

$x^2 - 3x + 2 = -x + 5$

$x^2 - 2x - 3 = 0$

$(x - 3)(x + 1) = 0$

$x - 3 = 0$ or $x + 1 = 0$

$x = 3$ or $x = -1$

$y = -(3) + 5, y = 2,$

$y = -(-1) + 5, y = 6$

$(-1, 6), (3, 2)$

3. hyperbola and a parabola

solve the parabola equation for x^2

and substitute into the hyperbola equation

$y = x^2 - 4,\ x^2 = y + 4;$

$\frac{y+4}{4} - \frac{y^2}{9} = 1,\ 36\left(\frac{y+4}{4} - \frac{y^2}{9}\right) = 36,$

$9y + 36 - 4y^2 = 36,\ -4y^2 + 9y = 0$

$y(-4y + 9) = 0,\ y = 0,$ or

$-4y + 9 = 0,\ -4y = -9,\ y = \frac{9}{4};$

$x^2 = 0 + 4, x = \pm 2$

$x^2 = \frac{9}{4} + 4,\ x^2 = \frac{25}{4},\ x = \pm \frac{5}{2}$

$(\pm\frac{5}{2}, \frac{9}{4}), (\pm 2, 0)$

5. Both equations are parabolas, set equal to each other and solve.

$5x^2 + 2x = -x^2 + x + 2,$

$6x^2 + x - 2 = 0,$

$(3x + 2)(2x - 1) = 0$

$3x + 2 = 0, 3x = -2, x = -\frac{2}{3};$

$2x - 1 = 0, 2x = 1, x = \frac{1}{2};$

$y = 5(-\frac{2}{3})^2 + 2(-\frac{2}{3}),$

$y = \frac{20}{9} - \frac{4}{3}, y = \frac{8}{9}$

$y = 5(\frac{1}{2})^2 + 2(\frac{1}{2}),$

$y = \frac{5}{4} + 1, y = \frac{9}{4};$

$(-\frac{2}{3}, \frac{8}{9}), (\frac{1}{2}, \frac{9}{4})$

7. Equations are an ellipse and a parabola. Substitute one into the other.

$y = x^2 - 1,\ x^2 = y + 1$

$y + 1 + \frac{y^2}{4} = 1$

$4y + 4 + y^2 = 4, y^2 + 4y = 0$

$y(y + 4) = 0$

$y = 0$ or $y + 4 = 0, y = -4$

$0 = x^2 - 1,\ x^2 = 1, x = \pm 1$

$-4 = x^2 - 1, x^2 = -3$ not possible

discard extraneous answer. $(\pm 1, 0)$

9. circle and a parabola substitute one into the other

$y = x^2 - 4,\ x^2 = y + 4;$

$y + 4 + y^2 = 16,$

$y^2 + y - 12 = 0,$

$(y + 4)(y - 3) = 0;$

$y + 4 = 0, y = -4;$

$y - 3 = 0,\ y = 3;$

$-4 = x^2 - 4, x^2 = 0,\ x = 0;$

$3 = x^2 - 4, x^2 = 7, x = \pm\sqrt{7},$

$(0, -4), (\pm\sqrt{7}, 3)$

11. Equations are both parabolas set equal to each other and solve

$x^2 - 6 = -x^2 + 2$

$2x^2 = 8,\ x^2 = 4$

$x = \pm 2$

$y = 4 - 6,\ y = -2$

$(\pm 2, -2)$

13. ellipse and a parabola.

$y = x^2 - 1,\ x^2 = y + 1;$

$\frac{y+1}{4} + y^2 = 1,$

$y + 1 + 4y^2 = 4,$

$4y^2 + y - 3 = 0,$

$(4y - 3)(y + 1) = 0,$

$y + 1 = 0, y = -1,$

$4y - 3 = 0, 4y = 3,\ y = 0.75;\ -1 = x^2 - 1, x^2 = 0, x = 0,$

$0.75 = x^2 - 1, x^2 = 1.75, x = \pm\sqrt{1.75};\ (0, -1), (\pm\sqrt{1.75} \approx \pm 1.323, 0.75)$

15. Equations are a circle and a hyperbola Use elimination, add equations.

$(x^2 + x^2) + (y^2 - y^2) = 4 + 4$

$2x^2 = 8; x^2 = 4, x = \pm 2$

$4 + y^2 = 4, y^2 = 0, y = 0$

$(\pm 2, 0)$

17. hyperbola and a line.

$x + y = 4,\ y = 4 - x;\ 4 - x = \frac{1}{x},$

$4x - x^2 = 1,\ x^2 - 4x + 1 = 0,$

$x = \dfrac{-(-4) \pm \sqrt{(-4)^2 - 4(1)(1)}}{2(1)} = \dfrac{4 \pm \sqrt{12}}{2},$

$x = \dfrac{4 \pm 2\sqrt{3}}{2},\ x = 2 \pm \sqrt{3};$

$y = 4 - (2 + \sqrt{3}),\ y = 2 - \sqrt{3},$

$y = 4 - (2 - \sqrt{3}),\ y = 2 + \sqrt{3};$

$(2 \pm \sqrt{3},\ 2 \pm \sqrt{3}) \approx (3{,}732,\ 0.268),\ (0.268,\ 3.732)$

19. Equations are a circle and an ellipse

$x^2 = 9 - y^2,\ \dfrac{9 - y^2}{16} + \dfrac{y^2}{4} = 1,$

$9 - y^2 + 4y^2 = 16,\ 3y^2 = 7,\ y^2 = \frac{7}{3}$

$y = \pm\sqrt{\frac{7}{3}},\ y = \pm\dfrac{\sqrt{7}}{\sqrt{3}},\ y = \pm\dfrac{\sqrt{21}}{3}$

$x^2 = 9 - \frac{7}{3},\ x^2 = \frac{20}{3},\ x = \pm\sqrt{\frac{20}{3}},$

$x = \pm\dfrac{2\sqrt{5}}{\sqrt{3}},\ x = \pm\dfrac{2\sqrt{15}}{3};$

$\left(\pm\dfrac{2\sqrt{15}}{3},\ \pm\dfrac{\sqrt{21}}{3}\right)$

21. A possible window is Xmin = -1.1, Xmax = 0.1, Ymin = -1, Ymax = 0.1

23.

$0 = -2(-1)^2 + 2,$	$(-1)^2 + (0 - 2)^2 = 4$	$(1)^2 + (0 - 2)^2 = 4$
$0 = -2 + 2,$	$1 + 4 = 4,$	$1 + 4 = 4,$
$0 = 0$	$5 \neq 4$	$5 \neq 4$

The points are not solutions, they do not satisfy both equations.

25. $3^x = 2 - 3^x$

$2 \cdot 3^x = 2,$

$3^x = 1,\ x = 0;$

$y = 3^0,\ y = 1;$

$(0, 1)$

27. $2^{x+2} = 4 + 2^x;\ 2^2 \cdot 2^x = 4 + 2^x$

$4 \cdot 2^x - 2^x = 4;\ 3 \cdot 2^x = 4,\ 2^x = \frac{4}{3}$

$\log 2^x = \log \frac{4}{3},\ x\log 2 = \log \frac{4}{3}$

$x = \dfrac{\log \frac{4}{3}}{\log 2},\ x \approx 0.415$

$y \approx 4 + 2^{0.415},\ y \approx 5.333$

$(0.415,\ 5.333)$

29. $\log x = 3 - \log x,\ 2\log x = 3,$

$\log x = 1.5,\ 10^{1.5} = x,\ x \approx 31.6;$

$y = \log 10^{1.5},\ y = 1.5$

$(31.6,\ 1.5)$

31. $\log_4 x = 2 - \log_4 x,\ 2\log_4 x = 2$

$\log_4 x = 1,\ 4^1 = x,\ x = 4;$

$y = \log_4 4,\ y = 1$

$(4, 1)$

33. $2^{x+2} = 2^2 \cdot 2^x = 4 \cdot 2^x \quad 4 \cdot \frac{1}{3} = \frac{4}{3}$

35. $3^{x+1} = 3^1 \cdot 3^x = 3 \cdot 3^x \quad 3 \cdot 2 = 6$

37. $2^{x-1} = 2^{-1} \cdot 2^x = \frac{1}{2} \cdot 2^x \quad \frac{1}{2} \cdot 5 = \frac{5}{2}$

39. $4^{x-2} = 4^{-2} \cdot 4^x = \frac{1}{16} \cdot 4^x \quad \frac{1}{16} \cdot 3 = \frac{3}{16}$

41. $\log_2 x^2 = 1,\ 2^1 = x^2,\ x = \pm\sqrt{2}$, The positive root is the same as the original solution, the negative root must be discarded.

Exercises 10.3

1. $x < 0$ and $y > 0$ is in quadrant 2

3. $x < 0$ and $y < 0$ is in quadrant 3

5. $y < 0$ is in quadrants 3 and 4

7. $x < 0$ is in quadrants 2 and 3

9. In quadrant 2 $x < 0$ and $y > 0$

11. In quadrants 2 and 4 $x < 0$ and $y > 0$ or $x > 0$ and $y < 0$

13. $300 \le y \le 2800$ and $x \ge 0$

15. $y \le x$ and $x \le 0$

17.

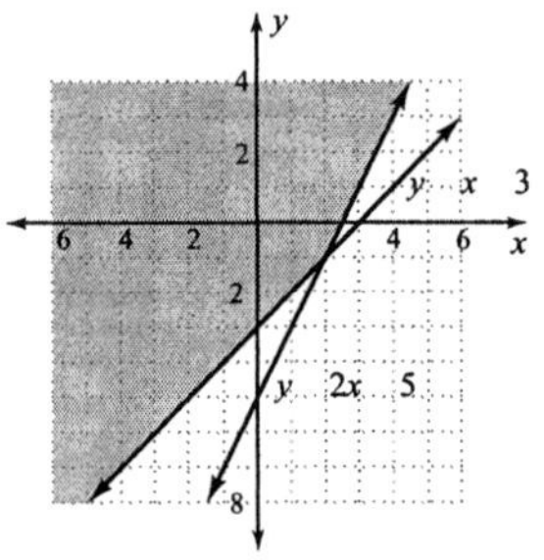

19.

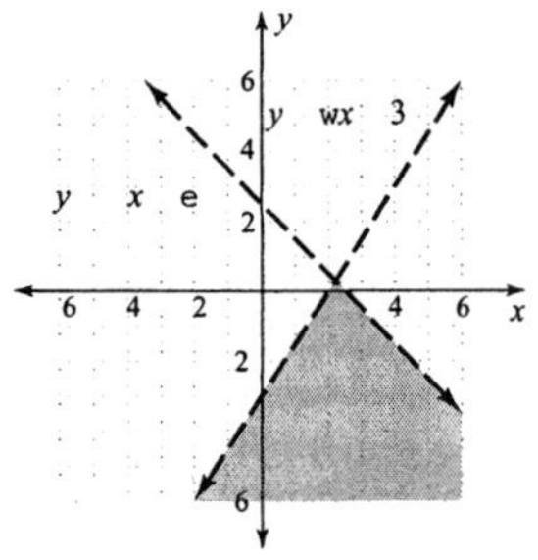

21.

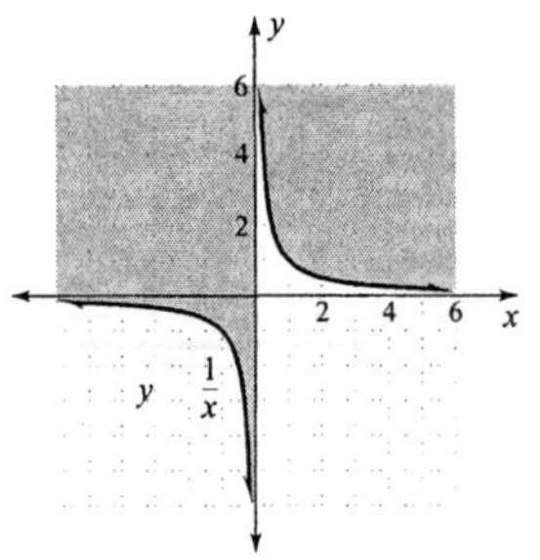

23.

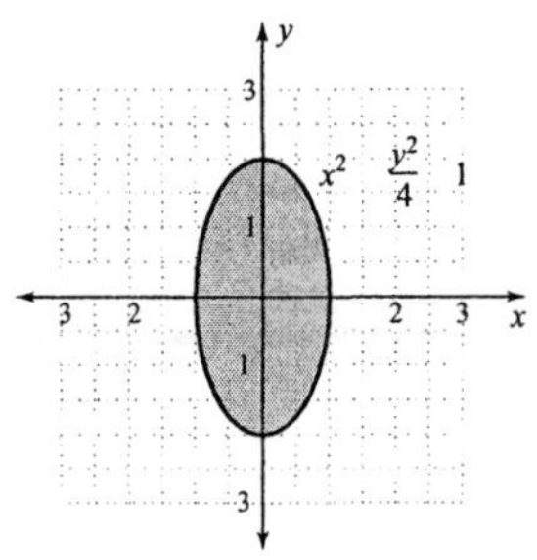

25.

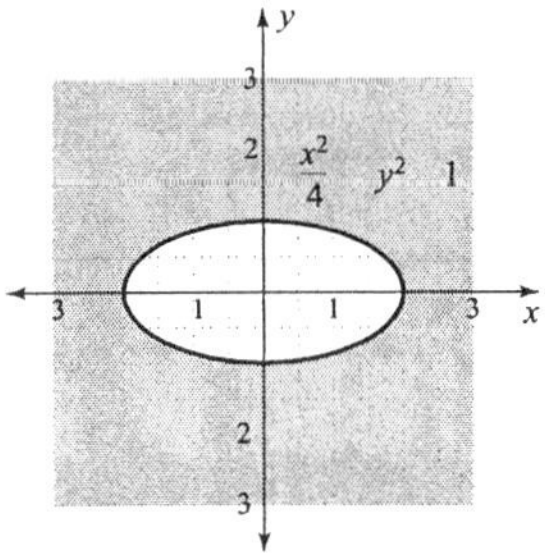

27.

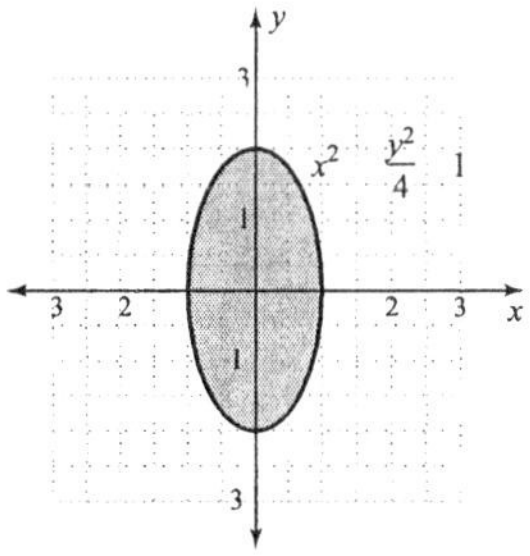

29. $y \le x^2$; graph **c**

31. $y \ge 0$ and $y \le -x^2 + 3$; graph **f**

33. $y \ge 0$ and $y \ge x^2 - 3$; graph **a**

35. x^2 is always ≥ 0, so if $y \ge x^2$, $y \ge 0$

37.

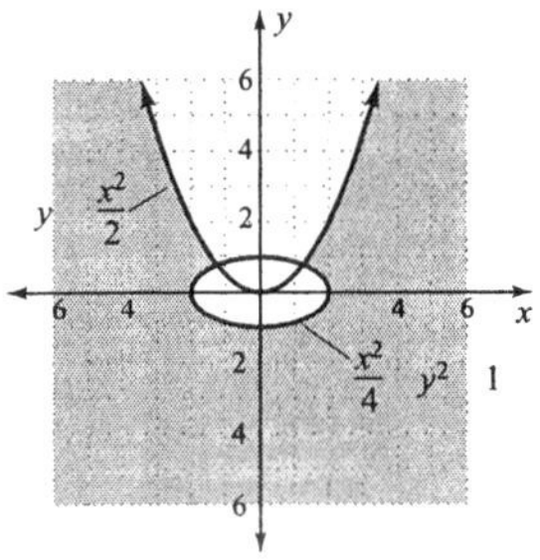

39.

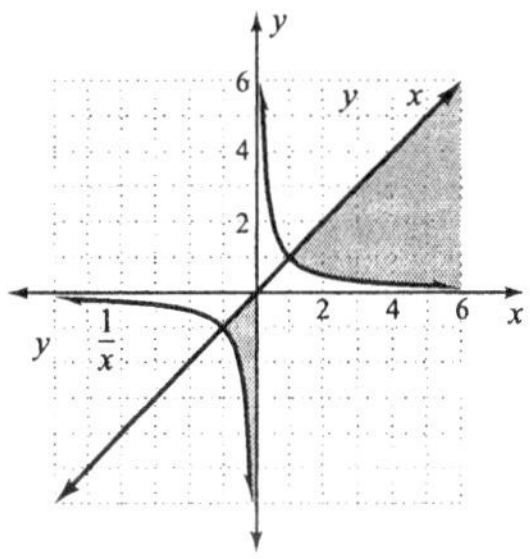

41.

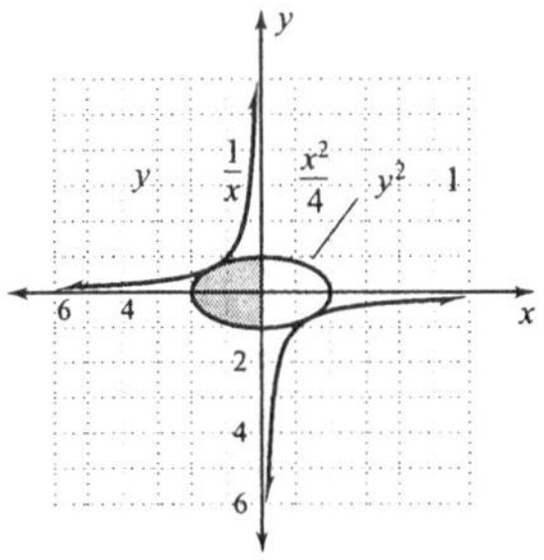

43. x = # of student tickets; y = # of reg. tickets

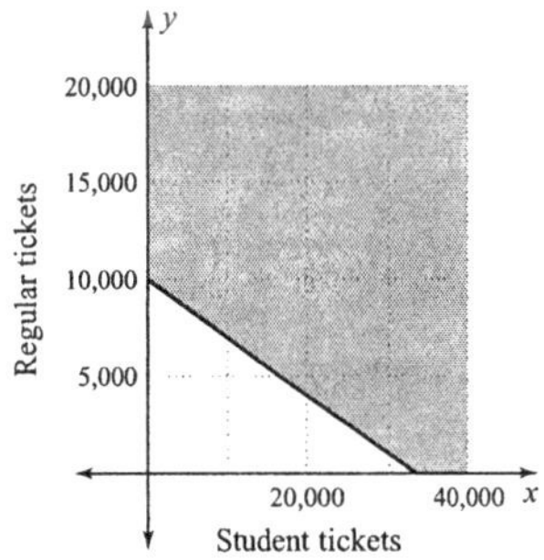

$15x + 50y \ge 500{,}000$

45. x = # of units IV; y = # of units fluid

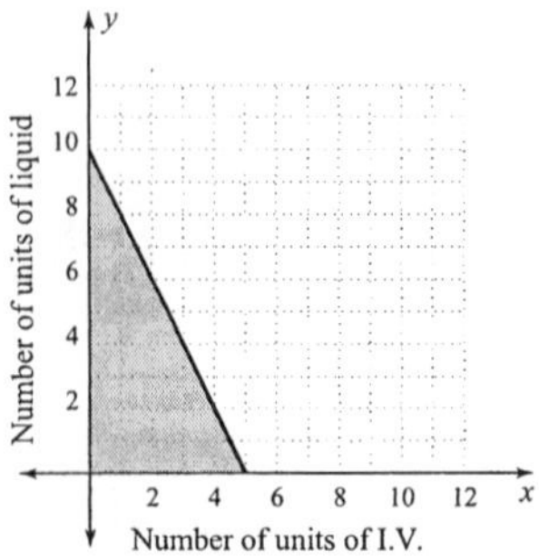

$500x + 250y \le 2500$

47. Solid boundaries are included in the solution, dashed boundaries are not.

Exercises 10.4

1. Matrix A has one row and two columns, it is a 1 x 2 matrix

3. Matrix F has 3 rows and 1 column, it is a 3 x 1 matrix.

5. $[2\ \ 3] + [2\ \ 4] = [2+2\ \ \ 3+4] = [4\ \ 7]$

7. $$\begin{bmatrix}1\\2\\3\end{bmatrix}+\begin{bmatrix}4\\5\\6\end{bmatrix}=\begin{bmatrix}1+4\\2+5\\3+6\end{bmatrix}=\begin{bmatrix}5\\7\\9\end{bmatrix}$$

9. $$\begin{bmatrix}6&7&8\\9&10&11\\12&13&14\end{bmatrix}-\begin{bmatrix}1&2&3\\4&5&6\\7&8&9\end{bmatrix}=\begin{bmatrix}6-1&7-2&8-3\\9-4&10-5&11-6\\12-7&13-8&14-9\end{bmatrix}=\begin{bmatrix}5&5&5\\5&5&5\\5&5&5\end{bmatrix}$$

11. $$\begin{bmatrix}1&3\\-1&-2\end{bmatrix}\begin{bmatrix}x\\y\end{bmatrix}=\begin{bmatrix}x+3y\\-x-2y\end{bmatrix}$$

13. $$\begin{bmatrix}1&3\\-1&-2\end{bmatrix}\begin{bmatrix}2\\-1\end{bmatrix}=\begin{bmatrix}-1\\0\end{bmatrix}$$

15. $$\begin{bmatrix}4&-2\\1&2\end{bmatrix}\begin{bmatrix}2\\-1\end{bmatrix}=\begin{bmatrix}10\\0\end{bmatrix}$$

17. $$\begin{bmatrix}1&3\\-1&-2\end{bmatrix}\begin{bmatrix}2&3\\4&5\end{bmatrix}=\begin{bmatrix}2+12&3+15\\-2-8&-3-10\end{bmatrix}=\begin{bmatrix}14&18\\-10&-13\end{bmatrix}$$

19. $\begin{bmatrix} 2 & 3 \\ 4 & 5 \end{bmatrix}\begin{bmatrix} 1 & 3 \\ -1 & -2 \end{bmatrix} = \begin{bmatrix} 2-3 & 6-6 \\ 4-5 & 12-10 \end{bmatrix} = \begin{bmatrix} -1 & 0 \\ -1 & 2 \end{bmatrix}$

21. $\begin{bmatrix} 4 & -2 \\ 1 & 2 \end{bmatrix}\begin{bmatrix} -2 & -1 \\ 1 & 2 \end{bmatrix} = \begin{bmatrix} -8-2 & -4-4 \\ -2+2 & -1+4 \end{bmatrix} = \begin{bmatrix} -10 & -9 \\ 0 & 3 \end{bmatrix}$

23. $\begin{bmatrix} -2 & -1 \\ 1 & 2 \end{bmatrix}\begin{bmatrix} 4 & -2 \\ 1 & 2 \end{bmatrix} = \begin{bmatrix} -8-1 & 4-2 \\ 4+2 & -2+4 \end{bmatrix} = \begin{bmatrix} -9 & 2 \\ 6 & 2 \end{bmatrix}$

25. det[A] = –2 – (–3) = 1

27. det[C] = 8 – (–2) = 10

29. det[A] = 1 (from exercise 25)

$[A]^{-1} = \begin{bmatrix} \frac{-2}{1} & \frac{-3}{1} \\ \frac{-(-1)}{1} & \frac{1}{1} \end{bmatrix} = \begin{bmatrix} -2 & -3 \\ 1 & 1 \end{bmatrix}$

31. det[C] = 10 (from exercise 27)

$[C]^{-1} = \begin{bmatrix} \frac{2}{10} & \frac{-(-2)}{10} \\ \frac{-1}{10} & \frac{4}{10} \end{bmatrix} = \begin{bmatrix} 0.2 & 0.2 \\ -0.1 & 0.4 \end{bmatrix}$

33. a. $[A][A]^{-1} = \begin{bmatrix} 1 & 0 \\ 0 & 1 \end{bmatrix}$

b. $[A]^{-1}[A] = \begin{bmatrix} 1 & 0 \\ 0 & 1 \end{bmatrix}$

35. $\begin{bmatrix} 1 & 2 \\ 4 & 4 \end{bmatrix}\begin{bmatrix} x \\ y \end{bmatrix} = \begin{bmatrix} 5 \\ 6 \end{bmatrix}$, $x = -2\ y = 3.5$

37. $\begin{bmatrix} 1 & 2 \\ 2 & 4 \end{bmatrix}\begin{bmatrix} x \\ y \end{bmatrix} = \begin{bmatrix} 5 \\ 6 \end{bmatrix}$, det[A] = 0, parallel lines.

39. $\begin{bmatrix} 1 & 2 \\ 0 & 4 \end{bmatrix}\begin{bmatrix} x \\ y \end{bmatrix} = \begin{bmatrix} 5 \\ 6 \end{bmatrix}$, $x = 2, y = 1.5$

41. $\begin{bmatrix} 1 & 2 \\ -2 & 4 \end{bmatrix}\begin{bmatrix} x \\ y \end{bmatrix} = \begin{bmatrix} 5 \\ 6 \end{bmatrix}$, $x = 1, y = 2$

43. The exercises all have 1x + 2y = 5 as one equation, and ax + 4y = 6 as the other. Changing *a* changes the slope of the graph. The y-intercept remains constant.

45. 5x = 28.2 + 2y, 5x – 2y = 28.2

$\begin{bmatrix} 5 & -2 \\ 3 & 5 \end{bmatrix}\begin{bmatrix} x \\ y \end{bmatrix} = \begin{bmatrix} 28.2 \\ -5.4 \end{bmatrix}$,

$x = 4.2, y = -3.6$

47. 3y + 6 = 2x, 2x – 3y = 6, –2x + 3y = 12

$\begin{bmatrix} 2 & -3 \\ -2 & 3 \end{bmatrix}\begin{bmatrix} x \\ y \end{bmatrix} = \begin{bmatrix} 6 \\ 12 \end{bmatrix}$, det[A] = 0;

parallel lines

49. $2x - 3y = 7$

$$\begin{bmatrix} 1 & -1.5 \\ 2 & -3 \end{bmatrix}\begin{bmatrix} x \\ y \end{bmatrix} = \begin{bmatrix} 3.5 \\ 7 \end{bmatrix}, \text{ det[A]} = 0;$$

$x - 1.5y = 3.5,\ x = 1.5y + 3.5$

$2x - 7 = 3y,\ 2x = 3y + 7, x = 1.5y + 3.5$

Coincident lines

51. a. det[E] = 0

b. $[E]^{-1} = \begin{bmatrix} \frac{2}{0} & \frac{-(-1)}{0} \\ \frac{-(-4)}{0} & \frac{2}{0} \end{bmatrix}$, zero denominators

c. matrix with no inverse

Exercises 10.5

1. $$\begin{bmatrix} 3 & 2 & 2 \\ 4 & -3 & -1 \\ 2 & 4 & 1 \end{bmatrix}\begin{bmatrix} x \\ y \\ z \end{bmatrix} = \begin{bmatrix} -2 \\ -16 \\ 7 \end{bmatrix}$$

(−2, 3, −1)

3. $$\begin{bmatrix} 2 & 3 & -1 \\ 4 & -2 & 2 \\ 1 & 4 & -\frac{1}{2} \end{bmatrix}\begin{bmatrix} x \\ y \\ z \end{bmatrix} = \begin{bmatrix} 8 \\ -6 \\ 9 \end{bmatrix}$$

$(\frac{1}{4}, 2, -1\frac{1}{2})$

5. determinant = 0
add first 2 equations
0 = 10; inconsistent
no solution

7. determinant = 0
add first 2 equations
x + y − z = 0, same as last equation
dependent; infinite # of solutions

9. $$\begin{bmatrix} 2 & 3 & 4 \\ -1 & 2 & 5 \\ -1 & 1 & -3 \end{bmatrix}\begin{bmatrix} x \\ y \\ z \end{bmatrix} = \begin{bmatrix} 13 \\ 4 \\ -11 \end{bmatrix}$$

(4, −1, 2)

11. $$\begin{bmatrix} 0 & 0 & 1 \\ 2^2 & 2 & 1 \\ 1^2 & 1 & 1 \end{bmatrix}\begin{bmatrix} a \\ b \\ c \end{bmatrix} = \begin{bmatrix} 3 \\ 1 \\ 0 \end{bmatrix}$$

$y = 2x^2 - 5x + 3$

13. $$\begin{bmatrix} (-2)^3 & (-2)^2 & -2 & 1 \\ (-1)^3 & (-1)^2 & -1 & 1 \\ 0 & 0 & 0 & 1 \\ 1^3 & 1^2 & 1 & 1 \end{bmatrix}\begin{bmatrix} a \\ b \\ c \\ d \end{bmatrix} = \begin{bmatrix} -27 \\ -7 \\ -1 \\ 3 \end{bmatrix}$$

$y = 2x^3 - x^2 + 3x - 1$

15. $$\begin{bmatrix} (-2)^4 & (-2)^3 & (-2)^2 & -2 & 1 \\ (-1)^4 & (-1)^3 & (-1)^2 & -1 & 1 \\ 0 & 0 & 0 & 0 & 1 \\ 1^4 & 1^3 & 1^2 & 1 & 1 \\ 2^4 & 2^3 & 2^2 & 2 & 1 \end{bmatrix}\begin{bmatrix} a \\ b \\ c \\ d \\ e \end{bmatrix} = \begin{bmatrix} -7 \\ -9 \\ -5 \\ -1 \\ 45 \end{bmatrix}$$

$y = 2x^4 + 3x^3 - 2x^2 + x - 5$

17. 2nd equation changes to s – m + 0 = 0 $\begin{bmatrix} 1 & 1 & 1 \\ 1 & -1 & 0 \\ 0 & 0 & 1 \end{bmatrix}\begin{bmatrix} s \\ m \\ l \end{bmatrix} = \begin{bmatrix} 620 \\ 0 \\ 20 \end{bmatrix}$

19. a. [A] is 1 × 3; [B] is 3 × 4; product will be 1 × 4

b. [C] = [\$4800 \$5100 \$5400 \$6000]

21. a. small = 4.5 oz ÷ 16 oz per lb = $\frac{4.5}{16}$ lb; medium = 12 oz ÷ 16 oz per lb = 0.75 lb

large = 3 lbs

b. enter weight as matrix [F], recall [B] from exercise 19 and multiply

$$\begin{bmatrix} \frac{4.5}{16} & 0.75 & 3 \end{bmatrix} \cdot \begin{bmatrix} 500 & 450 & 400 & 300 \\ 100 & 150 & 200 & 300 \\ 20 & 20 & 20 & 20 \end{bmatrix} = \begin{bmatrix} 275.6 & 299.1 & 322.5 & 369.4 \end{bmatrix}$$

c. multiply result from above by \$3.50, store as [E]; [\$965 \$1047 \$1129 \$1293]

23. [\$4800 \$5100 \$5400 \$6000] – [\$522 \$507 \$492 \$462] –
[\$965 \$1047 \$1129 \$1293] = [\$3313 \$3546 \$3779 \$4245]

25. The 3 equations are: $-\frac{3}{100} = 2a(0) + b$ (slope at K) $b = -\frac{3}{100}$

$\frac{5}{100} = 2a(2000) + b$ (slope at L) substitute b, solve for a; $a = \frac{1}{50{,}000}$

$700 = a(0)^2 + b(0) + c$ (parabola through K); c = 700

equation for road grade: $y = \frac{1}{50{,}000}x^2 - \frac{3}{100}x + 700$

could also be solved using matrix equation $\begin{bmatrix} 0 & 1 & 0 \\ 4000 & 1 & 0 \\ 0 & 0 & 1 \end{bmatrix}\begin{bmatrix} a \\ b \\ c \end{bmatrix} = \begin{bmatrix} -\frac{3}{100} \\ \frac{5}{100} \\ 700 \end{bmatrix}$

27. $-\frac{4}{100} = 2a(0) + b$ (slope at M); $0a + 1b + 0c = -\frac{1}{25}$

$\frac{2}{100} = 2a(2400) + b$ (slope at N) $4800a + 1b + 0c = \frac{1}{50}$

$1200 = 2a(2400)^2 + b(2400) + c$ (parabola through N) $2(2400)^2 + 2400b + 1c = 1200$

$\begin{bmatrix} 0 & 1 & 0 \\ 4800 & 1 & 0 \\ 2(2400)^2 & 2400 & 1 \end{bmatrix}\begin{bmatrix} a \\ b \\ c \end{bmatrix} = \begin{bmatrix} -\frac{1}{25} \\ \frac{1}{50} \\ 1200 \end{bmatrix}$ $y = \frac{1}{80{,}000}x^2 - \frac{1}{25}x + 1224$

Chapter 10 Review

1. opening to the left indicates $x = ay^2$ equation; $-3 = a(1)^2$; $a = -3$: $x = -3y^2$

3. $\dfrac{x^2}{3^2} + \dfrac{y^2}{2^2} = 1$; $\dfrac{x^2}{9} + \dfrac{y^2}{4} = 1$

5. parabola, $x^2 = y - 5$

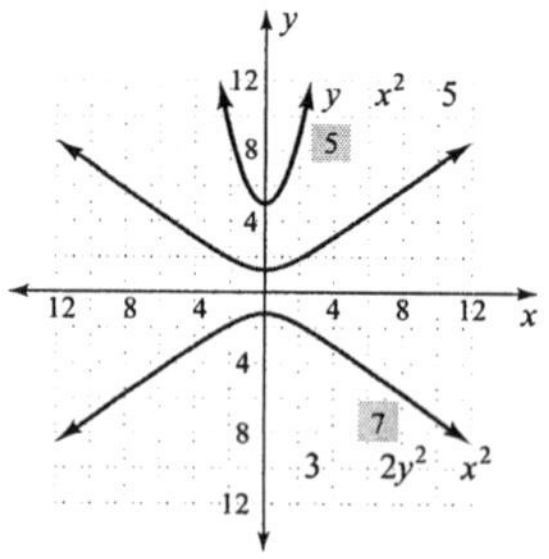

7. divide both sides by –3;

$\dfrac{2y^2}{3} - \dfrac{x^2}{3} = 1$;

hyperbola; $y^2 = \frac{1}{2}x^2 + \frac{3}{2}$

9. circle, $y = \pm\sqrt{8 - x^2}$

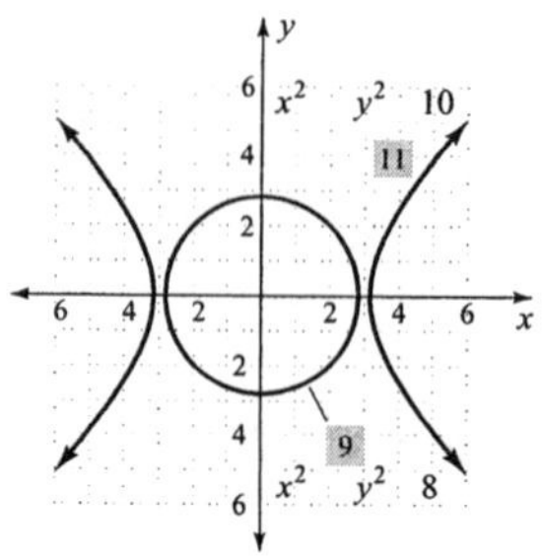

11. divide both sides by 10; $\dfrac{x^2}{10} - \dfrac{y^2}{10} = 1$;

hyperbola; $x = \pm\sqrt{10 + y^2}$

13. $y = \pm\sqrt{2 - \frac{x^2}{2}}$

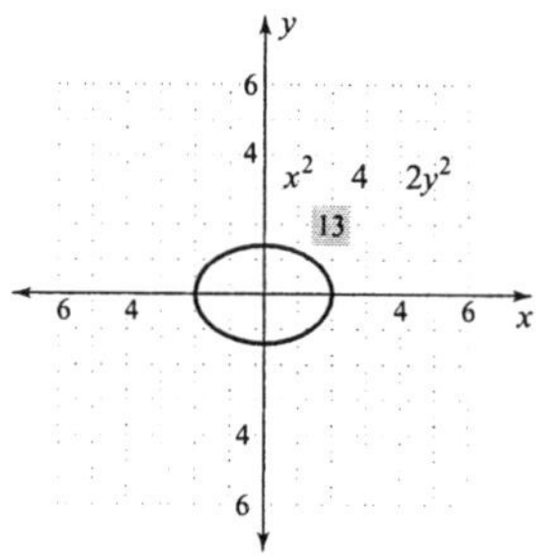

15. parabola and line

$x^2 + x - 2 = -x - 3$;

$x^2 + 2x + 1 = 0$

$(x + 1)^2 = 0$; $x = -1$

$-(-1) - 3 = -2$

$(-1, -2)$

17. parabola and line

$x^2 - 4x - 1 = -x + 3$

$x^2 - 3x - 4 = 0$

$(x - 4)(x + 1) = 0$; $x = 4$ or $x = -1$

$-4 + 3 = -1$; $-(-1) + 3 = 4$

$(4, -1), (-1, 4)$

19. circle and parabola

$x^2 = y - 1;\ \ y - 1 + y^2 = 16$

$y^2 + y - 17 = 0$

$$y = \frac{-1 \pm \sqrt{1^2 - 4(1)(-17)}}{2(1)}$$

$y \approx -4.653$ or $y \approx 3.653$

$x^2 = -4.653 - 1$ (not possible)

$x^2 = 3.653 - 1 = 2.653$

$x = \pm\sqrt{2.653} \approx \pm 1.629$

$(\pm 1.629,\ 3.653)$

21. parabolas

$x^2 - x = -x^2 + 3$

$2x^2 - x - 3 = 0$

$(2x - 3)(x + 1) = 0$

$x = \frac{3}{2}$ or $x = -1$

$(\frac{3}{2})^2 - \frac{3}{2} = \frac{3}{4}$

$(-1)^2 - (-1) = 2$

$(-1, 2), (\frac{3}{2}, \frac{3}{4})$

23. parabolas

$x^2 + 1 = x^2 - 5$

$1 = -5$; inconsistent

{ } or $\varnothing$

25. $2^x = 4 - 2^x$

$2 \cdot 2^x = 4$

$2^{x+1} = 2^2$

$x + 1 = 2;\ \ x = 1$

$y = 2^1;\ \ y = 2$

27. $\log x = 3 - 2\log x$

$3\log x = 3$

$\log x^3 = 3$

$10^3 = x^3;\ \ x = 10$

$y = \log 10;\ \ y = 1$

29.

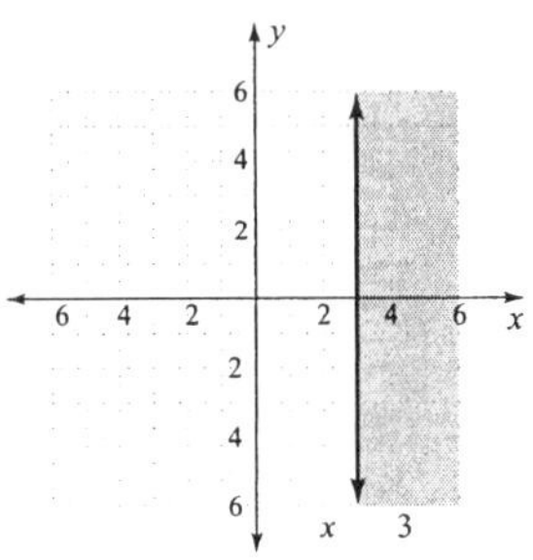

31.

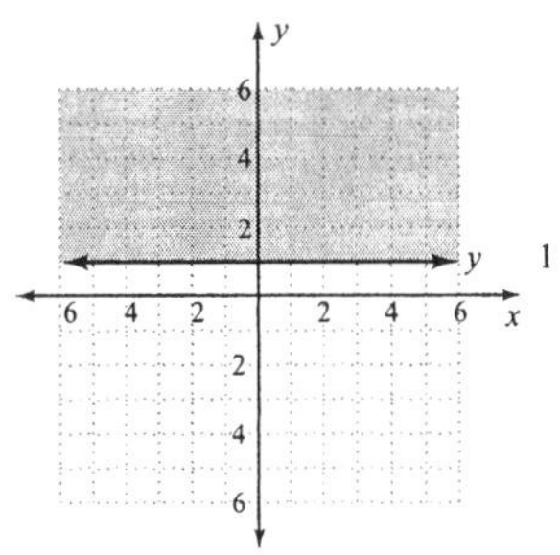

33.

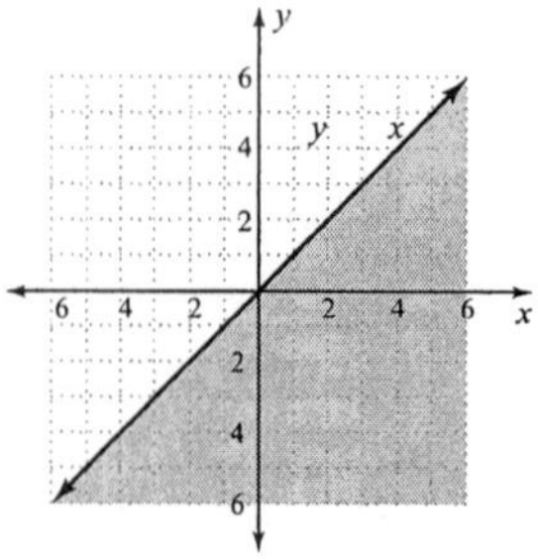

35.

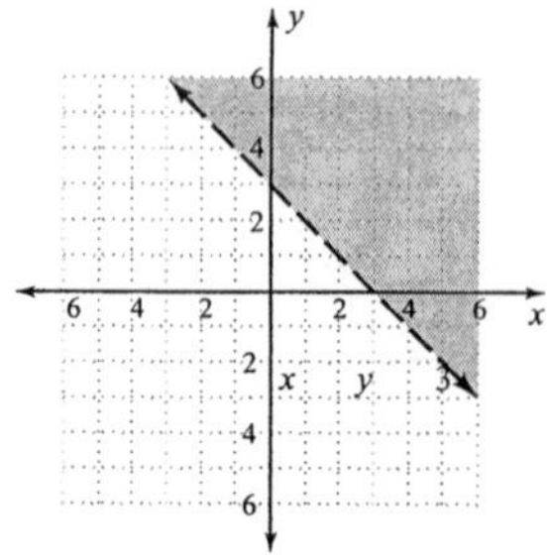

37.

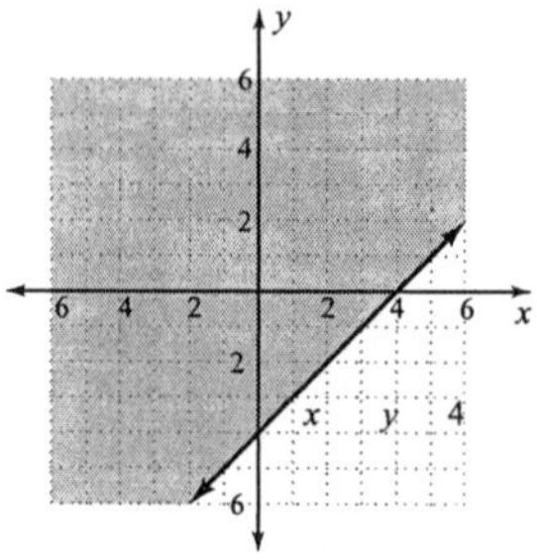

39.

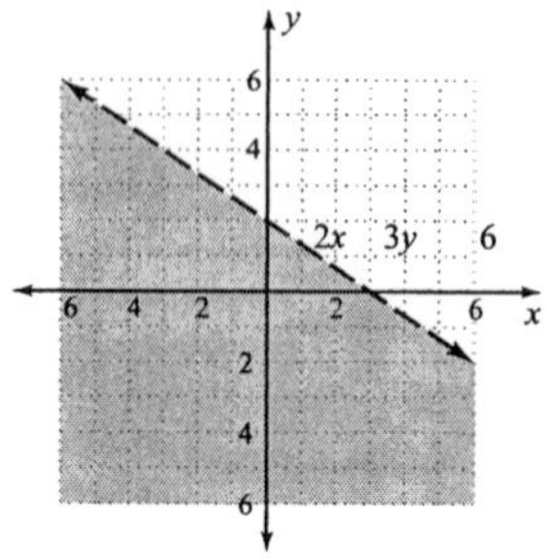

41.

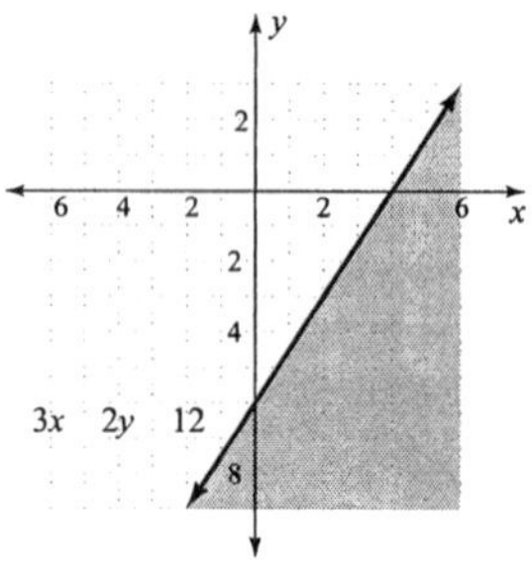

43.

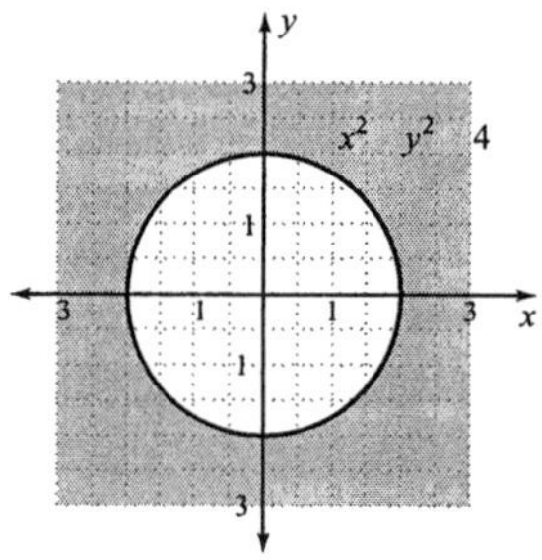

45.

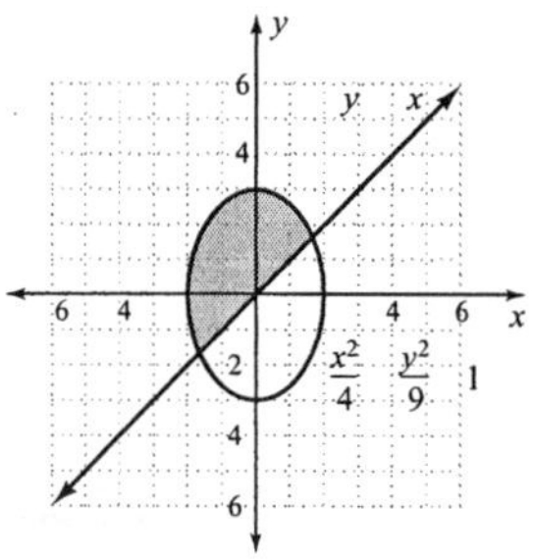

47.

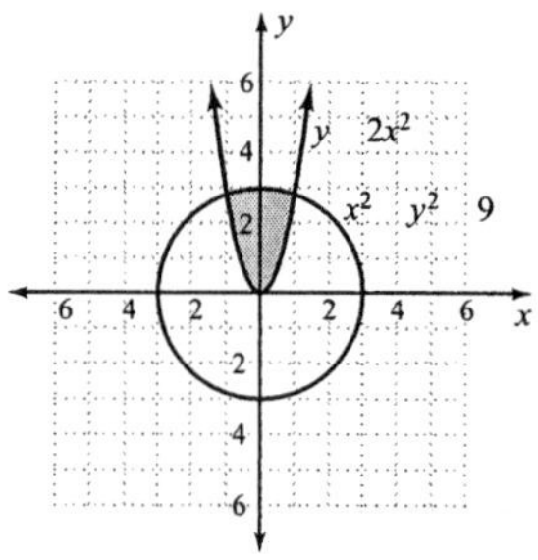

49. a. y is both < 0 and $< -x^2 + 4$
graph 3

b. y is > 0 and $> -x^2 + 4$
graph 1

c. y is < 0 and $> -x^2 + 4$
graph 4

d. y is > 0 and $< -x^2 + 4$
graph 2

51. $[A] + [B] = \begin{bmatrix} 2 & 4 \\ 1 & 3 \end{bmatrix} + \begin{bmatrix} 1 & 2 \\ 4 & 3 \end{bmatrix} = \begin{bmatrix} 2+1 & 4+2 \\ 1+4 & 3+3 \end{bmatrix} = \begin{bmatrix} 3 & 6 \\ 5 & 6 \end{bmatrix}$

53. $[A] \cdot [B] = \begin{bmatrix} 2 & 4 \\ 1 & 3 \end{bmatrix} \cdot \begin{bmatrix} 1 & 2 \\ 4 & 3 \end{bmatrix} = \begin{bmatrix} 2\cdot1+4\cdot4 & 2\cdot2+4\cdot3 \\ 1\cdot1+3\cdot4 & 1\cdot2+3\cdot3 \end{bmatrix} = \begin{bmatrix} 18 & 16 \\ 13 & 11 \end{bmatrix}$

55. $[C] \cdot [A] = \begin{bmatrix} 1.5 & -2 \\ -0.5 & 1 \end{bmatrix} \cdot \begin{bmatrix} 2 & 4 \\ 1 & 3 \end{bmatrix} = \begin{bmatrix} 1.5\cdot2+(-2)\cdot1 & 1.5\cdot4+(-2)\cdot3 \\ -0.5\cdot2+1\cdot1 & -0.5\cdot4+1\cdot3 \end{bmatrix} = \begin{bmatrix} 1 & 0 \\ 0 & 1 \end{bmatrix}$

57. $-2x + 3y = 12$
$2x - y = 4$
$\begin{bmatrix} -2 & 3 \\ 2 & -1 \end{bmatrix} \cdot \begin{bmatrix} x \\ y \end{bmatrix} = \begin{bmatrix} 12 \\ 4 \end{bmatrix}$
$x = 6, y = 8$

59. $2x - y = 13$
$x + 2y = -6$
$\begin{bmatrix} 2 & -1 \\ 1 & 2 \end{bmatrix} \cdot \begin{bmatrix} x \\ y \end{bmatrix} = \begin{bmatrix} 13 \\ -6 \end{bmatrix}$
$x = 4, y = -5$

61. $2x + y = 5$
$4x + 2y = 10$
determinant = 0
divide 2nd equation by 2
infinitely many solutions
coincident lines

63. $2x - 3y - 4z = -4$
$3x - 2y + z = -17$
$4x - y - 3z = -3$
$\begin{bmatrix} 2 & -3 & -4 \\ 3 & -2 & 1 \\ 4 & -1 & -3 \end{bmatrix} \cdot \begin{bmatrix} x \\ y \\ z \end{bmatrix} = \begin{bmatrix} -4 \\ -17 \\ -3 \end{bmatrix}$
$x = -2, y = 4, z = -3$

65. $4x + 2y - 3z = 5$
$-3x - 3y + 2z = 7$
$x - y - z = -2$

determinant = 0

add first 2 equations;
$x - y - z = 2$; inconsistent with 3rd equation
no solution

67. $-0.03 = 2a(0) + b;\ b = -0.03$ [note: 2800 to the left of (2800, 300) is (0, 300)]

$0.04 = 2a(2800) + (-0.03);\ a = \dfrac{0.07}{5600} = \dfrac{1}{80{,}000}$

$300 = \frac{1}{80{,}000}(2800)^2 - 0.03(2800) + c;\ c = 286$

69. $\begin{bmatrix} 3 & -1 \\ -6 & 2 \end{bmatrix} \cdot \begin{bmatrix} x \\ y \end{bmatrix} = \begin{bmatrix} 4 \\ -8 \end{bmatrix}$; $\det[A] = 0$;

$-2(3x - y) = (-2)4$

$-6x + 2y = -8$; coincident lines; infinitely many solutions;

$y = 3x - 4$; possible points $(0, -4)$; $(1, -1)$; $(2, 2)$

71. $a(0)^2 + b(0) + c = -2$
$a(1)^2 + b(1) + c = -1$
$a(2)^2 + b(2) + c = -8$

$$\begin{bmatrix} 0 & 0 & 1 \\ 1 & 1 & 1 \\ 2^2 & 2 & 1 \end{bmatrix} \cdot \begin{bmatrix} a \\ b \\ c \end{bmatrix} = \begin{bmatrix} -2 \\ -1 \\ -8 \end{bmatrix}$$

$y = -4x^2 + 5x - 2$

73. $a(-1)^3 + b(-1)^2 + c(-1) + d = 0$
$a(1)^3 + b(1)^2 + c(1) + d = 2$
$a(2)^3 + b(2)^2 + c(2) + d = 0$
$a(3)^3 + b(3)^2 + c(3) + d = 4$

$$\begin{bmatrix} (-1)^3 & (-1)^2 & -1 & 1 \\ 1^3 & 1^2 & 1 & 1 \\ 2^3 & 2^2 & 2 & 1 \\ 3^3 & 3^2 & 3 & 1 \end{bmatrix} \cdot \begin{bmatrix} a \\ b \\ c \\ d \end{bmatrix} = \begin{bmatrix} 0 \\ 2 \\ 0 \\ 4 \end{bmatrix}$$

$y = x^3 - 3x^2 + 4$

75. $a(-2)^4 + b(-2)^3 + c(-2)^2 + d(-2) + e = 8$
$a(-1)^4 + b(-1)^3 + c(-1)^2 + d(-1) + e = 1$
$a(0)^4 + b(0)^3 + c(0)^2 + d(0) + e = 2$
$a(2)^4 + b(2)^3 + c(2)^2 + d(2) + e = 4$
$a(3)^4 + b(3)^3 + c(3)^2 + d(3) + e = 53$

$$\begin{bmatrix} (-2)^4 & (-2)^3 & (-2)^2 & -2 & 1 \\ (-1)^4 & (-1)^3 & (-1)^2 & -1 & 1 \\ 0 & 0 & 0 & 0 & 1 \\ 2^4 & 2^3 & 2^2 & 2 & 1 \\ 3^4 & 3^3 & 3^2 & 3 & 1 \end{bmatrix} \cdot \begin{bmatrix} a \\ b \\ c \\ d \\ e \end{bmatrix} = \begin{bmatrix} 8 \\ 1 \\ 2 \\ 4 \\ 53 \end{bmatrix}$$

$y = x^4 - 3x^2 - x + 2$

Chapter 10 Test

1. $x^2 + y^2 = 16^2,\ x^2 + y^2 = 256$

2. $\dfrac{x^2}{1^2} + \dfrac{y^2}{4^2} = x^2 + \dfrac{y^2}{16} = 1$

3. Opening to the left indicates $x = ay^2$ equation; $-4 = a(1)^2$, $a = -4$; $x = -4y^2$

4. No y-intercepts indicates $\dfrac{x^2}{a^2} - \dfrac{y^2}{b^2} = 1$ equation; $\dfrac{x^2}{3^2} - \dfrac{y^2}{2^2} = \dfrac{x^2}{9} - \dfrac{y^2}{4} = 1$

5. parabola and line

$3x^2 - 4x - 5 = x - 3$

$3x^2 - 5x - 2 = 0$

$(3x + 1)(x - 2) = 0$

$x = -\frac{1}{3}$ or $x = 2$

$y = -\frac{1}{3} - 3 = -3\frac{1}{3} = -\frac{10}{3}$

$y = 2 - 3 = -1$

$(-\frac{1}{3}, -\frac{10}{3}), (2, -1)$

6. parabola and line

$x^2 - 3x + 2 = 0.5x - 2$

$x^2 - 3.5x + 4 = 0$

$$\frac{-(-3.5) \pm \sqrt{(-3.5)^2 - 4(1)(4)}}{2(1)} = \frac{3.5 \pm \sqrt{-3.75}}{2}$$

not a real number, no solution

7. circle and hyperbola

$x^2 + y^2 = -x^2 + y^2$

$2x^2 = 0;\ x = 0$

$0 + y^2 = 4;\ y = \pm 2$

$(0, \pm 2)$

8. parabola and circle

$x^2 = 4 - y^2;\ y = 4 - y^2 + 1$

$y^2 + y - 5 = 0$

$$\frac{-1 \pm \sqrt{1^2 - 4(1)(-5)}}{2(1)} = \frac{-1 \pm \sqrt{21}}{2}$$

$y = 1.791$ or $y = -2.791$

$-2.791 = x^2 + 1$ (not possible, discard soln.)

$1.791 = x^2 + 1;\ x^2 = 0.791;$

$x = \pm\sqrt{0.791} \approx \pm 0.890$

$(\pm 0.890, 1.791)$

9.

$3^x = 5 - 3^x$

$2(3^x) = 5$

$3^x = 2.5$

$\log 3^x = \log 2.5$

$x\log 3 = \log 2.5$

$x = \dfrac{\log 2.5}{\log 3} \approx 0.834$

$y = 3^{0.834} \approx 2.5$

$(0.834, 2.5)$

10.

$\log_3 x = 4 - \log_3 x$

$2\log_3 x = 4$

$\log_3 x^2 = 4$

$3^4 = x^2$

$81 = x^2$; $x = \pm 9$; (discard negative)

$y = \log_3 9$; $3^y = 9$; $3^y = 3^2$; $y = 2$

$(9, 2)$

11.

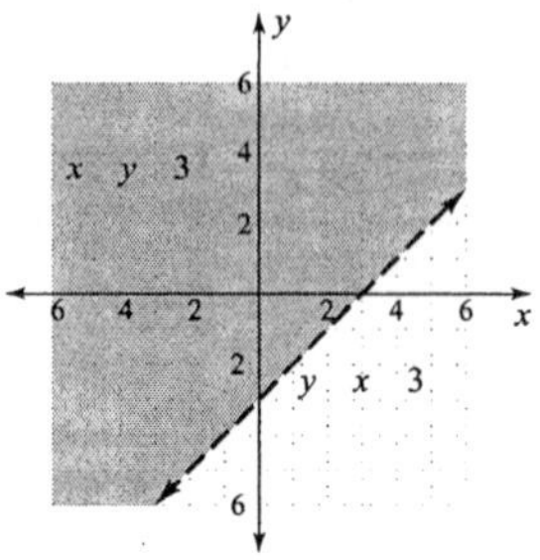

12.

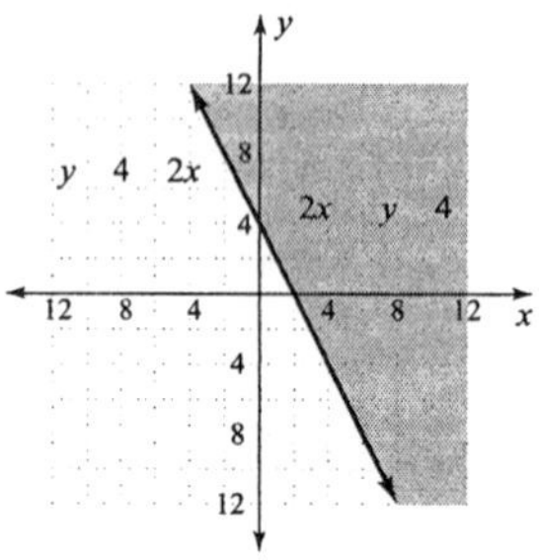

13.

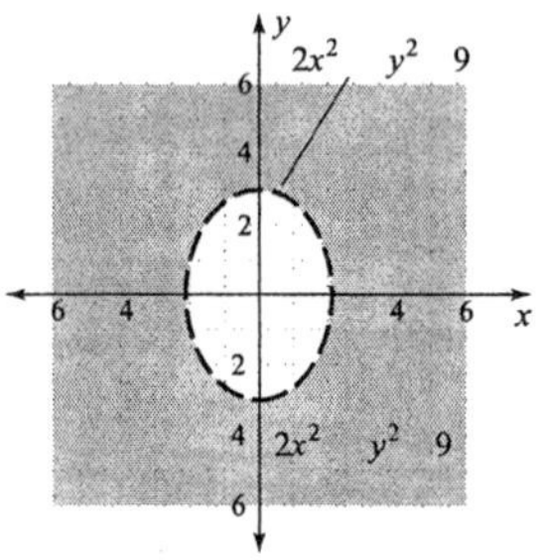

14.

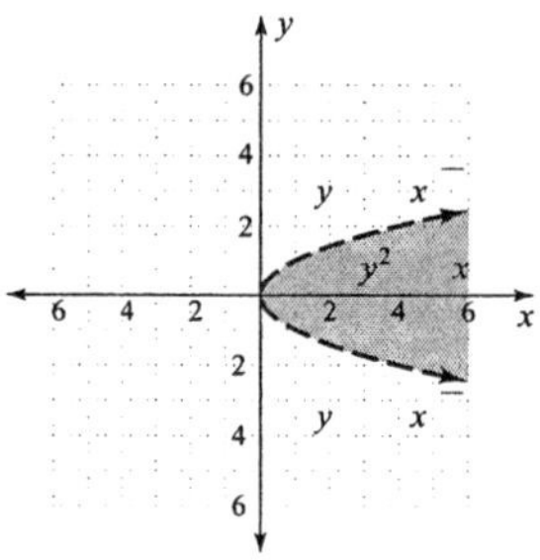

15.

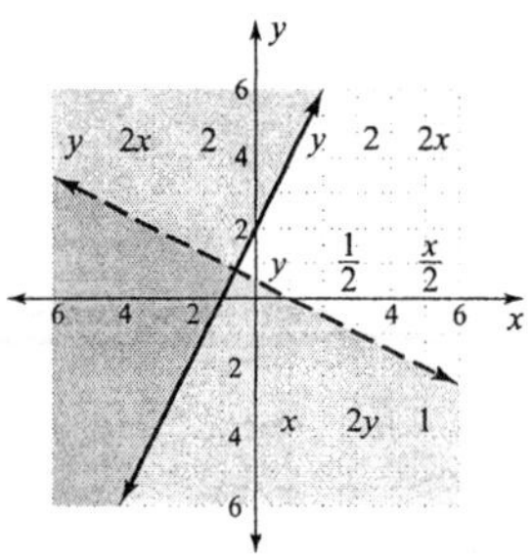

16.

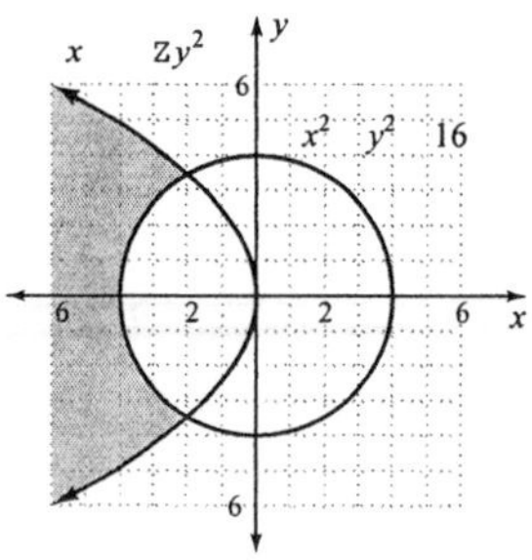

17. $-0.03 = 2a(0) + b$; $b = -0.03 = -\frac{3}{100}$

$0.06 = 2a(3200) + (-0.03)$; $a = \dfrac{0.09}{6400} = \dfrac{9}{640{,}000}$

$1500 = \frac{9}{640{,}000}(0)^2 - 0.03(0) + c$; $c = 1500$

18. a. $[A]\cdot[B] = \begin{bmatrix} 3 & -5 \\ -1 & 2 \end{bmatrix}\cdot\begin{bmatrix} 2 & 5 \\ 1 & 3 \end{bmatrix} = \begin{bmatrix} 3(2)+(-5)(1) & 3(5)+(-5)(3) \\ -1(2)+2(1) & -1(5)+2(3) \end{bmatrix} = \begin{bmatrix} 1 & 0 \\ 0 & 1 \end{bmatrix}$

b. $[A]^{-1} = \begin{bmatrix} 2 & 5 \\ 1 & 3 \end{bmatrix}$; (notice answer to **a**, $[A]\cdot[B]$ gave identity matrix, $[B] = [A]^{-1}$)

c. $\det[B] = (2)(3) - (5)(1) = 6 - 5 = 1$

19. $2x - 3y = 6$

$4x + 5y = -32$

$\begin{bmatrix} 2 & -3 \\ 4 & 5 \end{bmatrix}\cdot\begin{bmatrix} x \\ y \end{bmatrix} = \begin{bmatrix} 6 \\ -32 \end{bmatrix}$

$x = -3,\ y = -4$

20. $\begin{bmatrix} 3 & -2 & 1 \\ 2 & -3 & 4 \\ 1 & 0 & -1 \end{bmatrix}\cdot\begin{bmatrix} x \\ y \\ z \end{bmatrix} = \begin{bmatrix} 4 \\ 0 \\ 2 \end{bmatrix}$

$\det[A] = 0$

$-3(3x - 2y + z) = -3(4)$

$2(2x - 3y + 4z) = 2(0)$

$-9x + 6y - 3z = -12$

$4x - 6y + 8z = 0$ (add equations)

$-5x + 5z = 0$ (divide both sides by -5)

$x - z = 0$, inconsistent with third equation

21. $\begin{bmatrix} 1 & -2 & 1 \\ 2 & -3 & 0 \\ 1 & 0 & -1 \end{bmatrix}\cdot\begin{bmatrix} x \\ y \\ z \end{bmatrix} = \begin{bmatrix} 8 \\ 11 \\ 2 \end{bmatrix}$

$x = 4,\ y = -1,\ z = 2$

Exercises 11.1

1. a. GG, GB, BG, BB

b. GGG, GGB, GBG, GBB, BGG, BGB, BBG, BBB

c. GGGG, GGGB, GGBG, GGBB, GBGG, GBGB, GBBG, GBBB, BGGG, BGGB, BGBG, BGBB, BBGG, BBGB, BBBG, BBBB

3. Let Sergio Garcia = G, Vijay Singh = S and Tiger Woods = W
GSW, GWS, SWG, SGW, WSG, WGS

5.

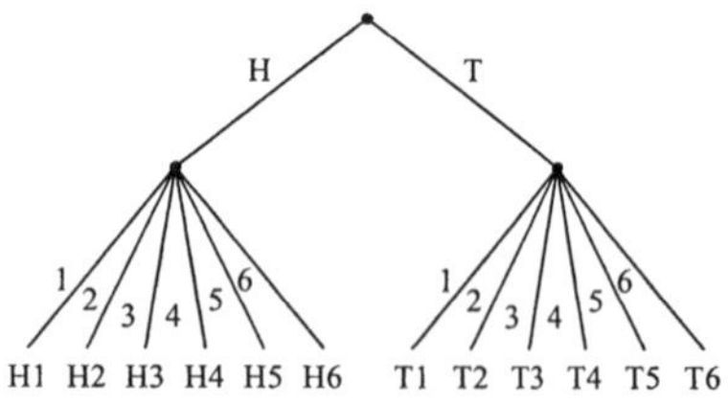

7.

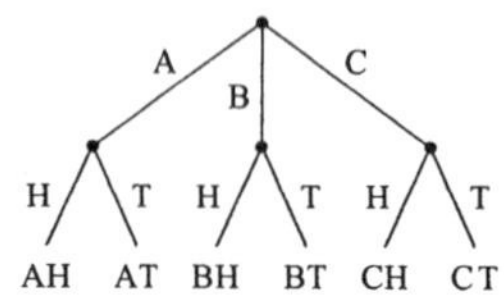

9. a. 5 children = $2^5 = 32$

b. 6 children = $2^6 = 64$

c. 10 children = $2^{10} = 1024$

d. n children = 2^n

11. There are two choices each time (B or G, H or T). The number of outcomes and the samples spaces (if H replaces B and T replaces G) are the same.

13. 10 choices for each of 9 places; 10^9

15. 4 choices for each of 20 questions; 4^{20}

17. a. 40 choices for each of 3 places; 40^3

b. 1 less choice for each place; $40 \cdot 39 \cdot 38 = 59{,}280$

19. $5 + 3 = 8$ choices

21. a. exactly 3 boys = 4; exactly 2 boys = 6; $4 + 6 = 10$

b. only 4 children, 3 girls and 2 boys are not possible

c. exactly 1 girl = 4; exactly 1 boy = 4; $4 + 4 = 8$

d. exactly 3 girls and exactly 1 boy = 4

23. a. $2 \cdot 26 \cdot 26 = 1352$

b. $2 \cdot 26 \cdot 26 \cdot 26 = 35{,}152$

c. $1352 + 35{,}152 = 36{,}504$

25. a. $27 \cdot 15 = 405$
b. $27 + 15 = 42$
c. $27 \cdot 26 = 702$
d. $15 \cdot 14 = 210$

27. a. $5 + 12 = 17$
b. $15 \cdot 3 = 45$
c. $(12 \cdot 11 \cdot 10) \cdot 15 = 19{,}800$
d. $12 + 15 = 27$
e. $15 \cdot 14 \cdot 13 = 2730$
f. $5 \cdot (12 + 3) = 75$

29. a. 6
b. one 2 = 10; two 2's = 1; 10 + 1 = 11
c. 2
d. a 2 = 10; a 3 = 10; 10 + 10 = 20
e. one 3 = 10; two 3's = 1; 10 + 1 = 11

31. To expand the listing to outcomes for 3 dice; put a 1 behind each of the 36 outcomes, then put a 2 behind each of the original 36 outcomes and repeat for 3, 4, 5 and 6. The total will be $6 \cdot 36 = 216$.

33. a. 8 for the first number, 2 for the middle and 10 for the last; $8 \cdot 2 \cdot 10 = 160$
b. still 8 for the first number; $8 \cdot 10 \cdot 10 = 800$
c. $8 \cdot 10 \cdot 10 \cdot 8 \cdot 10 \cdot 10 \cdot 10 \cdot 10 \cdot 10 \cdot 10 = 8^2 \cdot 10^8$
d. 555 occurs only once in the three digit prefix; break up the phone number into area code, prefix, number: $(8 \cdot 10 \cdot 10) \cdot (8 \cdot 10 \cdot 10 - 1) \cdot (10 \cdot 10 \cdot 10 \cdot 10) = 8 \cdot 10^2 \cdot 799 \cdot 10^4 = 8 \cdot 799 \cdot 10^6$
e. Answers will vary.

Exercises 11.2

1. a. $8! = 8 \cdot 7 \cdot 6 \cdot 5 \cdot 4 \cdot 3 \cdot 2 \cdot 1 = 40{,}320$
b. $10! = 10 \cdot 9 \cdot 8 \cdot 7 \cdot 6 \cdot 5 \cdot 4 \cdot 3 \cdot 2 \cdot 1 = 3{,}628,800$
c. $0! = 1$
d. $(8 + 2)! = 10! = 3{,}628{,}800$
e. $8! + 2! = 40{,}320 + 2 \cdot 1 = 40{,}322$
f. $(8 - 2)! = 6! = 6 \cdot 5 \cdot 4 \cdot 3 \cdot 2 \cdot 1 = 720$

3. a. false, missing the last 3 factors
b. true
c. true
d. false, $0! = 1$
e. false, $3! = 6$, $2! = 2$; $3!(2!) = 6(2) = 12$
f. true; use associative property of multiplication and rearrange factors;
$6! = 6 \cdot 5 \cdot 4 \cdot 3 \cdot 2 \cdot 1 = (5 \cdot 2) \cdot (6 \cdot 3) \cdot (4 \cdot 1) = 10 \cdot 18 \cdot 4 = 10 \cdot 9 \cdot 2 \cdot 4 = 10 \cdot 9 \cdot 8$

5. a. $5! = 5 \cdot 4 \cdot 3 \cdot 2 \cdot 1 = 120;\ 2! = 2 \cdot 1 = 2;\ 120 \cdot 2 = 240$

b. $3! = 3 \cdot 2 \cdot 1 = 6;\ 1! = 1;\ 6 \cdot 1 = 6$

c. $0! = 1;\ 4! = 4 \cdot 3 \cdot 2 \cdot 1 = 24;\ 1 \cdot 24 = 24$

d. $\dfrac{6!}{(6-2)!(2!)} = \dfrac{6 \cdot 5 \cdot 4!}{(4!)(2!)} = \dfrac{30}{2} = 15$

e. $\dfrac{8!}{(8-2)!(2!)} = \dfrac{8 \cdot 7 \cdot 6!}{(6!)(2!)} = \dfrac{56}{2} = 28$

f. $0! + 2! = 1 + 2 = 3$

7. After each is used there is one less, factorial; $6! = 720$

9. Repeats are allowed, multiplication principal; 7 choices for 7 days; $7^7 = 823{,}543$

11. After each is used there is one less, factorial; $7! = 5040$

13. Set contains 4 elements; ${}_4P_3$

15. Set of 4 elements, 2 at a time; ${}_4P_2$: ab, ac, ad, ba, bc, bd, ca, cb, cd, da, db, dc

17. a. ${}_5P_4$, $n = 5, r = 4;\ \dfrac{5!}{(5-4)!} = \dfrac{5 \cdot 4 \cdot 3 \cdot 2 \cdot 1}{1} = 120$

b. ${}_5P_5$, $n = 5, r = 5;\ \dfrac{5!}{(5-5)!} = \dfrac{5 \cdot 4 \cdot 3 \cdot 2 \cdot 1}{0!} = \dfrac{120}{1} = 120$

c. ${}_{12}P_1$, $n = 12, r = 1;\ \dfrac{12!}{(12-1)!} = \dfrac{12 \cdot 11!}{11!} = 12$

19. a. ${}_5P_3$, $n = 5, r = 3;\ \dfrac{5!}{(5-3)!} = \dfrac{5 \cdot 4 \cdot 3 \cdot 2!}{2!} = 60$

b. ${}_{52}P_3$, $n = 52, r = 3;\ \dfrac{52!}{(52-3)!} = \dfrac{52 \cdot 51 \cdot 50 \cdot 49!}{49!} = 132{,}600$

c. ${}_{12}P_8$, $n = 12, r = 8;\ \dfrac{12!}{(12-8)!} = \dfrac{12 \cdot 11 \cdot 10 \cdot 9 \cdot 8 \cdot 7 \cdot 6 \cdot 5 \cdot 4!}{4!} = 19{,}958{,}400$

21. a. $n = 5, r = 3;\ {}_5P_3 = \dfrac{5!}{(5-3)!} = \dfrac{5 \cdot 4 \cdot 3 \cdot 2!}{2!} = 60$

b. 5 choices in each of 3 places; $5^3 = 125$

23. a. 8 choices both times; $8^2 = 64$

b. $n = 8, r = 2;\ {}_8P_2 = \dfrac{8!}{(8-2)!} = \dfrac{8 \cdot 7 \cdot 6!}{6!} = 56$

25. $n = 15, r = 3;\ {}_{15}P_3 = \dfrac{15!}{(15-3)!} = \dfrac{15\cdot 14\cdot 13\cdot 12!}{12!} = 2730$

27. $n = 4, r = 3;\ {}_4C_3$, wxy, wxz, wyz, xyz

29. $n = 3, r = 2;\ {}_3C_2$, bc, bd, cd

31. a. ${}_6C_4, n = 6, r = 4;\ \dfrac{6!}{(6-4)!(4!)} = \dfrac{6\cdot 5\cdot 4!}{(2!)(4!)} = \dfrac{30}{2} = 15$

b. ${}_6C_5, n = 6, r = 5;\ \dfrac{6!}{(6-5)!(5!)} = \dfrac{6\cdot 5!}{(1!)(5!)} = 6$

c. ${}_{12}C_8, n = 12, r = 8;\ \dfrac{12!}{(12-8)!(8!)} = \dfrac{12\cdot 11\cdot 10\cdot 9\cdot 8!}{(4!)(8!)} = \dfrac{11{,}880}{24} = 495$

33. a. ${}_8C_3, n = 8, r = 3;\ \dfrac{8!}{(8-3)!(3!)} = \dfrac{8\cdot 7\cdot 6\cdot 5!}{(5!)(3!)} = \dfrac{8\cdot 7\cdot 6}{6} = 56$

b. ${}_{52}C_2, n = 52, r = 2;\ \dfrac{52!}{(52-2)!(2!)} = \dfrac{52\cdot 51\cdot 50!}{(50!)(2!)} = \dfrac{2652}{2} = 1326$

c. ${}_{12}C_1, n = 12, r = 1;\ \dfrac{12!}{(12-1)!(1!)} = \dfrac{12\cdot 11!}{(11!)(1)} = 12$

35. The words "different sets" indicates a combination; $n = 48, r = 6$,

$${}_{48}C_6 = \frac{48!}{(48-6)!(6!)} = \frac{48\cdot 47\cdot 46\cdot 45\cdot 44\cdot 43\cdot 42!}{(42!)(6!)} = \frac{48\cdot 47\cdot 46\cdot 45\cdot 44\cdot 43}{6\cdot 5\cdot 4\cdot 3\cdot 2\cdot 1} = 12{,}271{,}512$$

37. The words "different sets" indicates a combination; $n = 10, r = 3$,

$${}_{10}C_3 = \frac{10!}{(10-3)!(3!)} = \frac{10\cdot 9\cdot 8\cdot 7!}{(7!)(3!)} = \frac{720}{6} = 120$$

39. a. ${}_{12}C_0$ has 1 solution; (3) *There is only one set with nothing in it, the empty set.*

b. ${}_{12}C_{12}$ has 1 solution; (2) *There is only one way to choose al the elements from the original set.*

c. ${}_{12}C_1$ has 12 solutions; (1) *There are n ways to select 1 object from a set containing n objects.*

41. a. With repetitions use multiplication principle (MP), 10 numbers for each of 4 places and 26 letters for each of 4 places or 36 choice for each of 4 places: $36^4 = 1{,}679{,}616$

b. One less choice each time; permutation; ${}_{36}P_4 = 1{,}413{,}720$

41. c. Order not important; combination; ${}_{36}C_4 = 58{,}905$

d. 4 choices taken 4 at a time, order matters (each student ID has an order) ${}_4P_4 = 24$

43. a. no repetitions, order is implied; permutation; ${}_8P_3 = 336$

b. no repetitions, order does not matter; combination; ${}_8C_3 = 56$

c. no repetition, order matters; permutation; ${}_8P_8 = 8! = 40{,}320$

d. repetition is possible; multiplication principle; $8^5 = 32{,}768$

45. a. without repetition, permutation; ${}_{26}P_6 = 165{,}765{,}600$

b. with repetition, multiplication principle; $26^6 = 308{,}915{,}776$

47. a. 1 item from 9 possible choices; combination or addition principal; $4 + 2 + 3 = 9$

${}_9C_1 = 9$

b. combinations and multiplication principle; $({}_4C_2)\cdot({}_2C_1)\cdot({}_3C_1) = 6\cdot 2\cdot 3 = 36$

49. a. no repetition, order does not matter; combination; ${}_{35}C_6 = 1{,}623{,}160$

b. combination and multiplication principle: $({}_3C_1)\cdot({}_{12}C_3)\cdot({}_{20}C_2) = 3\cdot 220\cdot 190 = 125{,}400$

51. a. no, order gives twice as many permutations.

b. yes, there is only one order for a single object.

53. a. $${}_nC_1 = \frac{n!}{(n-1)!(1!)} = \frac{n!}{(n-1)!(1)} = \frac{n!}{(n-1)!} = {}_nP_1$$

b. $${}_nP_1 = \frac{n!}{(n-1)!} = \frac{n\cdot(n-1)!}{(n-1)!} = n$$

c. $${}_nC_1 = \frac{n!}{(n-1)!(1!)} = \frac{n\cdot(n-1)!}{(n-1)!(1)} = n$$

d. $${}_nC_n = \frac{n!}{(n-n)!(n!)} = \frac{1}{0!} = 1;\quad {}_nP_n = \frac{n!}{(n-n)!} = \frac{n!}{0!} = n!;$$ they are not equal

e. $${}_nP_n = \frac{n!}{(n-n)!} = \frac{n!}{0!} = n!$$

f. $${}_nC_n = \frac{n!}{(n-n)!(n!)} = \frac{1}{0!} = 1$$

g. There is only one set with nothing in it; therefore ${}_nP_0 = {}_nC_o$

h. $${}_nP_0 = \frac{n!}{(n-0)!} = \frac{n!}{n!} = 1$$

i. $${}_nC_0 = \frac{n!}{(n-0)!(0!)} = \frac{n!}{n!(1)} = 1$$

Exercises 11.3

1. $\frac{2}{5}$

3. **a.** The first 9 numbers are single digits; P(a 2 digit number) = $\frac{8}{17}$

b. 1 is not prime; 2, 3, 5, 7, 11, 13, & 17 are; P(prime number) = $\frac{7}{17}$

c. 6 and 12 are divisible by 2 and 3; P(a number divisible by both 2 and 3) = $\frac{2}{17}$

d. 1, 4, 9, & 16 are perfect squares; P(a perfect square) = $\frac{4}{17}$

e. P(a number less than 8) = $\frac{7}{17}$

5. **a.** P(A) = $\frac{1}{11}$ **b.** P(B) = $\frac{2}{11}$

c. P(a vowel) = $\frac{5}{11}$ (y is a vowel here) **d.** P(also in the word BABAR) = $\frac{4}{11}$

e. P(also in the word BROTHER) = $\frac{4}{11}$

7. **a.** P(A and H) = $\frac{1}{6}$ **b.** P(outcome contains a T) = $\frac{1}{2}$ **c.** P(C and either H or T) = $\frac{1}{3}$

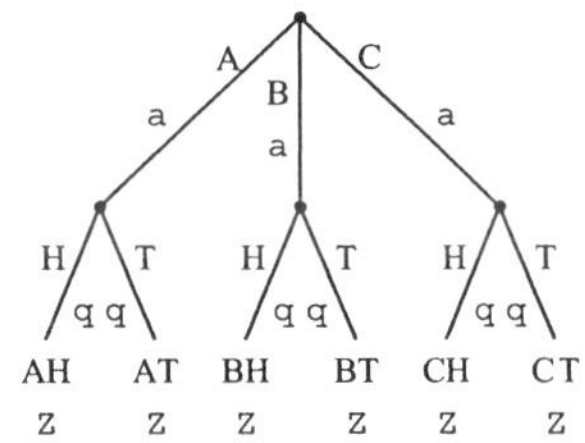

9. **a.** Addition Principle; P(cold) or P(hot) = P(cold) + P(hot) = $\frac{1}{2} + \frac{1}{2} = 1$

b. Multiplication Principle; P(boy and boy) = P(boy) · P(boy) = $\frac{1}{2} \cdot \frac{1}{2} = \frac{1}{4}$

c. Addition Principle; P(one) or P(two) = P(one) + P(two) = $\frac{1}{6} + \frac{1}{6} = \frac{2}{6} = \frac{1}{3}$

d. Multiplication Principle; P(three and head) = P(three) · P(head) = $\frac{1}{6} \cdot \frac{1}{2} = \frac{1}{12}$

e. events are not mutually exclusive; neither; from a tree diagram, 4 appears twice - 4H & 4T; heads appear 6 times - once with each number, one overlap at 4H, $\frac{7}{12}$

11.

Sample Space	Event (A)	Event (not A)	P(A)	P(not A)
Birth of child	Girl	**Boy**	**1/2**	**1/2**
BB, BG, GB, GG	**GG**	BB, BG, GB	**1/4**	**3/4**
BB. BG. GB, GG	At most 1 boy	**BB**	**3/4**	**1/4**
1, 2, 3, 4, 5	**2, 4**	1, 3, 5	2/5	**3/5**
x, y, z	**z**	x, y	1/3	**2/3**
Sum of 2 dice	**odd sum**	even sum	**1/2**	**1/2**

13. a. P(sum of 7) = $\frac{6}{36}$ so P(not sum of 7) = $\frac{30}{36} = \frac{5}{6}$

b. P(not sum of 1) = 1

c. P(not a sum of 2, 3, 11 or 12) = $\frac{30}{36} = \frac{5}{6}$

d. P(sum of 2 and sum of 9) = 0,

e. P(sum of 2 or sum of 9) = $\frac{5}{36}$

15. a. P(zero dots appear on at least one side) = $\frac{7}{28} = \frac{1}{4}$

b. P(six dots appear on at least one side) = $\frac{7}{28} = \frac{1}{4}$

c. P(sum of dots is 6) = $\frac{4}{28} = \frac{1}{7}$

d. P(sum of dots is even) = $\frac{16}{28} = \frac{4}{7}$

e. P(sum of dots is not a perfect square) = $\frac{21}{28} = \frac{3}{4}$

f. P(sum of 4 or sum of 5) = $\frac{6}{28} = \frac{3}{14}$

g. P(sum of 4 and sum of 5) = 0

17. a. $2^7 = 128$

b. ${}_7C_2 = 21$; P(exactly 2 tails) = $\frac{21}{128}$

c. There is only 1 way to get THHHHHT, P(THHHHHT) = $\frac{1}{128}$

d. At least 2 tails means 2, 3, 4, 5, 6, and 7 tails; ${}_7C_2 + {}_7C_3 + {}_7C_4 + {}_7C_5 + {}_7C_6 + {}_7C_7 =$ $21 + 35 + 35 + 21 + 7 + 1 = 120$; P(at least 2 tails) = $\frac{120}{128} = \frac{15}{16}$

e. P(exactly 2 tails or exactly 3 tails) = $\frac{21}{128} + \frac{35}{128} = \frac{56}{128} = \frac{7}{16}$

f. P(exactly 2 tails and exactly 3 tails) = 0

g. ${}_7C_0 = 1$; ${}_7C_1 = 7$; ${}_7C_2 = 21$; P(no tails or 1 tail or 2 tails) = $\frac{1+7+21}{128} = \frac{29}{128}$

19. There are $2^{12} = 4096$ possible outcomes, P(6 boys then 6 girls) = $\frac{1}{4096}$

21. a. Order is important; ${}_4P_4 = 24$

b. P(D, C, B, A) = $\frac{1}{24}$

23. a. Order is not important; ${}_4C_2 = 6$

b. P(C and D) = $\frac{1}{6}$

25. domain: any event;

range: any rational number p, $0 \le p \le 1$

27. P(a) = $\frac{1}{10,000}$

Mid-Chapter 11 Test

1. **a.**

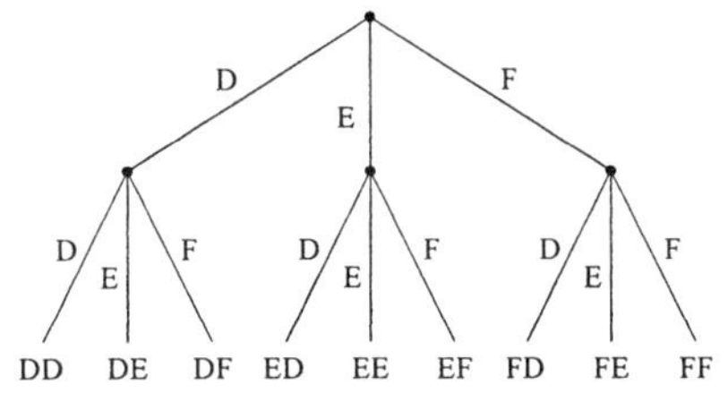

b. DD, DE, DF, ED, EE, EF, FD, FE, FF

2. **a.** Event (not A) = DD **b.** Event (not B) = DF and FD

3. **a.** $15 \cdot 9 = 135$ **b.** $15 + 9 = 24$

c. $9 \cdot 8 = 72$

4. $26 \cdot 10 \cdot 26 \cdot 10 \cdot 26 \cdot 10 = 17{,}576{,}000$

5. **a.** $8! = 8 \cdot 7 \cdot 6 \cdot 5 \cdot 4 \cdot 3 \cdot 2 \cdot 1 = 40{,}320$ **b.** ${}_4C_2 = \dfrac{4!}{(4-2)!(2!)} = \dfrac{4 \cdot 3 \cdot 2!}{(2!)(2 \cdot 1)} = \dfrac{12}{2} = 6$

c. ${}_6C_4 = \dfrac{6!}{(6-4)!(4!)} = \dfrac{6 \cdot 5 \cdot 4!}{(2!)(4!)} = \dfrac{30}{2} = 15$ **d.** ${}_6P_4 = \dfrac{6!}{(6-4)!} = \dfrac{6 \cdot 5 \cdot 4 \cdot 3 \cdot 2!}{2!} = 360$

e. ${}_{26}C_{25} = \dfrac{26!}{(26-25)!(25!)} = \dfrac{26 \cdot 25!}{1(25!)} = 26$ **e.** ${}_8P_1 = \dfrac{8!}{(8-1)!} = \dfrac{8 \cdot 7!}{7!} = 8$

6. order is not important; combination; ${}_8C_3 = 56$

7. order is not important; combination; ${}_8C_5 = 56$

8. numbered tickets implies order is important; permutation; ${}_8P_5 = 6720$

9. **a.** order is not important; combination; ${}_{52}C_6 = 20{,}358{,}520$

b. order is not important; combination; ${}_{52}C_{13} \approx 6.35 \times 10^{11}$

10. **a.** careful was I, careful I was, I was careful, I careful was, was I careful, was careful I

b. May vary but likely "I was careful"

11. a. P(odd number) = $\frac{4}{8} = \frac{1}{2}$; (1, 3, 5, 7) **b.** P(number less than 4) = $\frac{3}{8}$; (1, 2, 3)

c. P(5 or greater) = $\frac{4}{8} = \frac{1}{2}$; (5, 6, 7, 8)

d. P(an integral power of two) = $\frac{4}{8} = \frac{1}{2}$; ($2^0, 2^1, 2^2, 2^3$)

e. P(divisible by 3) = $\frac{2}{8} = \frac{1}{4}$; (3, 6) **f.** P(a perfect square) = $\frac{2}{8} = \frac{1}{4}$; (1, 4)

12. order not important; $\dfrac{1}{_{52}C_6} = \dfrac{1}{20{,}358{,}520}$

13. a. there are 10^4 possible combinations; P(guessing the right combination) = $\frac{1}{10{,}000}$

b. Mathematically the set of four numbers is a permutation because order counts.

14. P(not a) = 1 – P(a) = $1 - \frac{1}{10{,}000} = \frac{9{,}999}{10{,}000}$

Exercises 11.4

1. a. 4:5, $\frac{4}{5}$ **b.** $\frac{3}{7}$, 3 to 7 **c.** 3 to 2, 3:2

3. a. 6 possible, 2:4 = 1:2 **b.** $\dfrac{\frac{2}{6}}{\frac{4}{6}} = \dfrac{2}{6} \cdot \dfrac{6}{4} = \dfrac{2}{4} = \dfrac{1}{2}$

5. BB, BG, GB, GG **a.** 4 possible, 1:3 **b.** $\dfrac{\frac{1}{4}}{\frac{3}{4}} = \dfrac{1}{4} \cdot \dfrac{4}{3} = \dfrac{1}{3}$

7. a. 4 outcomes, HH, HT, TH, TT **b.** 1:3

c. 1:3 **d.** 2:2 = 1:1

9. a. $H_1H_{1/2}$, $H_1T_{1/2}$, $H_1H_{1/2}$, $H_1T_{1/2}$ **b.** 2:2 = 1:1

c. P(2 tails) = 0 **d.** 0:4

e. 4:0 **f.** P(at least one head) = 4

11. In *odds* the answer must be the ratio of two numbers, 1:1.

13. The answer must be the ratio of two numbers, 3:4.

15. a. No, 1:1 means P(A) = $\frac{1}{2}$; 1 of 2 possible outcomes **b.** 2:0

c. P(not A) = 0 **d.** 0:2 **e.** True

17. 1:1 odds is the same as $P(A) = \frac{1}{2}$

19. 2:3 odds is the same as $P(A) = \frac{2}{5}$

21. 2:1 odds is the same as $P(A) = \frac{2}{3}$

23. a. Total of $6^2 = 36$ outcomes; P(3 first die) $= \frac{6}{36}$; P(3 second die) $= \frac{6}{36}$, 1 overlap;

P(3 on either die) $= \frac{6}{36} + \frac{6}{36} - \frac{1}{36} = \frac{11}{36}$

b. P(2 on either die) $= \frac{11}{36}$, same as above

c. P(2 and 3) = P(2 first and 3 next) + P(3 first and 2 next) $= \frac{1}{36} + \frac{1}{36} = \frac{2}{36} = \frac{1}{18}$

d. P(2 or 3) = P(2) + P(3) – P(2 and 3) $= \frac{11}{36} + \frac{11}{36} - \frac{2}{36} = \frac{20}{36} = \frac{5}{9}$

e. P(doubles or 3) = P(doubles) + P(3) – P(doubles & 3) $= \frac{6}{36} + \frac{11}{36} - \frac{1}{36} = \frac{16}{36} = \frac{4}{9}$

25. a. Total of $6^2 = 36$ outcomes; P(sum of 2 or sum of 3) $= \frac{1}{36} + \frac{2}{36} = \frac{3}{36} = \frac{1}{12}$

b. P(doubles or sum of 7) $= \frac{6}{36} + \frac{6}{36} = \frac{12}{36} = \frac{1}{3}$

c. P(doubles or sum of 6) = P(doubles) + P(sum of 6) – P(doubles and sum of 6) =

$\frac{6}{36} + \frac{5}{36} - \frac{1}{36} = \frac{10}{36} = \frac{5}{18}$

d. P(3 or sum of 8) = P(3) + P(sum of 8) – P(3 and sum of 8) $= \frac{11}{36} + \frac{5}{36} - \frac{2}{36} = \frac{14}{36} = \frac{7}{18}$

27. a. P(diamond or a face card) = P(diamond) + P(face card) – P(diamond and face card) =

$\frac{13}{52} + \frac{12}{52} - \frac{3}{52} = \frac{22}{52}$

b. P(9 or club) = P(9) + P(club) – P(9 and club) $= \frac{4}{52} + \frac{13}{52} - \frac{1}{52} = \frac{16}{52} = \frac{4}{13}$

c. P(diamond or heart) $= \frac{13}{52} + \frac{13}{52} = \frac{26}{52} = \frac{1}{2}$

d. P(club or numbers 3, 4, 5 or 6) =

P(club) + P(3) + P(4) + P(5) + P(6) – P(club & 3, 4, 5 & 6) $= \frac{13}{52} + \frac{4}{52} + \frac{4}{52} + \frac{4}{52} + \frac{4}{52} - \frac{4}{52} = \frac{25}{52}$

29. a. 355 (science) + 933 (fiction) – 112 (science fiction) = 1176

b. P(science or fiction) $= \frac{1176}{10{,}001}$

31. a. 232 (cats) + 465 (licensed pets) – 207 (licensed cats) = 490

b. 232 – 207 = 25 not licensed cats; P(not licensed) $= \frac{25}{232}$

c. 465 – 207 = 258 dogs; P(dog) $= \frac{258}{490} = \frac{129}{245}$

d. P(cat) $= \frac{232}{490} = \frac{116}{245}$

33. P(A and B) = 0

35. a.

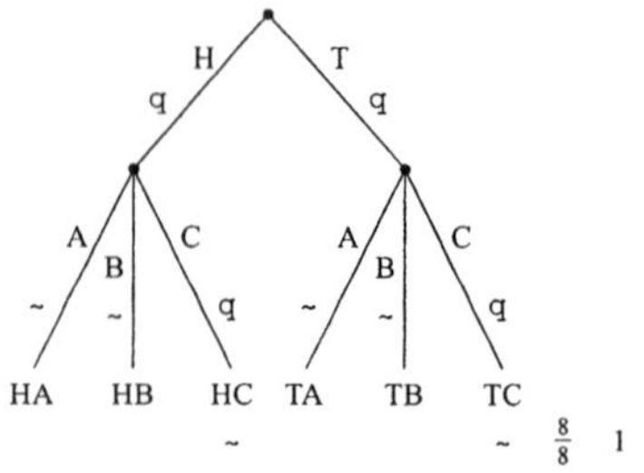

b. HA $\frac{1}{8}$, HB $\frac{1}{8}$, HC $\frac{1}{4}$, TA $\frac{1}{8}$, TB $\frac{1}{8}$, TC $\frac{1}{4}$

c. $P(\text{head}) = \frac{1}{8}+\frac{1}{8}+\frac{1}{4} = \frac{1}{8}+\frac{1}{8}+\frac{2}{8} = \frac{4}{8} = \frac{1}{2}$

d. $P(\text{B or tail}) = P(\text{B}) + P(\text{tail}) - P(\text{B and tail}) = \frac{1}{4} + \frac{1}{2} - \frac{1}{8} = \frac{5}{8}$

e. $P(\text{B or C}) = P(\text{B}) + P(\text{C}) = \frac{1}{4} + \frac{1}{2} = \frac{3}{4}$

f. $P(\text{B and head}) = \frac{1}{8}$

37. a.

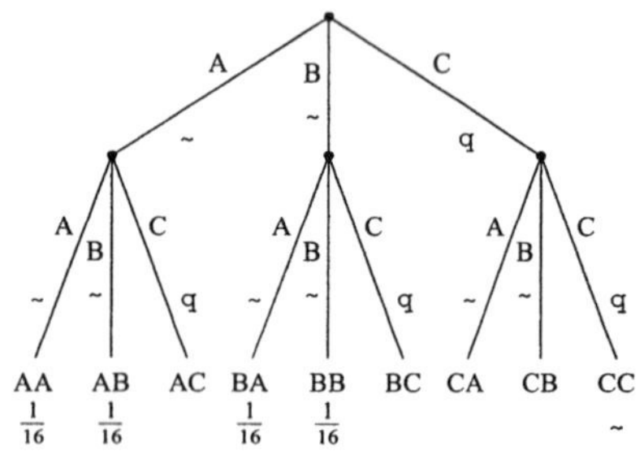

b. $P(\text{AA}) = \frac{1}{16}$

c. $P(\text{at least one A}) = \frac{1}{16}+\frac{1}{16}+\frac{1}{8}+\frac{1}{16}+\frac{1}{8} = \frac{7}{16}$

d. $P(\text{B and C}) = \frac{1}{8}+\frac{1}{8} = \frac{1}{4}$

e. $P(\text{at least one B or at least one C}) = P(\text{not AA}) = 1 - \frac{1}{16} = \frac{15}{16}$

f. $P(\text{at least one A or at least one B}) = P(\text{not CC}) = 1 - \frac{1}{4} = \frac{3}{4}$

39. a.

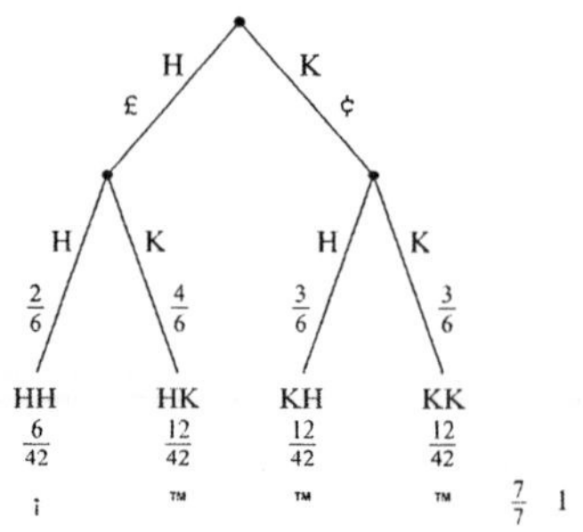

b. $P(\text{2 Kisses}) = \frac{2}{7}$

c. $P(\text{at least one Hug}) = P(\text{not 2 Kisses}) = 1 - \frac{2}{7} = \frac{5}{7}$

d. $P(\text{Hug and Kiss}) = \frac{2}{7} + \frac{2}{7} = \frac{4}{7}$

e. $P(\text{at least one Kiss or a Hug and a Kiss}) = P(\text{not 2 Hugs}) = 1 - \frac{1}{7} = \frac{6}{7}$

41. a.

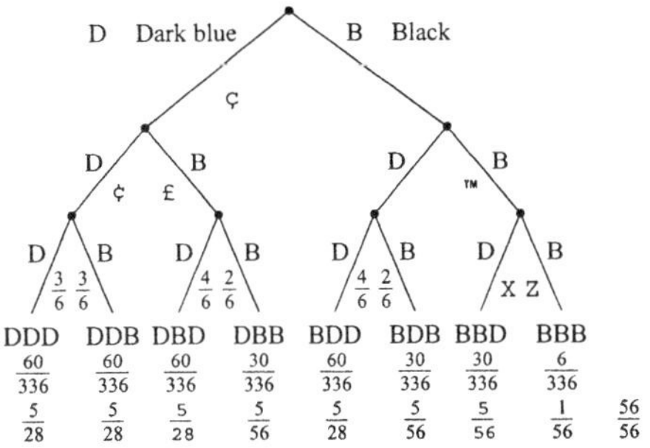

b. P(2 dark blue) = $\frac{5}{28}+\frac{5}{28}+\frac{5}{28}+\frac{5}{28}=\frac{20}{28}=\frac{5}{7}$

c. P(2 black) = P(not 2 dark blue) = 1 - $\frac{5}{7}=\frac{2}{7}$

d. P(at least 2 alike) = 1

e. 3 socks in only 2 colors must always have one pair of matching colors

43. P(red 3 and odd green) = $\frac{1}{12}$

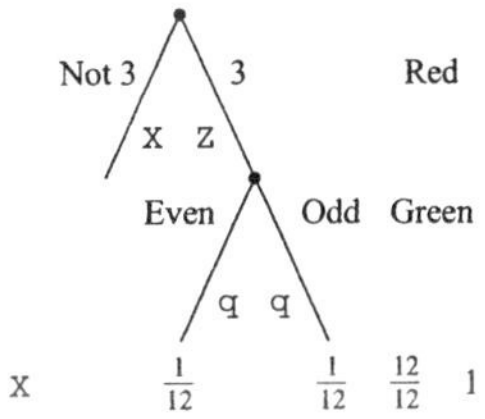

45. P(2 sixes) = $\frac{1}{169}$

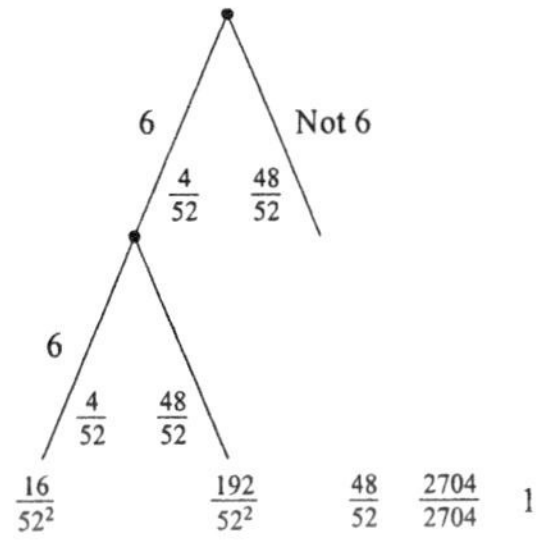

47. a.

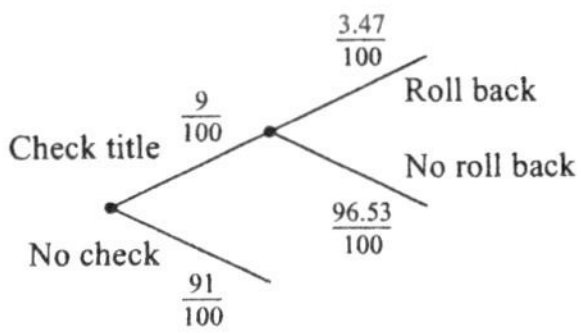

b. P(roll back) = $\frac{9}{100}\cdot\frac{3.47}{100}\approx\frac{31}{10{,}000}$

49. a.

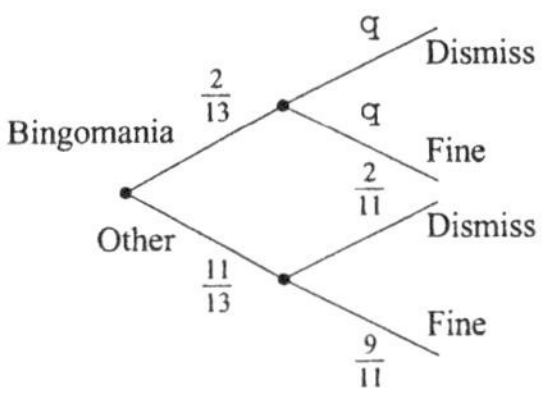

b. P(citation and fine) = $\frac{2}{13}\cdot\frac{1}{2}=\frac{1}{13}$

c. P(others citation and fine) = $\frac{11}{13}\cdot\frac{9}{11}=\frac{9}{13}$

51. a. Yes, the two-stage is reasonable if the survey first asked about being tired and then asked those people if they were exhausted.

b. $\frac{1}{2}$ are tired; $\frac{1}{5}$ are tired and exhausted; $\frac{1}{2}$ · P(tired and exhausted) = $\frac{1}{5}$; P = $\frac{2}{5}$

53. The friend forgot the overlap with (3, 3). The probability is $\frac{6}{36}+\frac{6}{36}-\frac{1}{36}=\frac{11}{36}$

55. $a+b=1,\ a=1-b;$
$c+d=1$
$ac+ad=a$ (multiply both sides by a)
$ac+ad=1-b$ (substitute for a)
$ac+ad+b=1$ (add b to both sides)

Exercises 11.5

1.

1 1
1 2 1
1 3 3 1
1 4 6 4 1
1 5 10 10 5 1
1 6 15 20 15 6 1
1 7 21 35 35 21 6 1
1 8 28 56 70 56 28 8 1
1 9 36 84 126 126 84 36 9 1
1 10 45 120 210 252 210 120 45 10 1

3. $1+7+21+35+35+21+7+1=128=2^7$
$1+8+28+56+70+56+28+8+1=256=2^8$

5. Listing is left to the student, there is 1 birth order with all girls, 5 with 4 girls, 10 with 3 girls, 10 with 2 girls, 5 with 1 girl and 1 with 0 girls.

7. $\frac{12!}{0!(12-0)!}=1,\ \frac{12!}{1!(12-1)!}=12,\ \frac{12!}{2!(12-2)!}=66,\ \frac{12!}{3!(12-3)!}=220,\ \frac{12!}{4!(12-4)!}=495,$
$\frac{12!}{5!(12-5)!}=792,\ \frac{12!}{6!(12-6)!}=924,\ \frac{12!}{7!(12-7)!}=792,$ 495, 220, 66, 12, 1

9. **a.** row 1, 7; fourth term = 35 **b.** row 1, 6; third term = 15
c. row 1, 4; third term = 6

11. a. ${}_{20}C_1 = 20$ (row 1, 20; second term)
b. ${}_{15}C_{15} = 1$ (row 1, 15; last term)
c. ${}_{20}C_0 = 1$ (row 1, 20; first term)
d. ${}_{16}C_{15} = 16$ (row 1, 16; next to last term)

13. The first two terms are 1 and *n*, and the last two terms are *n* and 1

15. a. ${}_3C_0 = {}_3C_3$
b. ${}_8C_2 = {}_8C_6$
c. ${}_7C_3 = {}_7C_4$
d. ${}_nC_b = {}_nC_{n-b}$

17. For odd *n*, all numbers in the 1, *n* row repeat.

19. ${}_nC_a = {}_nC_{n-a}$ for all *n*.

21. The row beginning 1, 7, is really the 8th row

23. $2^5 = 32$, P(4 girls) = (second or next to last term in 1, 5 row) ÷ (total outcome) = $\frac{5}{32}$

25. $2^5 = 32$, P(5 boys) = (first or last term in 1, 5 row) ÷ (total outcome) = $\frac{1}{32}$

27. $2^6 = 64$, P(1 tail) = (second or next to last term in 1, 6 row) ÷ (total outcome) = $\frac{6}{64} = \frac{3}{32}$

29. $2^{12} = 4096$

31. $\frac{{}_{12}C_{10}}{2^{12}} = \frac{66}{4096} = \frac{33}{2048}$

33. $\frac{{}_{12}C_7}{2^{12}} = \frac{792}{4096} = \frac{99}{512}$

35. $\frac{{}_{12}C_2}{2^{12}} = \frac{66}{4096} = \frac{33}{2048}$

37. ${}_{12}C_{10} = {}_{12}C_2$

39. a. 10 ways from A to B
b. 20 ways from A to B
c. 21 ways from A to B
d. 66 ways from A to B

Chapter 11 Review Exercises

1. No, not all months have 31 days.

3. a. $2^1 = 2$
b. $2^2 = 4$
c. $2^3 = 8$
d. $2^4 = 16$

5. a. $7 \cdot 6 \cdot 5 \cdot 4 \cdot 3 \cdot 2 \cdot 1 = 5040$

b. ${}_7P_2 = \frac{7!}{(7-2)!} = \frac{7 \cdot 6 \cdot 5!}{5!} = 42$

c. ${}_6P_4 = \frac{6!}{(6-4)!} = \frac{6 \cdot 5 \cdot 4 \cdot 3 \cdot 2!}{2!} = 360$

d. ${}_9P_4 = \frac{9!}{(9-4)!} = \frac{9 \cdot 8 \cdot 7 \cdot 6 \cdot 5!}{5!} = 3024$

e. ${}_{20}C_{20} = 1$

f. ${}_8C_5 = \frac{8!}{5!(8-5)!} = \frac{8 \cdot 7 \cdot 6 \cdot 5!}{5! \cdot 3 \cdot 2 \cdot 1} = 56$

7. a. (original) Hard to see the dark side is. To see the dark side is hard. The dark side is hard to see. Hard the dark side is to see. Hard the dark side to see is. Hard to see is the dark side.

b. ${}_7P_7 = \frac{7!}{(7-7)!} = \frac{7 \cdot 6 \cdot 5 \cdot 4 \cdot 3 \cdot 2 \cdot 1}{0!} = 5040$

9. ${}_3P_3 = 6$; I love you. I you love. You I love. You love I. Love you I. Love I you.

11. a. ${}_{15}P_4 = 32{,}760$

b. ${}_{15}C_3 = 455$

13. a. ${}_{10}C_3 = 120$

b. ${}_{10}P_{10} = 3{,}628{,}800$

c. ${}_{10}P_3 = 720$

d. ${}_{10}C_4 = 210$

e. MP only $10^5 = 100{,}000$

15. a. $8 + 12 = 20$

b. $8 \cdot 12 = 96$

c. ${}_8C_2 = 28$

d. ${}_{12}P_2 = 132$

17.

Sample Space	Event A	Event not A	P(A)	P(not A)
HH, HT, TH, TT	**HT, TH**	HH, TT	**1/2**	**1/2**
HH, HT, TH, TT	TT	**HH, HT, TH**	**1/4**	**3/4**
1, 2, 3, 4, 5	even numbers	**odd numbers**	**2/5**	**3/5**
red, white, blue	**white, blue**	red	**1/3**	**2/3**
a, b, c, d	letters also in cat	**b, d**	**1/2**	**1/2**
a, b, c, d	**a, b, c**	letters also in dog	**3/4**	**1/4**

19. a. AP; $P(1) + P(2) + P(3) + P(4) + P(5) + P(6) = \frac{1}{6} + \frac{1}{6} + \frac{1}{6} + \frac{1}{6} + \frac{1}{6} + \frac{1}{6} = 1$

b. MP; $P(\text{girl}) \cdot P(\text{boy}) = \frac{1}{2} \cdot \frac{1}{2} = \frac{1}{4}$

21. P(tail or 3) = P(tail) + P(3) – P(tail and 3) = $\frac{6}{12} + \frac{2}{12} - \frac{1}{12} = \frac{7}{12}$

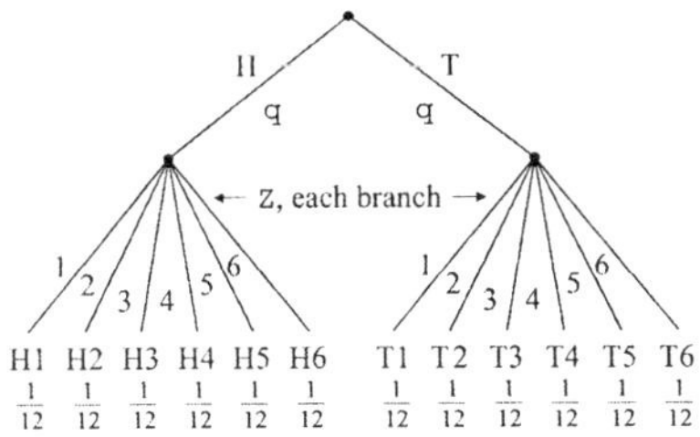

23. a. 10^3 possible variations only 1 will match P(match) = $\frac{1}{1{,}000}$

b. ${}_{10}C_3 = 120$ possible combinations of the same numbers, P(same numbers) = $\frac{1}{120}$

c. No, not the same.

25. a. total outcomes = 6(red) · 6(green) = 36 **b.** 6 doubles; odds for doubles = 6:30

c. 6 (red 3 + any green) + 6 (green 3 + any red) – 1 (3 red & 3 green) = 11 possible
odd for at least one three = 11:(36 – 11) = 11:25

d. odd against at least one three = 25:11 (opposite of odds for)

27. a. overlap with 6, 1 and 1, 6; pairs with 6 or sum of 7 = pairs with 6 + pairs with sum of 7 = 6 *(6 on first)* + 6 *(6 on second)* – 1 *(6 on both)* + 6 *(sum of 7)* – 2 *(overlap)* = 15 pairs

b. overlap with 5,1 and 1, 5; pairs with 5 or sum of 6 = pairs with 5 + pairs with sum of 6 = 6 *(5 on first)* + 6 *(5 on second)* – 1 *(5 on both)* + 5 *(sum of 6)* – 2 *(overlap)* = 14 pairs

c. overlap; pairs with 4 or 6 = pairs with 4 + pairs with 6 = 6 *(4 on first)* + 6 *(4 on second)* – 1 *(4 on both)* + 6 *(6 on first)* + 6 *(6 on second)* – 1 *(6 on both)* – 2 *(4, 6 and 6, 4)* = 20

d. no overlap; P(6 or sum of 6) = P(6) + P(sum of 6) = $\frac{11}{36} + \frac{5}{36} = \frac{16}{36}$ *(Note: the set containing 6 on either die does have an overlap but not with the set making a sum of 6.)*

e. overlap, 2, 2; P(double or sum of 4) = P(dbl) + P(sum of 4) = $\frac{6}{36} + \frac{3}{36} - \frac{1}{36} = \frac{8}{36} = \frac{2}{9}$

f. no overlap (all doubles are even sums); P(double or odd sum) = P(dbl) + P(odd sum) = $\frac{6}{36} + \frac{18}{36} = \frac{24}{36} = \frac{2}{3}$

29. a.

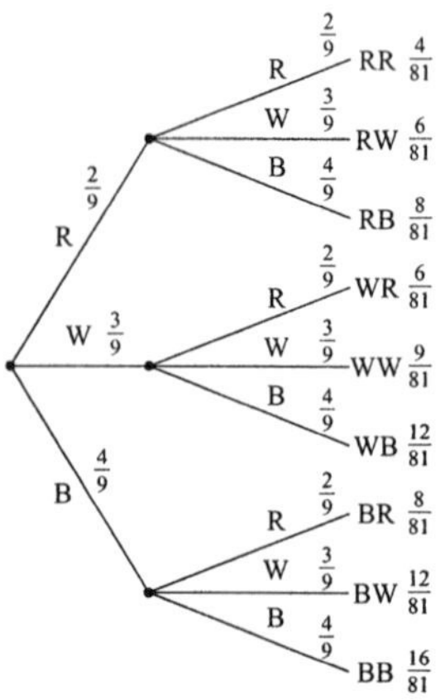

b. P(red and white) = $\frac{6}{81}+\frac{6}{81}=\frac{12}{81}=\frac{4}{27}$

c. P(red or white) = $\frac{2}{9}+\frac{3}{9}+\frac{8}{81}+\frac{12}{81}=\frac{65}{81}$

d. P(not BB) = P(red or white) = $\frac{65}{81}$

e. P(not blue) = $\frac{4}{81}+\frac{6}{81}+\frac{6}{81}+\frac{9}{81}=\frac{25}{81}$

31. a.

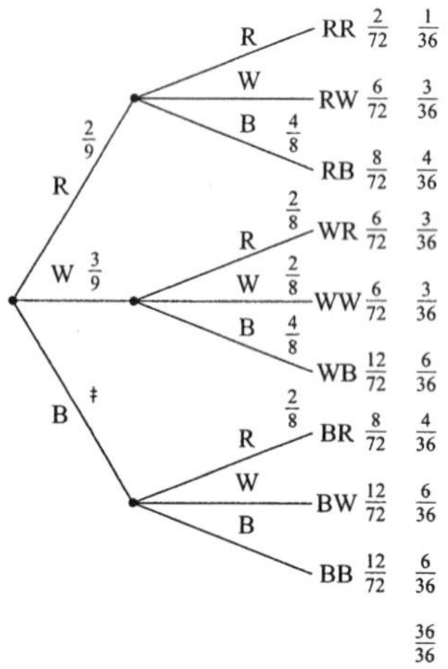

b. P(red and white) = $\frac{3}{36}+\frac{3}{36}=\frac{6}{36}=\frac{1}{6}$

c. P(red or white) = $\frac{2}{9}+\frac{3}{9}+\frac{4}{36}+\frac{6}{36}=\frac{30}{36}=\frac{5}{6}$

d. P(not BB) = P(red or white) = $\frac{5}{6}$

e. P(not blue) = $\frac{1}{36}+\frac{3}{36}+\frac{3}{36}+\frac{3}{36}=\frac{10}{36}=\frac{5}{18}$

33. a.

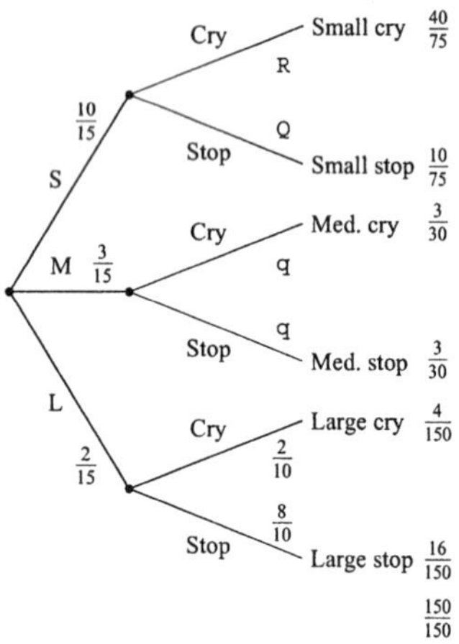

b. P(large and stops crying) = $\frac{16}{150}=\frac{8}{75}$

c. P(small and crying) = $\frac{40}{75}$

d. P(stops crying) = $\frac{10}{75}+\frac{3}{30}+\frac{16}{150}=\frac{51}{150}=\frac{17}{50}$

35. P(3 and odd) = $\frac{3}{36}=\frac{1}{12}$

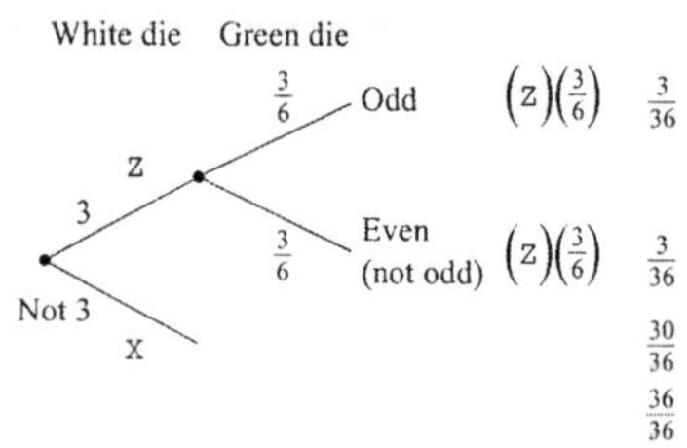

37. P(5 twice) = $\frac{16}{2704}=\frac{1}{169}$

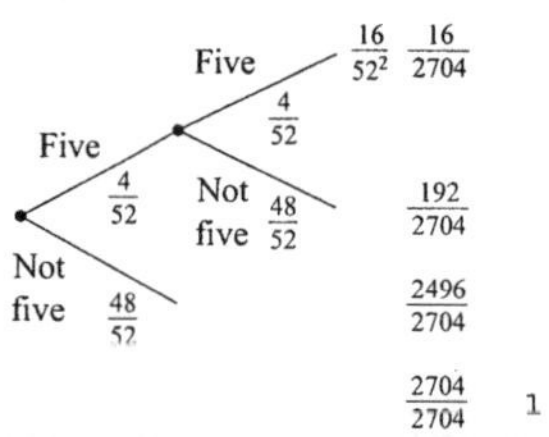

41. a. 1 6 14 22 14 6 1

1 7 20 36 36 20 7 1

b. $1 + 7 + 20 + 36 + 36 + 20 + 7 + 1 = 128$, yes $128 = 2^7$

c. from the fake row ${}_6C_4 = 14$

d. from the fake row, total outcomes = 64; P(3 boys and 3 girls) = $\frac{{}_6C_3}{64} = \frac{22}{64} = \frac{11}{32}$

e. 1 is the number of ways of obtaining 0 items from a set of n items.

43. ${}_aC_b = \frac{a!}{(a-b)!b!}$; $1, a, \frac{a!}{(a-2)!2!}, \ldots, \frac{a!}{(a-2)!2!}, a, 1$

Chapter 11 Test

1. a.

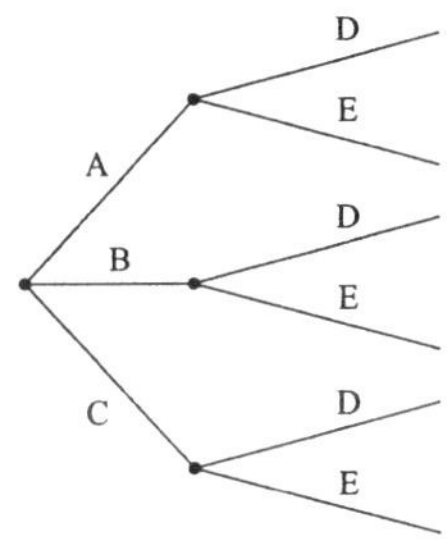

b. AD, AE, BD, BE, CD, CE

c. Event not M is AD or BD

d. Event not N is BD or BE

e. P(A or E) = P(A) + P(E) – P(A & E) = $\frac{2}{6} + \frac{3}{6} - \frac{1}{6} = \frac{4}{6} = \frac{2}{3}$

f. P(A & E) = $\frac{1}{6}$

2. a. $10 + 15 = 25$

b. $10 \cdot 15 = 150$

c. $15 \cdot 14 = 210$

d. P(novel) = $\frac{15}{25} = \frac{3}{5}$

e. P(novel or mystery) = 1

f. P(novel and novel) = $\frac{1}{210}$

3. a. $10 \cdot 25 \cdot 25 = 6250$

b. Assume 5 numbers; $10^5 = 100{,}000$

4. ${}_nP_0$, ${}_nC_0$, and ${}_nC_n$ equal one, ${}_nC_1$ and ${}_nC_{n-1}$ equal n

a. ${}_{52}C_1 = 52$

b. ${}_{52}P_{52} = \frac{52!}{0!} \approx 8.07 \times 10^{67}$

c. ${}_{52}C_{52} = 1$

d. ${}_{52}P_0 = 1$

e. ${}_{52}C_0 = 1$

f. ${}_{52}C_{51} = 52$

5. a. combination; ${}_{10}C_7$
b. permutation; ${}_{10}P_5$
c. multiplication principal; 10^5
d. permutation; ${}_{10}P_7$
e. combination; ${}_{10}C_3$

6. a. $P(I) = \frac{3}{8}$
b. $P(L \text{ or } I) = \frac{2}{8} + \frac{3}{8} = \frac{5}{8}$
c. $P(\text{letter also in SNOW}) = \frac{3}{8}$

7. a. $P(\text{red}) = \frac{1}{6}$
b. $P(\text{green}) = \frac{2}{6} = \frac{1}{3}$
c. $P(\text{not white}) = 1 - \frac{3}{6} = \frac{1}{2}$
d. $P(\text{red or green}) = \frac{1}{6} + \frac{2}{6} = \frac{3}{6} = \frac{1}{2}$
e. $P(\text{blue}) = 0$

8. a.

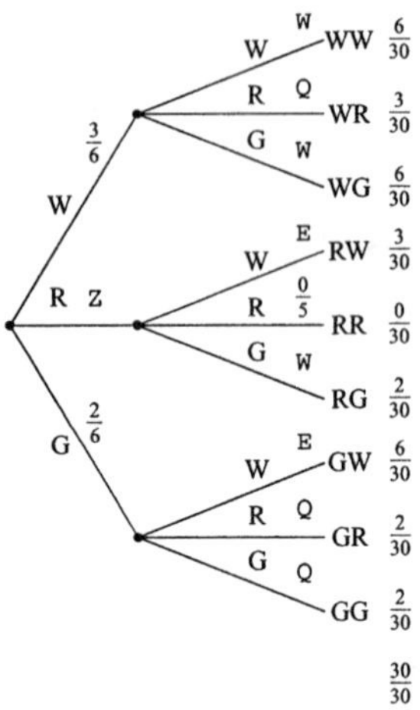

b. $P(\text{red on either draw}) = \frac{1}{6} + \frac{3}{30} + \frac{2}{30} = \frac{10}{30} = \frac{1}{3}$

c. $P(\text{red on both draws}) = 0$

d. P(green on either or the outcome WG) = P(green on either) + P(WG) – P(WG) *overlap* =
$\left(\frac{2}{6} + \frac{6}{30} + \frac{2}{30}\right) + \frac{6}{30} - \frac{6}{30} = \frac{18}{30} = \frac{3}{5}$

9. possible combinations = ${}_{30}C_5$; $P(\text{match}) = \dfrac{1}{{}_{30}C_5} = \dfrac{1}{142{,}506}$

10. a. 1 7 20 36 36 20 7 1
1 8 27 56 72 56 27 8 1

b. $1 + 8 + 27 + 56 + 72 + 56 + 27 + 8 + 1 = 256 = 2^8$

c. from the fake row; ${}_7C_5 = 20$

d. original fake row totals to 128; $P(\text{3 boys and 4 girls}) = \frac{36}{128} = \frac{9}{32}$

e. The row represents 0, 1, 2, 3, ... boys; the 4th term in the row is 36.

f. The row totals to $128 = 2^7$

Part 1: Short Answers

1. $-(-2)^2 = -4$

3. $\dfrac{xy^2}{x^2y} = \dfrac{xyy}{xxy} = \dfrac{y}{x}$

5. $x - 6 + 3(2 - x) = 8$
$x - 6 + 6 - 3x = 8$
$-2x = 8$
$x = -4$

7. The -4 in $-4xyz$ is the **coefficient** of xyz.

9. $f(-2) = 3(-2) + 4$
$f(-2) = -6 + 4 = -2$

11. The graphs of $y = 2x - 3$ and $y = -2x + 3$ are **intersecting** lines.

13. In $y = 1.5x$, $m = \frac{3}{2}$; perpendicular line has negative reciprocal slope, $m = -\frac{2}{3}$; $y = -\frac{2}{3}x$

15. $m = \dfrac{y_2 - y_1}{x_2 - x_1}$

17. $y = 79 - x$; $2x - 3(79 - x) = 28$
$2x - 237 + 3x = 28$; $5x = 265$
$x = 53$; $y = 79 - 53 = 26$

19. The result when solving a system with parallel graphs is **false, such as 0 = 4.**

21. Let w = Williamsport score and h = Lock Haven score
$w + h = 23$, $w = 23 - h$ $w - h = 9$, $w = 9 + h$
$23 - h = 9 + h$; $2h = 14$; $h = 7$
$w = 23 - 7 = 16$

23. $x + 7 \le 4x - 8$
$7 + 8 \le 4x - x$
$15 \le 3x$
$5 \le x$ or $x \ge 5$

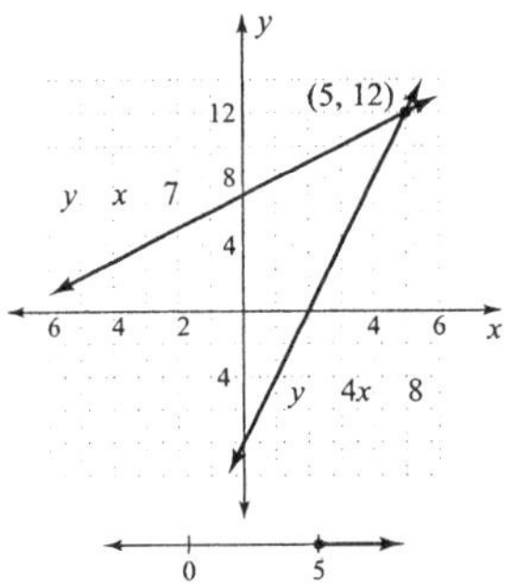

25. The graph of a quadratic eqn. is a parabola.

27. $f(a + b) = 2(a + b)^2$
$f(a + b) = 2(a^2 + 2ab + b^2)$
$f(a + b) = 2a^2 + 4ab + 2b^2$

29. $6x^2 - 5x - 4 = 6x^2 + 3x - 8x - 4 =$
$3x(2x + 1) - 4(2x + 1) =$
$(3x - 4)(2x + 1)$

31. $8x^3 - 27 = (2x)^3 - 3^3 =$
$(2x - 3)[(2x)^2 + 3(2x) + 3^2]$
$(2x - 3)(4x^2 + 6x + 9)$

33. multiply both sides by 3; $4x^2 = 45x + 36$
$4x^2 - 45x - 36 = 0$; $4x^2 - 48x + 3x - 36 = 0$
$4x(x - 12) + 3(x - 12) = 0$
$(x - 12)(4x + 3) = 0$; $x = 12$ or $x = -\frac{3}{4}$

35. $a^2 + b^2 = c^2$

37. $(x + \frac{5}{2})^2 = x^2 + 5x + \frac{25}{4}$

39. Possible answers: π, e, $\sqrt{2}$, $\sqrt{3}$

41. One method is to graph the equation and find the x-intercepts.
Factor: $5x^2 - x - 4 = (5x + 4)(x - 1) = 0$ $x = -\frac{4}{5}$ or $x = 1$

Quadratic formula: $x = \dfrac{-(-1) \pm \sqrt{(-1)^2 - 4(5)(-4)}}{2(5)} = \dfrac{1 \pm \sqrt{81}}{10} = \dfrac{1 \pm 9}{10}$; $\{-0.8, 1\}$

43. The graph opens up (+a); equation will be ≤ 0 between the x-intercepts; $-2 \leq x \leq 0.4$.

45. $(3 + 2i)(3 - 2i) = 9 - 6i + 6i - 4i^2 = 9 - 4(-1) = 9 + 4 = 13$

47. conjugate of $-3 - 9i$ is $-3 + 9i$

49. The vertex tells the **minimum** or **maximum** value of a quadratic function.

51. The expression $b^2 - 4ac$ is called the **discriminant**.

53. Additive inverses add to 0; $-x + 2$

55. $y = k/x$ is an equation for **inverse** variation.

57. Compare the ratios of the coordinates, x to y or y to x. If the function passes through the origin the ratios will be equal.

59. $\dfrac{4x^2}{28x} = \dfrac{4xx}{4 \cdot 7x} = \dfrac{x}{7}$

61. $\dfrac{5x^2 + 15x}{6x + 18} = \dfrac{5x(x + 3)}{6(x + 3)} = \dfrac{5x}{6}$

63. $\dfrac{x+1}{x-3}+\dfrac{x}{x^2-9}=\dfrac{(x+3)(x+1)}{(x+3)(x-3)}+\dfrac{x}{(x+3)(x-3)}=\dfrac{x^2+4x+3+x}{(x+3)(x-3)}=\dfrac{x^2+5x+3}{(x+3)(x-3)}$

65. $\dfrac{x^2-8x+12}{x^2-1}\div\dfrac{2x-x^2}{x^2-5x-6}=\dfrac{(x-6)(x-2)}{(x+1)(x-1)}\cdot\dfrac{(x-6)(x+1)}{x(2-x)}=\dfrac{-(x-6)^2}{x(x-1)}$

67.

$$
\begin{array}{r}
x^2-x+2+\frac{6}{x-1} \\
x-1\overline{)x^3-2x^2+3x+4} \\
\underline{-(x^3-x^2)} \\
-x^2+3x \\
\underline{-(-x^2+x)} \\
2x+4 \\
\underline{-(2x-2)} \\
+6
\end{array}
$$

69. defined for all $\Re$, $x \neq -1$

$\dfrac{x+1}{8}=\dfrac{2}{x+1}$

$8(x+1)\dfrac{x+1}{8}=\dfrac{2}{x+1}(x+1)8$

$(x+1)^2=16$

$|x+1|=4;\ \ x+1=\pm 4$

$x=-5\ or\ x=3$

71. defined for all $\Re$, $x \neq 5$

$\dfrac{x}{2}=\dfrac{3}{5-x}$

$(5-x)x=6$

$5x-x^2=6;\ x^2-5x+6=0$

$(x-3)(x-2)=0;\ \{2, 3\}$

73. defined for all $\Re$, $x \neq 3$

$\dfrac{x}{3}=\dfrac{3-x}{x-3}=\dfrac{-(x-3)}{(x-3)}$

$\dfrac{x}{3}=-1;\ x=-3$

75. $\sqrt[3]{8x^6y^{12}}=(8x^6y^{12})^{1/3}=(2^3)^{1/3}(x^6)^{1/3}(y^{12})^{1/3}=2^{3/3}x^{6/3}y^{12/3}=2x^2y^4$

77. $0.000\,000\,567=5.67\times 10^{-7}$

79. $d=\sqrt{(x_2-x_1)^2+(y_2-y_1)^2}$

81. $3^0=1;\ y=0$

83. $(3x+2)^2=9x^2+12x+4$

85. a. $x^{1.5}=x^{3/2}=\sqrt{x^3}\ or\ (\sqrt{x})^3$

b. $x^{n/m}=(\sqrt[m]{x})^n\ or\ \sqrt[m]{x^n}$

c. $x^{1/2}(y^2)^{1/4}=x^{1/2}y^{1/2}=\sqrt{x}\sqrt{y}=\sqrt{xy}$

87. a. $\sqrt[3]{a^3b^4c^5} = \sqrt[3]{a^3b^3c^3} \cdot \sqrt[3]{bc^2} = abc\sqrt[3]{bc^2}$ **b.** $\sqrt[4]{w^4x^5y^8z^{12}} = |wxz^3|y^2\sqrt[4]{x}$

89. If $a^x = a^y$ and $a > 0$, $a \neq 1$, then <u>**x = y**</u>.

91. $16^x = 8$
$(2^4)^x = 2^3$
$4x = 3$
$x = \frac{3}{4}$

93. $0.01^x = 100$
$(10^{-2})^x = 10^2$
$-2x = 2$
$x = -1$

95. $9^{x+1} = \frac{1}{27}$
$(3^2)^{x+1} = 3^{-3}$
$2x + 2 = -3$
$2x = -5,\ x = -\frac{5}{2}$

97. $10^y = x$

99. $b^c = a$

101. $\dfrac{\log_{10} a}{\log_{10} b}$

103. base = 2

105. When we write a logarithm without a base, we mean log base <u>**10.**</u>

107. Calculating y = antilog 2.3 means raising <u>**10**</u> to the power of 2.3.

109. $2^{x+1} = 19$
$\log 2^{x+1} = \log 19$
$(x+1)\log 2 = \log 19$
$x + 1 = \dfrac{\log 19}{\log 2}$
$x = \dfrac{\log 19}{\log 2} - 1$
$x \approx 3.248$

111. $10^x = 17$
$\log 17 = x$
$x \approx 1.230$

113. $\log_2 x = 1$
$2^1 = x;\ x = 2$

115. $\log_{10} 2 = x;\ x \approx 0.301$

117. $\log_{10} 1000 = x;\ x = 3$

119. $\log_9 27 = x$
$9^x = 27$
$(3^2)^x = 3^3$
$2x = 3;\ x = 1.5$

121. $\ln x = -1$
$e^{-1} = x$
$x \approx 0.368$

123. $\ln x = e;\ e^e = x;\ x \approx 15.154$

125. $2500 = 1000e^{0.08t}$

$2.5 = e^{0.08t}$

$\ln 2.5 = 0.08t$

$\ln 2.5 \div 0.08 = t$

$t \approx 11.5$ yr

127. circle and hyperbola

$x^2 = 11 - y^2$

$(11 - y^2) - y^2 = 7$

$11 - 2y^2 = 7$

$2y^2 = 4;\ y^2 = 2;\ y = \pm\sqrt{2}$

$x^2 + 2 = 11;\ x^2 = 9;\ x = \pm\sqrt{3}$

129. $\begin{bmatrix} 2 & 3 & -1 \\ 3 & -4 & 1 \\ 4 & -1 & 2 \end{bmatrix}\begin{bmatrix} x \\ y \\ z \end{bmatrix} = \begin{bmatrix} 10.7 \\ 4.2 \\ 19.6 \end{bmatrix}$; x = 3.5, y = 2.6, z = 4.1

131.a. $P(E) = \frac{2}{11}$

b. $P(\text{vowel or N}) = \frac{5}{11} + \frac{2}{11} = \frac{7}{11}$

c. $P(\text{N \& N no replacement}) = \frac{2}{11} \cdot \frac{1}{10} = \frac{2}{110}$

d. $P(\text{N \& N with replacement}) = \frac{2}{11} \cdot \frac{2}{11} = \frac{4}{121}$

Part 2: Extended Problems

1. a. 0.8(200 – 100) = 80; 0.8(200 – 150) = 40; 0.8(200 – 200) = 0;

0.08(200 – 250) = –40 (not possible)

b. y = a + 0.8(200 – a) for a < 200; y = a

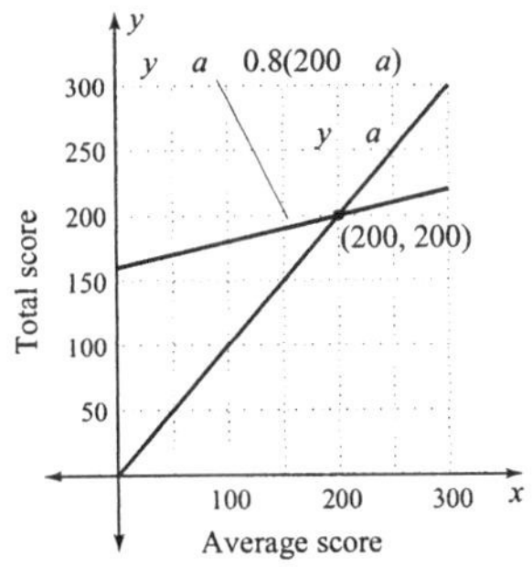

3. Substitution back into the first equation created a system of equations in which the first line was coincident with itself.

Following the students method, solve the first equation for x and substitute into the second equation.

$2x - 3y = 18;\ 2x = 3y + 18;\ x = 1.5y + 9$

$5(1.5y + 9) + 2y = 26;\ 7.5y + 45 + 2y = 26;\ 9.5y = -19;\ y = -2$

$x = 1.5(-2) + 9;\ x = -3 + 9 = 6$

5. convert weights to ounces: 13 pounds · 16 oz/lb = 208 oz; 208 + 15 = 223 oz total
2 lbs · 16 oz/lb = 32 oz; 32 + 4.5 = 36.5 oz.

$t + p + s = 11;$ $\quad t = 11 - p - s$

$17t + 16p + 36.5s = 223;$ $\quad 17(11 - p - s) + 16p + 36.5s = 223$

$0.50t + 1.62p + 2.86s = 13.58;$ $\quad 0.50(11 - p - s) + 1.62p + 2.86s = 13.58$

$187 - 17p - 17s + 16p + 36.5s = 223;$ $\quad -p + 19.5s = 36;\ p = 19.5s - 36$

$5.5 - 0.5p - 0.5s + 1.62p + 2.86s = 13.58;$ $\quad 1.12p + 2.36s = 8.08$

$1.12(19.5s - 36) + 2.36s = 8.08;$ $\quad 21.84s - 40.32 + 2.36s = 8.08$

$24.20s = 48.40;\ s = 2$ $\quad p = 19.5(2) - 36;\ p = 3$

$t = 11 - 3 - 2;\ t = 6$

6 cans of tomatoes, 3 pizza sauce, 2 salsa

7. points are x-intercepts; $a = \Re$; $a(x - 3)(x + 2) = 0$

9. $2a = 32$, $a = 16$; $-b = -6$, $b = 6$; $-4(a)(c) = 64$, $-4(16)(c) = 64$, $c = -1$

$16t^2 + 6t - 1 = 0;\ t = \dfrac{-6 \pm 10}{32}$; $\{-0.5, 0.125\}$; graph as 2 horizontal axis intercepts

11. $2a = 10$, $a = 5$; $-b = 8$, $b = -8$; $-4(5)c = -80$, $c = 4$

$5x^2 - 8x + 4 = 0;\ x = \dfrac{8 \pm \sqrt{-16}}{10} = \dfrac{8 \pm 4i}{10} = \dfrac{4}{5} \pm \dfrac{2}{5}i$; graph has no x-axis intercepts

13. $x^2 - 3x - 10$ is an expression; $y = x^2 - 3x - 10$ is an equation with a parabolic graph and an infinite number of solutions; $0 = x^2 - 3x - 10$ is an equation whose two solutions, $x = -2$ and $x = 5$, are the x-intercepts for the second equation.

15. In $y = ax^2 + bx + c$, c is the vertical axis intercept.

17. $(-3 - 2i)^2 + 6(-3 - 2i) + 13 = 9 + 12i - 4 - 18 - 12i + 13 = 0$, yes it is a solution
the other solution is the conjugate; $-3 + 2i$

19. linear: $y = 1.2755x - 0.53$; quadratic: $y = 0.02375x^2 + 0.7055x + 2.7$
The data appears linear so the linear regression is logical. Shorter panels might cost relatively more so the positive y-intercept of the quadratic equation makes sense.

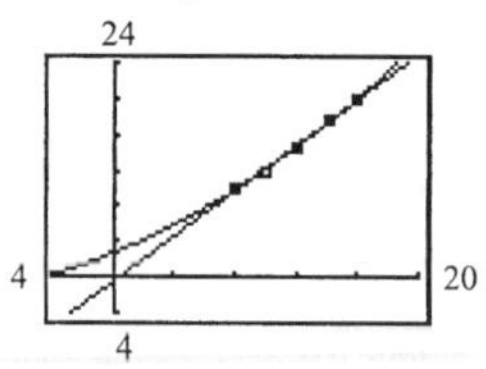

21. $\dfrac{y+\frac{1}{y}}{y-\frac{1}{y}} = \dfrac{\frac{y^2+1}{y}}{\frac{y^2-1}{y}} = \dfrac{y^2+1}{y} \cdot \dfrac{y}{y^2-1} = \dfrac{y^2+1}{y^2-1}$

23. $\dfrac{x}{2} + \dfrac{3}{x} = \dfrac{x^2}{2x} + \dfrac{6}{2x} = \dfrac{x^2+6}{2x}$; reciprocal $= \dfrac{2x}{x^2+6}$

25. a.
$A = \frac{1}{2}h(a+b)$

$2A = h(a+b)$

$\dfrac{2A}{h} = a+b$

$\dfrac{2A}{h} - a = b$

b.
$A = \frac{1}{2}h(a+b)$

$\dfrac{A}{\frac{1}{2}h} = a+b$

$\dfrac{A}{\frac{1}{2}h} - a = b$

c.
$A = \frac{1}{2}h(a+b)$

$A = \frac{1}{2}ah + \frac{1}{2}bh$

$A - \frac{1}{2}ah = \frac{1}{2}bh$

$\dfrac{A-\frac{1}{2}ah}{\frac{1}{2}h} = b$

d.
$A = \frac{1}{2}h(a+b)$

$2A = ah + bh$

$2A - ah = bh$

$\dfrac{2A-ah}{h} = b$

All are correct; part d contains no fractions in the numerator or denominator and is a single term.

27.
$\sqrt{3x-11} = 1$

$3x - 11 = 1^2$

$3x = 12$

$x = 4$

29.
$\sqrt{x-9} = 9 - \sqrt{x}$

$x - 9 = (9-\sqrt{x})^2 = 81 - 18\sqrt{x} + x$

$18\sqrt{x} = 90$

$18^2 x = 90^2$; $324x = 8100$; $x = 25$

31.
$2000 = 1000(1 + \mathrm{r})^{10}$
$2 = (1 + \mathrm{r})^{10}$
$2^{1/10} = 1 + \mathrm{r}$
$\mathrm{r} = 2^{1/10} - 1 \approx 7.2\%$

$2000 = 1000(1 + \mathrm{r})^{8}$
$2 = (1 + \mathrm{r})^{8}$
$2^{1/8} = 1 + \mathrm{r}$
$\mathrm{r} = 2^{1/8} - 1 \approx 9.1\%$

$2000 = 1000(1 + \mathrm{r})^{43/8}$
$2 = (1 + \mathrm{r})^{43/8}$
$2^{8/43} = 1 + \mathrm{r}$
$\mathrm{r} = 2^{8/43} - 1 \approx 13.8\%$

33. a. Geometric; constant ratio = $\frac{1}{3}$

b. $a_n = 27(\frac{1}{3})^{n-1} = 27(\frac{1}{3})^n(\frac{1}{3})^{-1} = 27(3)(\frac{1}{3})^n = 81(\frac{1}{3})^n$; $y = 81(\frac{1}{3})^x$

d. $S_{10} = \dfrac{27(1 - \frac{1}{3}^{10})}{1 - \frac{1}{3}} = \dfrac{27(0.999983)}{\frac{2}{3}} = 40.4993$

e. $S_\infty = \dfrac{27}{1 - \frac{1}{3}} = 40.5$

35. a. Quadratic; constant second differences = 4

c. From quadratic regression; $y = 2x^2 - x$

37. a. Exponential regression has best fit, $r \approx 0.9793$; $y = 164.6(1.208)^x$

b. $1.208 - 1 = 0.208$ or 20.8%

c. With 1985 = 1, 2001 = 17; $164.6(1.208)^{17} \approx 4088$

39. Set up matrix.

$$\begin{bmatrix} 2 & -4 & 1 \\ 3 & 2 & -1 \\ 1 & 2 & -1 \end{bmatrix}\begin{bmatrix} a \\ b \\ c \end{bmatrix} = \begin{bmatrix} 13 \\ 6 \\ 0 \end{bmatrix}, \ a = 3, b = -2, c = -1$$

41. At $x = 0$, $m = -0.05$; $-0.05 = 2a(0) + b$; $b = -0.05 = -\frac{1}{20}$

At $x = 3200$, $m = 0.04$; $0.04 = 2a(3200) + (-0.05)$; $0.09 = 6400a$; $a =$

$1000 = \frac{9}{640{,}000}(3200)^2 - \frac{1}{20}(3200) + c$; $c = 1016$

$y = \frac{9}{640{,}000}x^2 - \frac{1}{20}x + 1016$

43. a. cake and ice cream: $6 \cdot 10 = 60$

b. cake or ice cream: $6 + 10 = 16$

c. cake and 2 (different) ice cream flavors: $6 \cdot 10 \cdot 9 = 540$

d. three (different) ice cream flavors: $10 \cdot 9 \cdot 8 = 720$

45. ${}_xC_n = {}_xC_5$ when $n = x - 5$

Appendix 2

1. **a.** $0.30 \cdot 100 = 30\%$ **b.** $0.20 \cdot 100 = 20\%$ **c.** $0.075 \cdot 100 = 7.5\%$
d. $0.0005 \cdot 100 = 0.05\%$ **e.** $2.0 \cdot 100 = 200\%$ **f.** $5.0 \cdot 100 = 500\%$
g. $20 \cdot 100 = 2000\%$ **h.** $100 \cdot 100 = 10{,}000\%$ **i.** $2000 \cdot 100 = 200{,}000\%$

3. **a.** $15\% = \frac{15}{100} = \frac{3}{20} = 0.15$ **b.** $80\% = \frac{80}{100} = \frac{4}{5} = 0.80$

c. $0.25\% = \frac{0.25}{100} = \frac{25}{10{,}000} = \frac{1}{400} = 0.0025$ **d.** $10\% = \frac{10}{100} = \frac{1}{10} = 0.10$

e. $1000\% = \frac{1000}{100} = 10$ **f.** $17.5\% = \frac{17.5}{100} = \frac{175}{1000} = \frac{7}{40} = 0.175$

g. $0.15\% = \frac{0.15}{100} = \frac{15}{10{,}000} = \frac{3}{2000} = 0.0015$

h. $35\% = \frac{35}{100} = \frac{7}{20} = 0.35$ **i.** $200\% = \frac{200}{100} = 2$

5. Slope gives change per year, percent change describes the total change. Without time stated, percent change can be misleading.

7. Dividing changes the fraction to a decimal, moving the decimal 2 places to the right is multiplying by 100. Placing a percent sign means dividing by 100, net effect is multiplication by 1.

Appendix 3

1. Constant difference is –2, find 20th term: $a_{20} = 18 + (20 - 1)(-2) = -20$
$S_{20} = \frac{20}{2}(18 + (-20)) = -20$

3. Constant difference is 8, $a_{20} = 10 + (20 - 1)(8) = 162$; $S_{20} = \frac{20}{2}(10 + 162) = 1720$

5. Sequence is not arithmetic

7. Not arithmetic; pairs have different sums.

9. Constant difference is 1;
$a_n = 47 + (n - 1)(1)$;
$113 = 47 + (n - 1)$
$66 = n - 1$
$n = 67$ seats

11. Constant difference is 4;
$a_n = 1896 + (n - 1)4$;
$2000 = 1896 + (n - 1)4$
$104 = 4(n - 1)$; $26 = n - 1$; $n = 27$ games
games cancelled during WWI and WWII

13. Constant difference is 3;

$a_n = 1 + (n - 1)3;$

$145 = 1 + 3(n - 1)$

$144 = 3(n - 1);\ 48 = n - 1;$

$n = 49$ tests

17. Constant ratio is 3;

$$S_{10} = \frac{9(1-3^{10})}{1-3} = 265{,}716$$

infinite sum not appropriate, $r > 1$

19. Constant ratio is $\frac{1}{2}$;

$$S_{10} = \frac{64[1-(\frac{1}{2})^{10}]}{1-\frac{1}{2}} = 127.875$$

$$S_{\infty} = \frac{64}{1-\frac{1}{2}} = 128$$

21. Constant ratio is 3;

$$S_{10} = \frac{3(1-3^{10})}{1-3} = 88{,}572$$

infinite sum not appropriate, $r > 1$

23. $S_{20} = \dfrac{108[1-(\frac{2}{3})^{20}]}{1-\frac{2}{3}} \approx 323.9$ in.

25. $S_{20} = \dfrac{135[1-(\frac{2}{3})^{20}]}{1-\frac{2}{3}} \approx 404.88$ in.

27. Constant ratio is $\frac{1}{2}$; $S_{\infty} = \dfrac{1}{1-\frac{1}{2}} = 2$

29. $S_{\infty} = \dfrac{18}{1-\frac{5}{6}} = 108$ ft.